High-Performance Super-capacitors: Preparation and Application

High-Performance Super-capacitors: Preparation and Application

Guest Editor

Xin Chen

MDPI

Basel • Beijing • Wuhan • Barcelona • Belgrade • Novi Sad • Cluj • Manchester

Guest Editor
Xin Chen
School of Materials Science
and Engineering
East China University of
Science and Technology
Shanghai
China

Editorial Office
MDPI AG
Grosspeteranlage 5
4052 Basel, Switzerland

This is a reprint of the Special Issue, published open access by the journal *Batteries* (ISSN 2313-0105), freely accessible at: https://www.mdpi.com/journal/batteries/special_issues/J3792YH2XN.

For citation purposes, cite each article independently as indicated on the article page online and as indicated below:

Lastname, A.A.; Lastname, B.B. Article Title. *Journal Name* **Year**, *Volume Number*, Page Range.

ISBN 978-3-7258-6566-6 (Hbk)
ISBN 978-3-7258-6567-3 (PDF)
https://doi.org/10.3390/books978-3-7258-6567-3

Contents

About the Editor

Xin Chen

Xin Chen is a Professor (Ph.D), School of Materials Science and Engineering, East China University of Science and Technology in China. He obtained his B.Sc. in Modern Applied Physics at Tsinghua University, and M.Sc. and Ph.D. in Physics at University of Houston. He has worked at the State's Key Open Laboratory of Surface Physics in Chinese Academy of Sciences, University of Houston, Peking University, and the University of Illinois at Urbana-Champaign, and he is now a Professor at East China University of Science and Technology. Dr. Chen has published over 80 papers in peer reviewed international journals, which have been cited some 2000 times. He has chaired several international conferences; held plenary, keynote, and invited talks in many international conferences; received seven U.S. patents and eight China patents; and edited/co-edited three books. He was awarded the Sigma Xi Research Achievement Award, USA, in 2000; NASA Space Act Award, USA, in 2007; and First Prize of the Book Award for Excellent Publications from China Petroleum and Chemical Industry in 2018. Currently Dr. Chen's research focuses on supercapacitor materials, in situ liquid cell electron microscopy, and other nano functional materials. He is currently a Council Member of Shanghai Micrology Society.

Preface

Thanks to MDPI for giving me the chance to edit this Reprint, and thanks to all of the authors who have kindly contributed to its success. Supercapacitors continue to attract extensive research interest from all over the world, owing to the advantages of their relatively high power/energy density properties, high safety, long cycling lifetimes, etc., which are envisioned to be important for solving the world's energy and environment problems. However, to successfully compete with the existing energy storage technologies, the performance of supercapacitor materials, especially their overall power/energy density performance, needs to be continuously improved. Thus, the development of high-performance supercapacitor materials is a timely and critical task of the field. In addition to this collection, future projects will be carried out on this topic. Experts in the field are invited to continue joining this effort, and we hope that all subsequent collections will be very successful.

Xin Chen
Guest Editor

batteries

MDPI

Article

Secondary High-Temperature Treatment of Porous Carbons for High-Performance Supercapacitors

Weihao Chi, Guanwen Wang, Zhipeng Qiu, Qiqi Li, Zheng Xu, Zhiyuan Li, Bin Qi, Ke Cao, Chunlei Chi *, Tong Wei and Zhuangjun Fan *

School of Materials Science and Engineering, China University of Petroleum (East China), Qingdao 266580, China
* Correspondence: chichunleiccl4@163.com (C.C.); fanzhj666@163.com (Z.F.)

Abstract: Supercapacitors are extensively used in urban rail transit, electric vehicles, renewable energy storage, electronic products, and the military industry due to its long cycle life and high power density. Porous carbon materials are regarded as promising anode materials for supercapacitors due to their high specific surface areas and well-developed pore structures. However, the over-developed pore structure often results in poor conductivity and reduced cycle stability due to the destruction of a carbon skeleton. Herein, we introduce an advanced strategy for preparing porous carbon with high specific surface areas (3333 $m^2\ g^{-1}$), high electrical conductivity (68.6 S m^{-1}), and fast ion transport channels through secondary high-temperature carbonization treatment. As a result, the fabricated porous carbon anode delivers a high specific capacitance (199.2 F g^{-1} at 1 A g^{-1}) and outstanding rate performance (136.3 F g^{-1} at 20 A g^{-1}) in organic electrolyte. Furthermore, the assembled symmetrical supercapacitor achieves an energy density of 43.2 Wh kg^{-1} at 625.0 W kg^{-1}, highlighting the potential of a secondary high-temperature carbonization strategy in practical applications.

Keywords: supercapacitors; porous carbon; secondary high-temperature carbonization

1. Introduction

Environmental challenges and the restricted availability of fossil fuels have promoted the great development and application of environmentally friendly renewable energy, such as solar energy, geothermal energy, wind energy, tidal energy, and hydrogen energy [1–5]. However, sustainable energy is inherently intermittent and unevenly distributed in terms of time and space. Thus, energy storage devices (ESDs) with a high energy density, long cycle life, and power density are eagerly demanded. Among various ESDs, supercapacitors (SCs) have become increasingly important as a short-term energy storage buffer due to their high power densities, long lifespans, great safety, and adaptability to temperature [6–8]. Nevertheless, the limited energy density hinders the further application of SCs.

Based on the previous research, enlarging the voltage window and increasing capacitance are both effective to improve the energy density of SCs [9,10]. Chang et al. [11] reported a frame-filled C/C composite with coal tar pitch and natural flake graphite as raw materials, showing both a high specific surface area and electrical conductivity. When tested in the $TEABF_4$ in the PC electrolyte (3 V) and medium pure $EMIMBF_4$ electrolyte (3.5 V), the C/C composite exhibited both high gravimetric and volumetric capacitances. Moreover, it displayed a high energy density as high as 66.3 Wh kg^{-1}. Similarly, Leng et al. [12] synthesized a three-dimensional carbon framework (3DCF) under H_2/Ar at high temperature, with a high specific surface area and good conductive network, exhibiting both ultra-fast and high charge storage. Dead leaves of Ginkgo biloba were used as the precursor to prepare activated carbon by using different activating agents [13]; the carbon activated by KOH showed the highest specific capacitance of 374 F g^{-1} at a current density of 0.5 A g^{-1}. Further, the energy density can reach 9.2 Wh kg^{-1} at a power density of 48 W kg^{-1}. Compared to SCs with the aqueous electrolyte working within 1 V, organic electrolyte-based SCs provide a wider voltage range of 2.5–2.7 V, contributing to the higher energy density of SCs [14]. However, organic electrolytes always consist of large-sized ions, resulting in sluggish ion diffusion [15]. Therefore, the preparation of electrode materials with both plentiful energy storage active sites and rapid transport rates for large-size ions is crucial for the higher energy density of SCs. Improving specific surface areas (SSAs) of electrode materials is one effective method to enhance the ion transport rate because it can expose more active sites for the rapid physical adsorption of TEA^+ ions [16,17]. In addition, according to the definition equation of capacitance: $C = \varepsilon_r \varepsilon_0 A/d$, the increase in SSAs can help increase the capacitance.

Porous carbons with high SSAs and hierarchical pore structures are the most common material used for the electrode material of commercial SCs [18]. Due to the physical adsorption and desorption, charge is stored in the Helmholtz bilayer at the electrolyte–electrode interface, resulting in the rapid ion storage capability. Moreover, porous carbon materials can generally be produced by a variety of synthesis methods, with a wide range of raw material sources and relatively low costs. Early strategies focused on exposing larger SSAs to further increase the energy storage active sites and enhance capacitance [19]. Nevertheless, the excessive increasing of SSAs and porosity will probably block the formation of continuous electron pathways, restricting the rapid charge transport and lowering the electric conductivity of active materials. Therefore, balancing the relationship among the SSAs, conductivity, and cycling stability of porous carbon has become the current research focus.

To address these issues, we propose an optimized porous carbon with the secondary high-temperature treatment strategy after the activation of phenolic resin. On the one hand, the preferential removal of oxygen-containing functional groups inside the pores at high temperatures can enlarge the transport channels and reduce the diffusion barriers of ions. On the other hand, the repeated carbonization treatment promotes the formation of more micropores and enhanced SSAs, resulting in more energy storage active sites for ion storage. In addition, the released oxygen impurities lead to the formation of a coherent conductive carbon skeleton with effectively enhanced electric and ionic conductivity. As a result, the synthesized porous carbon exhibits excellent specific capacitance (199.2 F g^{-1} at 1 A g^{-1}) and rate performance (136.3 F g^{-1} at 20 A g^{-1}). When assembled as the symmetrical SCs, it also displays a high energy density of 43.2 Wh kg^{-1} at a power density of 625.0 W kg^{-1}, demonstrating significant potential for large-scale application.

2. Materials and Methods

2.1. Materials

The materials used in this experiment included phenolic resin (chemistry), concentrated hydrochloric acid (HCl, AR), polytetrafluoroethylene (60 wt%), ethyl alcohol (AR), potassium hydroxide (AR), N_2, foamed nickel, and ketjen black. The heat-resistant phenolic resin (phenolic resin foam) is formed by the phenolic resin as the matrix and hexamethylenetetramine as the curing agent, which exhibits the internal linear or branched-chain polymer chains connected by covalent bonds, resulting in a network polymer structure. The electrolyte used in the system is 1 mol L^{-1} $TEABF_4$ in PC. The phenolic resin foam was ball-ground into powder before being used. Other reagents for experiments are purchased and used directly.

2.2. Preparation of High-Temperature Treated Porous Carbons (HTCs)

The high-temperature treated porous carbons (HTCs) were synthesized by a secondary high-temperature treatment strategy. Firstly, 1 g of phenol-formaldehyde resin was added to the solution of KOH (3 g) in deionized water (20 mL). The mixture was stirred for 30 min to homogenous dispersal. Subsequently, the solution was dried in oven at 100 °C, then the obtained product was ground into powder. Afterward, the powder was carbonized at 900 °C for 2 h. After chemical activation, the obtained mixture was immersed in 1 mol L^{-1} HCl solution for 12 h to remove residual K components. After filtering and drying, the sample was carbonized for the second time at 1000 or 1100 °C for 2 h. The resulting samples were named HTC-1000 and HTC-1100, respectively. In addition, the sample without secondary high-temperature treatment is named HTC-0.

2.3. Material Characterization

Scanning electron microscopy (SEM, Hitachi Regulus 8100, Hitachi, Japan) was used to characterize the morphology and structure of the HTCs. The X-ray diffraction (XRD, Bruker Focus D8, Bruker, Germany) was used to characterize the crystal structure of the HTCs. The HORIBA Scientific LabRAM HR Evolution (Gières, France) spectrometer was used to characterize Raman spectroscopy. The magnification of the objective lens is 50×, the focal length of the spectrometer is 800 mm, and the spectral resolution is 1 cm^{-1} with a wavelength of 532 nm. The ESCALAB 250 Xi spectrometer (Thermo Fisher, Waltham, MA, USA) was used to characterize the X-ray photoelectron spectroscopy (XPS) of the HTCs. The IRTRACER-100, (Shimadzu, Japan) was used to characterize the Fourier-transform infrared spectrum (FTIR) of the HTCs. The N_2 adsorption–desorption isotherms for HTCs were characterized by Quanta chrome IQ2 at −196 °C.

2.4. Electrochemical Measurements

The slurry was mixed with active materials (HTCs), conductive carbon black, and binder (PTFE) in a mass ratio of 75:20:5. Firstly, a small amount of anhydrous ethanol and a certain mass of diluted PTFE solution was added to the conductive carbon black and ultrasonicated homogeneously. Secondly, a certain mass of active substance was added and ultrasonicated again until they were mixed homogeneously, and the obtained homogeneously mixed slurry was put into a vacuum oven for drying. Thirdly, after weighing the material of ~2 mg cm^{-2}, ethanol was added onto the material by drops for ease of rolling the material into the sheet and the sheet was pressed into a stainless-steel mesh. Finally, the pressed electrode sheets were dried in oven at 80 °C. The prepared supercapacitor assembly was pressed into the cell in the order of the negative case, spacer, electrode sheet, diaphragm, electrode sheet, and positive case. Before pressing, the electrolyte was injected into the supercapacitor to ensure that the electrolyte fully saturates the electrodes. In addition, symmetrical SCs were assembled using two identical electrodes in the CR2032 coin cell with cellulose diaphragm TF4035 as the separator. The organic electrolyte was 1 mol L^{-1} $TEABF_4$ in PC solution with a voltage window of 0–2.5 V. The CHI660E electrochemical workstation was used to conduct the electrochemical properties of cells at 25 °C. The rate range of galvanostatic charge-discharge (GCD) was from 1 to 20 A g^{-1}, and the current density of cyclic voltammetry (CV) was from 10 to 200 mV S^{-1}. The electrochemical impedance spectroscopy was measured at the frequency range of 0.01 to 100 kHz.

3. Results and Discussion

High-temperature treatment carbons (HTCs) were prepared by a secondary high-temperature treatment strategy after the KOH activation of phenolic resin (Figure 1). After high-temperature treatment, the order degree of amorphous carbon increased and the partial oxygen impurities were removed, resulting in excellent electrical conductivity and open ions transport channels, which is good for the rate capability of porous carbon materials [20]. Simultaneously, the pore volume and SSAs increased, which can introduce more electrochemical active sites and further increase specific capacitance for SCs [21–23]. A scanning electron microscopy (SEM) image of HTC-0 shows a block structure with a size of 20~40 μm (Figure 2a). After carbonizing at 1000 °C for the second time, HTC-1000 still consists of chunks of different blocks, which is consistent with HTC-0 (Figures 2b and S1). This means that HTC-0 has excellent structural stability and there was no chalking after the high-temperature treatment. The excellent strength and toughness make the active materials bond tightly and maintain integrity during the charge and discharge process. The corresponding EDS mappings shown in Figure 2d,f demonstrate the uniform distribution of C and O elements of HTC-1000 (Figure S2). Even when the secondary treatment temperature increases to 1100 °C, HTC-1100 still maintains the original bulk morphology, indicating the high stability of the macrostructure (Figure 2c). TEM and HRTEM images of HTC-1000 have been provided, as shown in Figure S3. It can be seen that a large number of micropores are evenly distributed in the carbon matrix, which is in accordance with the BET analysis. In addition, HTC-1000 shows partially formed graphite nanodomains, indicating that secondary high-temperature carbonization can improve the degree of order (circled in Figure S3b).

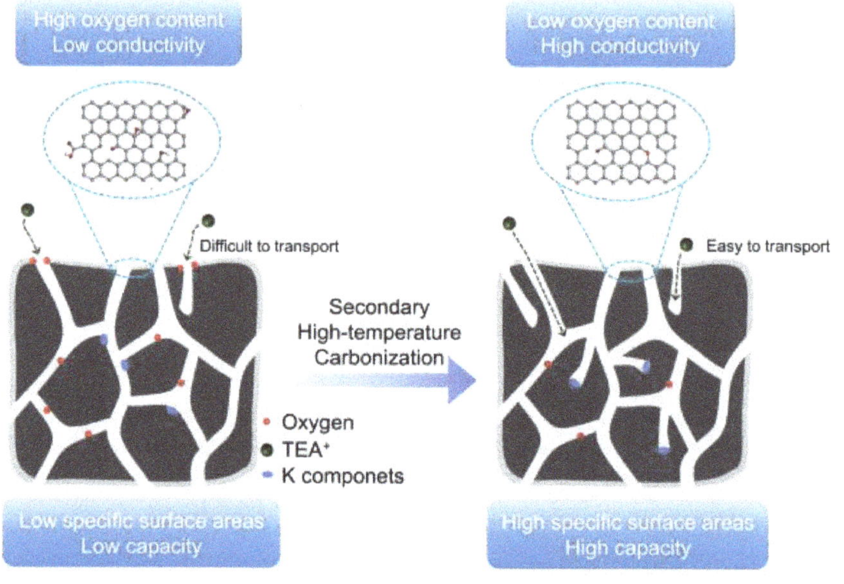

Figure 1. Schematic diagram of preparation process and structure evolution of HTC-1000.

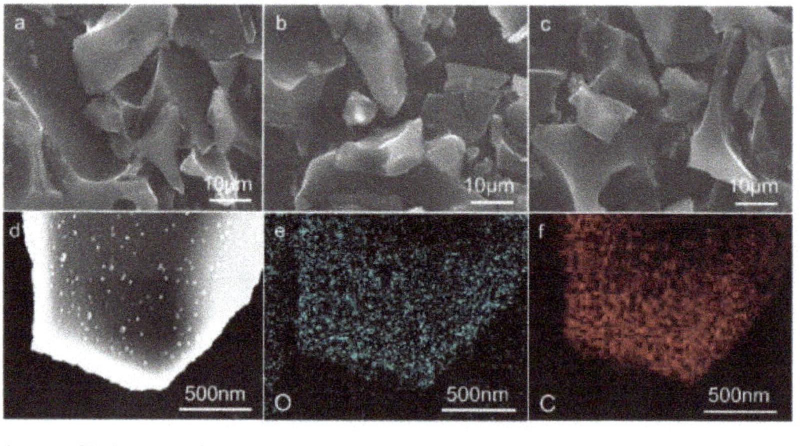

Figure 2. SEM images of (**a**) HTC-0, (**b**) HTC-1000, (**c**) HTC-1100, (**d**) SEM image of HTC-1000, (**e**) O mapping of HTC-1000, (**f**) C mapping of HTC-1000.

To explore the pore structure of the three samples, N_2 adsorption-desorption tests were performed. The vast gas uptake in the low-pressure region is dominated by the microporous structures while the adsorption generated in the pressure region of $0.4 < P/P_0 < 0.8$ is dominated by the mesoporous structures. Therefore, from N_2 adsorption–desorption isotherms, all three samples exhibit pore distribution with both mesoporous and microporous. Moreover, from the hysteresis loop in the N_2 adsorption curve, it is observed that the internal pore type of the material is a slit pore (Figure 3a). According to the Brunauer–Emmett–Teller (BET) method, the untreated HTC-0 shows an SSA of 3156 $m^2\ g^{-1}$, while after second high-temperature treatment, the SSA of HTC-1000 increased to 3333 $m^2\ g^{-1}$. The further increased SSA is due to the secondary activation of residual K components and the release of small gas molecules containing C, H, and O under high temperature [24,25]. The achieved higher SSAs could offer more energy storage active sites for TEA^+ ion adsorption, thus improving capacitance. Moreover, the increased SSAs can increase the effective adsorption area between the capacitor motor and the electrolyte, shorten the diffusion adsorption path of TEA^+ ions in the electrode material, and increase the diffusion speed of particles in the electrode material. However, while the secondary treatment temperature further increased to 1100 °C, the pores in HTC-1100 collapsed, leading to a decreased SSA of 2467 $m^2\ g^{-1}$. Consequently, excessively high temperatures would destroy the pore structure and lead to partial pore blockage, thus adversely affecting the specific capacitance and rate capability of the material. To further investigate the effect of the secondary treatment on the pore structure, the content of micropores and mesopores was analyzed in detail. According to Figure 3b, the total pore volume of HTC-1000 is 2.60 $cm^3\ g^{-1}$, which is much larger than the 2.00 $cm^3\ g^{-1}$ of HTC-0. Furthermore, compared with HTC-0, the micropore volume of HTC-1000 slightly decreased from 1.27 $cm^3\ g^{-1}$ to 1.24 $cm^3\ g^{-1}$ while the mesopore volume has an obvious increase from 0.75 $cm^3\ g^{-1}$ to 1.35 $cm^3\ g^{-1}$, suggesting that high-temperature treatment generates ordered mesoporous pores [26–28]. The expanded pores volume and micropores can effectively promote the transport of the TEA^+ ions and provide plenty of space to store the electrolyte [29,30]. However, the total pore volume of HTC-1100 decreased to 1.69 $cm^3\ g^{-1}$ with the micropore volume decreasing to 0.95 $cm^3\ g^{-1}$ and mesopore volume decreasing to 0.74 $cm^3\ g^{-1}$ due to the pore shrinkage and collapse caused by excessive high-temperature treatment (Table S1). Compared to aqueous systems, organic electrolyte ions are larger (the non-solvated TEA^+ ion sizes are 0.67 nm, respectively), requiring larger pore volumes and higher mesopore ratios to store and transport electrolyte ions [31–34]. The high-temperature treatment at 1000 °C significantly expands the pore volume and increases the mesopore ratio of the material without significant changes in morphology, providing HTC-1000 with an increased specific capacitance and improved rate performance under high current densities. The pore diameter distribution of the three samples is shown in Figure 3c. The micropore diameter is centered at 1.17 nm, which is beneficial for ion transport, while the mesoporous diameter is centered at 2.7–4 nm, which can form a repository for the electrolyte, quickly providing ions for energy storage during charging and discharging.

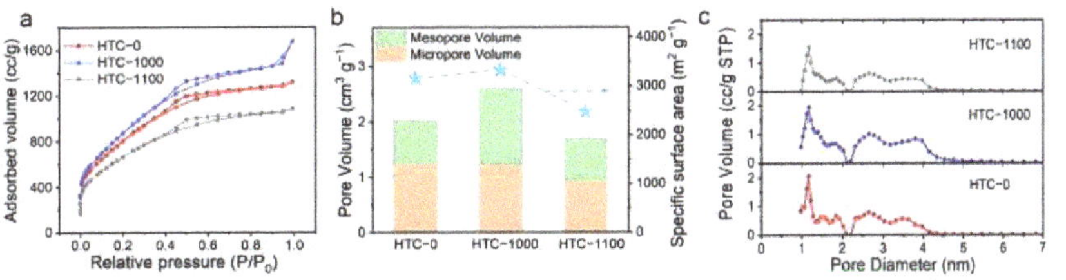

Figure 3. (**a**) N_2 adsorption-desorption isotherms, (**b**) porous structure evolution and specific surface area, and (**c**) pore size distributions of HTCs.

The X-ray diffraction (XRD) patterns of HTCs are shown in Figure 4a. The peaks in the low-angle scattering region of all samples indicate that the material HTCs have a porous structure [35–37]. Due to the non-graphitic disordered stacking and amorphous structure of the porous carbon material, the XRD pattern appears as a bread-like shape and the broad diffraction peaks at around 26° and 43° correspond to the reflections of the (002) and

(100) planes, respectively [38,39]. It can be observed from the pattern that the low-angle scattering intensity of HTC-1000 increases after secondary high-temperature treatment, indicating that HTC-1000 has a more developed porous structure. However, there is no obvious peak shift for HTC-0, HTC-1000, and HTC-1100, illustrating that secondary high-temperature treatment cannot affect the interlayer spacings of graphite nanodomains in these three samples [40]. The intensities of peaks in HTC-1000 and HTC-1100 are stronger than that of HTC-0, indicating an increase in the ordered degree, which contributed to the improvement of conductivity. The Raman spectra of HTCs are shown in Figure 4b. The Raman spectra of three samples showed D and G peaks located at 1350 and 1580 cm^{-1}. The Raman spectra were fitted with four peaks located at 1150, 1350, 1510, and 1580 cm^{-1}, which characterize the T band, D band, D′ band, and G band, respectively, where the D band and G band represent sp^3-type disordered carbon and sp^2-type graphitic carbon, respectively (Figure 4c and Figure S4). The area ratio of the D band and G band (A_D/A_G) can be used to characterize the defect content of the sample. After areas fitting, the A_D/A_G of HTC-0, HTC-1000, and HTC-1100 are 3.28, 2.87, and 2.76, respectively. The decrease in A_D/A_G ratios with an increasing treatment temperature demonstrates the improvement in the crystallinity and graphitization degree, which is identical to the XRD results [41]. As the internal crystal structure of the material becomes more ordered, more conjugated π electrons are formed, and electrons can move freely inside the material, leading to good electrical conductivity [42]. The increase in graphitization during high-temperature treatment produces a more stable crystal structure. This structural stabilization improves the chemical stability of HTC-1000, which results in better cycling stability. The Fourier-transform infrared spectrum of HTC-0 exhibits specific absorption peaks, including 700 cm^{-1} belonging to -CH in the aromatic ring, 1200 cm^{-1} caused by C=O, 1500 cm^{-1} belonging to $-CH_3$ connected to the aromatic ring, 1600 cm^{-1} caused by C=C vibration, 2850 cm^{-1} and 2950 cm^{-1} belonging to the $-CH_2$ peak connected to the aromatic ring, and 3700 cm^{-1} belonging to -OH (Figure 4d, Table S2). After secondary high-temperature treatment, the -OH peak and the C=O peak are significantly weakened in HTC-1000 and HTC-1100, suggesting that the unstable -OH and C=O are further released at high temperatures [43–45]. The removal of oxygen-containing functional groups is beneficial to the properties of porous carbon materials. Firstly, the removal of oxygen-containing functional groups promotes the formation of carbon six-membered rings, which is conducive to the construction of a continuous carbon conductive network and thus improves electrical conductivity. Secondly, since oxygen functional groups are highly reactive, removing them can avoid overmuch side reactions and improve the electrochemical stabilization of the material in SCs. Finally, the removal of oxygen-containing functional groups reduces the impediment for ion transport, creating abundant channels for rapid ion transport. As shown in Figure 4e, the total XPS spectra of HTCs all show O 1s peaks at 532.8 eV and C 1s peaks at 284.8 eV, and the C/O ratio of HTC-0, HTC-1000, and HTC-1100 are 16.1, 19.2, and 21.7, respectively, proving the decreased O content and increased C content after secondary high-temperature carbonization. The enhanced C content illustrates that the secondary high-temperature treatment can effectively remove the unstable oxygen-containing functional groups [46]. As shown in Figure 4f, high-resolution C 1s spectra can be fitted into three peaks centered at 287.9, 285.8, and 284.5 eV, corresponding to C=O, C-O, and C=C bonds, respectively. The ratios of C=C and C-O are 57% and 23% of HTC-0, respectively. After secondary high-temperature treatment, HTC-1000 shows an increased ratio of C=C (60%) and a decreased ratio of C-O (16%). The increase in the C=C bonding ratio indicates the formation of a more complete carbon conducting network in HTC-1000, while the decrease in the C-O bonding ratio suggests that the O impurities are effectively removed.

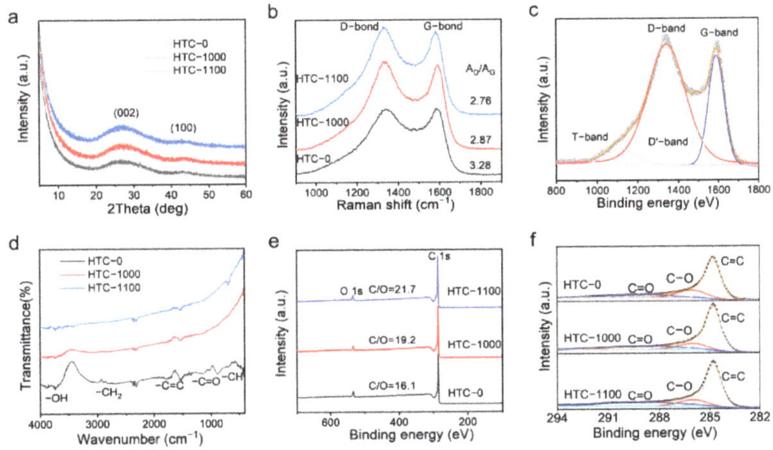

Figure 4. (**a**) XRD patterns and (**b**) Raman spectra of HTCs. (**c**) Raman fit plot of HTC-1000, (**d**) FTIR spectra, (**e**) XPS survey spectra, and (**f**) high-resolution XPS C 1s spectra of the HTCs.

To evaluate the electrochemical performance of HTCs, GCD, CV, rate performance, and cyclic stability, tests were conducted [47]. The active mass loadings of electrodes were 2.0 mg cm^{-2}. CV tests are carried out from 0 to 2.5 V at the scan rate of 10 mV s^{-1}, and the CV curves of HTC-0, HTC-1000, and HTC-1100 are all rectangular-shaped, indicating good double-layer capacitance behavior (Figure 5a). Compared with HTC-0 and HTC-1100, HTC-1000 shows a larger area at the same current, indicating a higher capacitance. The excellent energy storage properties of HTC-1000 benefited from its larger SSAs and more abundant pore distribution, which can provide sufficient active sites. Moreover, the CV tests were carried out at a series of scan rates of 20, 50, 100, and 200 mV s^{-1}. From Figure S5, the shape of CV curves is maintained well, indicating the good stability of HTC-1000. GCD tests were conducted and based on the equation (m is the active mass loading on the single electrode), and HTC-1000 exhibits a high capacitance of 199.2 F g^{-1} at 1 A g^{-1}, which is higher than the 170.4 F g^{-1} of HTC-0 (Figure 5b). When the current densities increase to 2, 3, 4, 5, and 10 A g^{-1}, HTC-1000 exhibits a high capacitance of 189.8, 182.9, 177.9, 173.6, and 158.2 F g^{-1}, respectively. In addition, the GCD curves of HTC-1000 exhibit a standard isosceles triangular shape, meaning that there is no redox reaction in HTC-1000, resulting in the fast charge-discharge capability [48]. As shown in Figure 5c, HTC-0 exhibits a specific capacitance of 170.4 F g^{-1} at 1 A g^{-1} and a specific capacitance of 107.5 F g^{-1} at 20 A g^{-1}, with a capacity retention ratio of 62.9%. After secondary high-temperature treatment, HTC-1000 displays a high specific capacitance of 199.2 F g^{-1} at 1 A g^{-1}. In addition, the capacity retention ratio of HTC-1000 at 20 A g^{-1} increases from 62.9% to 68.3%, with a specific capacitance of 136.3 F g^{-1} (Table S3). The improved capacitance and rate performance benefit from the increased SSAs, abundant mesoporous, and excellent conductivity, which can provide more energy storage sites and promote the transport kinetics of ions and the conduction of electrons [49,50]. In addition, the commonly used YP-50F was also tested for comparing cycling stability. Compared with YP-50F, which displays a specific capacitance of 90.0 F g^{-1} at 1 A g^{-1} and 26.0 F g^{-1} at 20 A g^{-1}, HTC-1000 has a significant advantage (Figure S6). As for HTC-1100, due to the excessively high temperature during secondary high-temperature treatment, some pores collapse and merge, resulting in a decreased specific capacitance of 168.0 F g^{-1} at 1 A g^{-1} and 80.0 F g^{-1} at 20 A g^{-1}. Furthermore, the HTC-1000//HTC-1000 symmetric SCs were also assembled and measured at 1, 5, 10, 20, 30, 40, 50, 70, and 100 A g^{-1}. According to the equation: $E = C(\Delta V)^2/(8 \times 3.6)$, $P = 3600E/\Delta t$, the HTC-1000//HTC-1000 symmetric SCs exhibit an excellent energy density of 43.2 Wh kg^{-1} at the power density of 625.0 W kg^{-1}, which is much higher than the other reported carbon material (Figure 5d, Table S5) [48,51–60]. Furthermore, the cycling performance of symmetric organic SCs assembled with HTC-1000 was tested at 5 A g^{-1}. As shown in Figure 5e, even after as long as 50,000 cycles, the symmetric supercapacitor still maintains a high capacitance of 179.0 F g^{-1}, with a high capacity retention ratio of 90%. The outstanding cycling stability of symmetric organic supercapacitors derives from the excellent structural stability of the electrode material and the few side reactions. SEM images of HTC-1000 after cycling for 50,000 cycles still maintain a blocky structure of 20–40 μm, which is in accordance with the pristine HTC-1000 (Figure S7). It shows that the electrode material can maintain the structural integrity during charging and discharging, thus improving the cycle stability. Under high pressure in the organic electrolyte, the -COOH groups and C-O groups are easily decomposed to produce H_2O, CO_2, CO, and other gas molecules, which can raise side reactions and affect electrode stability. Moreover, the oxygen-containing functional groups can react with the electrolyte and catalyze the decomposition of the electrolyte. Therefore, the removal of oxygen-containing functional groups can effectively hinder the occurrence of side reactions and enhance the cycling stability of electrode materials. In general, due to the excellent structural stability of the materials and the decrease in side reactions, symmetric organic supercapacitors assembled with HTC-1000 exhibit outstanding cycling stability.

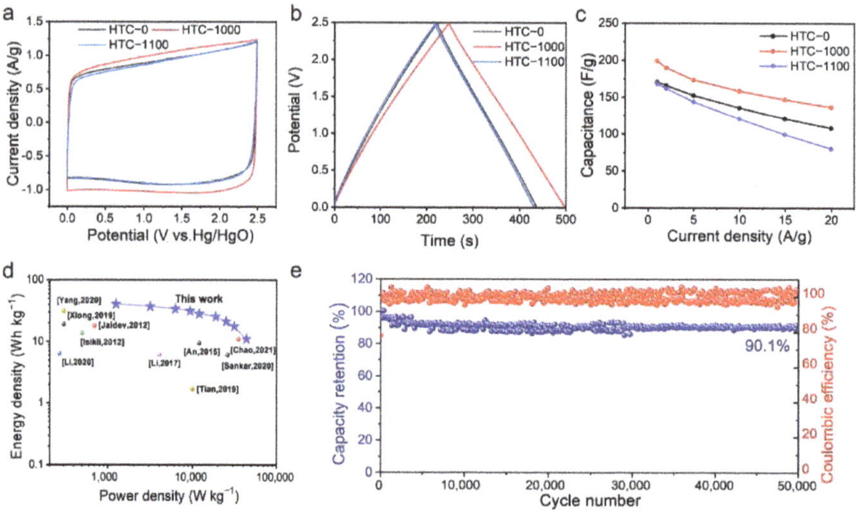

Figure 5. (**a**) CV curves at 10 mV s^{-1}, (**b**) galvanostatic charge−discharge curves at 1 A g^{-1}, (**c**) gravimetric capacitance, (**d**) Ragone plots [48,51–60], (**e**) Cycle performance with the mass loading of 2 mg cm^{-2} at 5 A g^{-1}.

Electrochemical impedance tests were conducted to explore the ionic and electronic conductivity. According to the Nyquist plots of HTCs, HTC-1000 treated at high temperature exhibits better electronic transport efficiency and conductivity compared with HTC-0 (Figure 6a) [61,62]. To further analyze the effects of secondary high-temperature treatment on ionic conductivity and electronic conductivity, the Nyquist plots have been fitted and the corresponding equivalent electrical circuit diagram is shown in Figure S8. The electrode resistances obtained from equivalent circuits by fitting the experimental data of HTCs electrodes are shown in Table S4. According to the fitted data, the Rct of HTC-1000 and HTC-1100 are 5.7 Ω and 2.5 Ω, which are much lower than that of HTC-0 16.2. It shows that secondary high-temperature treatment can effectively improve electronic conductivity. As shown in Figure S9, the slope between Z' and $\omega^{-1/2}$ represents the Warburg coefficient (σ_w). The slopes of HTC-0, HTC-1000, and HTC-1100 are 2.9, 2.7, and 4.1, respectively, indicating the highest diffusion coefficient of HTC-1000. This is because after secondary high-temperature treatment, the porosity of HTC-1000 is expanded and more mesoporous is formed. However, the high temperature of HTC-1100 leads to the collapse of pore structures, resulting in lower ion transport efficiency. To further verify the conductivity of the samples, we measured the I-V curves of the sample HTCs with an electrochemical workstation. From Figure 6b,c, it can be found that the conductivities of HTC-0, HTC-1000, and HTC-1100 are 58.4, 68.6, and 97.4 S m^{-1}, respectively, which is consistent with the R_{ct} results.

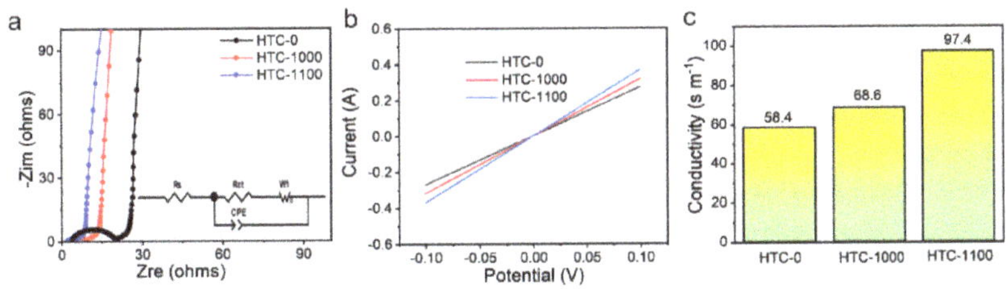

Figure 6. (**a**) Nyquist plots, (**b**) I–V curves, (**c**) conductivity of HTCs.

4. Conclusions

In conclusion, we have fabricated a kind of porous carbon with both high SSAs and excellent conductivity via the secondary high-temperature treatment strategy. Firstly, the enhanced SSAs and micropores of the porous carbon material provide more reactive active sites, leading to increased specific capacitance. Moreover, the more complete carbon six-numbered rings due to oxygen elimination effectively improve the conductivity of the material, which results in better rate performance. In addition, rapid ion transport has been achieved and benefited from the elimination of oxygen. As a result, the synthesized anode displays a high specific capacitance of 199.2 F g^{-1} at 1 A g^{-1}, and the assembled symmetrical supercapacitor has an outstanding energy density of 43.2 Wh kg^{-1} at a power density of 625.0 W kg^{-1}. The strategy can provide novel enlightenment for designing porous carbon in the energy storage field.

Supplementary Materials: The following supporting information can be downloaded at: https://www.mdpi.com/article/10.3390/batteries10010005/s1, Figure S1: SEM images of (a,b) HTC-0, (c,d) HTC-1000, (e,f) HTC-1100; Figure S2: The EDS spectrum of HTC-1000; Figure S3. (a) TEM images of HTC-1000 and (b) HRTEM images of HTC-1000; Figure S4: Raman spectra of (a) HTC-0, (b) HTC-1100; Figure S5: CV curves at different scan rates and galvanostatic charge–discharge curves at different current densities of the (a,b) HTC-0, (c,d) HTC-1000, (e,f) HTC-1100; Figure S6: Gravimetric capacitance of YP-50F; Figure S7: SEM images of after cycling HTC-1000; Figure S8: (a) Nyquist plots and (b) the corresponding equivalent electrical circuit; Figure S9: line plots between Z′ and $\omega^{-1/2}$. Table S1: Texture properties of HTCs measured by N_2 adsorption–desorption isotherms; Table S2. Peaks attribution in FTIR spectra of porous carbons; Table S3: Specific capacitance of HTC-0, HTC-1000, HTC-1100, and YP-50F; Table S4: Fitting experimental resistances of HTCs electrodes; Table S5: Electrochemical performance comparison between the HTC-1000//HTC-1000 symmetric device and other devices.

Author Contributions: Conceptualization, Z.F.; methodology, validation, writing—original draft preparation, W.C.; formal analysis, writing—review and editing, C.C.; writing—review and editing, G.W.; writing—review and editing, Z.Q.; software, Q.L.; resources, T.W.; Visualization, Z.X.; Visualization, Z.L.; Visualization, B.Q.; Visualization, K.C. All authors have read and agreed to the published version of the manuscript.

Funding: This research was funded by the National Natural Science Foundation of China, grant number 52202311, 51972342, 51872056, Shandong Provincial Natural Science Foundation, China, grant number ZR2020QE050, Fundamental Research Funds for the Central Universities, grant number

20CX06024A, Key Basic Research Projects of Natural Science Foundation of Shandong province, grant number ZR2019ZD51, Taishan Scholar Project of Shandong Province, grant number ts20190922.

Data Availability Statement: All created data are contained in the manuscript and the Supporting Information.

Conflicts of Interest: The authors declare no conflicts of interest.

References

1. Zhu, Y.W.; Murali, S.; Stoller, M.D.; Ganesh, K.J.; Cai, W.W.; Ferreira, P.J.; Pirkle, A.; Wallace, R.M.; Cychosz, K.A.; Thommes, M.; et al. Carbon-based supercapacitors produced by activation of graphene. *Science* **2011**, *332*, 1537–1541. [CrossRef] [PubMed]
2. Yu, Z.N.; Tetard, L.; Zhai, L.; Thomas, J. Supercapacitor electrode materials: Nanostructures from 0 to 3 dimensions. *Energy Environ. Sci.* **2015**, *8*, 702–730. [CrossRef]
3. Gogotsi, Y.; Nikitin, A.; Ye, H.H.; Zhou, W.; Fischer, J.E.; Yi, B.; Foley, H.C.; Barsoum, M.W. Nanoporous carbide-derived carbon with tunable pore size. *Nat. Mater.* **2003**, *2*, 591–594. [CrossRef] [PubMed]
4. Fan, Z.J.; Liu, Y.; Yan, J.; Ning, G.Q.; Wang, Q.; Wei, T.; Zhi, L.J.; Wei, F. Template-directed synthesis of pillared-porous carbon nanosheet architectures: High-performance electrode materials for supercapacitors. *Adv. Energy Mater.* **2012**, *2*, 419–424. [CrossRef]
5. Pang, Z.Y.; Li, G.S.; Xiong, X.L.; Ji, L.; Xu, Q.; Zou, X.L.; Lu, X.G. Molten salt synthesis of porous carbon and its application in supercapacitors: A review. *J. Energy Chem.* **2021**, *61*, 622–640. [CrossRef]
6. Tian, W.H.; Zhu, J.Y.; Dong, Y.; Zhao, J.; Li, J.; Guo, N.N.; Lin, H.; Zhang, S.; Jia, D.Z. Micelle-induced assembly of graphene quantum dots into conductive porous carbon for high rate supercapacitor electrodes at high mass loadings. *Carbon* **2020**, *161*, 89–96. [CrossRef]
7. Zhang, S.; Zhu, J.Y.; Qing, Y.; Wang, L.X.; Zhao, J.; Li, J.; Tian, W.H.; Jia, D.Z.; Fan, Z.J. Ultramicroporous carbons puzzled by graphene quantum dots: Integrated high gravimetric, volumetric, and areal capacitances for supercapacitors. *Adv. Funct. Mater.* **2018**, *28*, 10. [CrossRef]
8. Li, P.; Li, H.; Han, D.L.; Shang, T.X.; Deng, Y.Q.; Tao, Y.; Lv, W.; Yang, Q.H. Packing activated carbons into dense graphene network by capillarity for high volumetric performance supercapacitors. *Adv. Sci.* **2019**, *6*, 8. [CrossRef]
9. Li, Q.Q.; Jiang, Y.T.; Jiang, Z.M.; Zhu, J.Y.; Gan, X.M.; Qin, F.W.; Tang, T.T.; Luo, W.X.; Guo, N.N.; Liu, Z.; et al. Ultrafast pore-tailoring of dense microporous carbon for high volumetric performance supercapacitors in organic electrolyte. *Carbon* **2022**, *191*, 19–27. [CrossRef]
10. Wang, Q.; Yan, J.; Fan, Z.J. Carbon materials for high volumetric performance supercapacitors: Design, progress, challenges and opportunities. *Energy Environ. Sci.* **2016**, *9*, 729–762. [CrossRef]
11. Chang, P.P.; Wang, C.Y.; Kinumoto, T.; Tsumura, T.; Chen, M.M.; Toyoda, M. Frame-filling C/C composite for high-performance EDLCs with high withstanding voltage. *Carbon* **2018**, *131*, 184–192. [CrossRef]
12. Leng, C.Y.; Zhao, Z.B.; Song, Y.Z.; Sun, L.L.; Fan, Z.J.; Yang, Y.Z.; Liu, X.G.; Wang, X.Z.; Qiu, J.S. 3D carbon frameworks for ultrafast charge/discharge rate supercapacitors with high energy-power density. *Nano-Micro Lett.* **2021**, *13*, 11. [CrossRef] [PubMed]
13. Zhu, X.Q.; Yu, S.; Xu, K.T.; Zhang, Y.; Zhang, L.M.; Lou, G.B.; Wu, Y.T.; Zhu, E.H.; Chen, H.; Shen, Z.H.; et al. Sustainable activated carbons from dead ginkgo leaves for supercapacitor electrode active materials. *Chem. Eng. Sci.* **2018**, *181*, 36–45. [CrossRef]
14. Gomibuchi, E.; Ichikawa, T.; Kimura, K.; Isobe, S.; Nabeta, K.; Fujii, H. Electrode properties of a double layer capacitor of nano-structured graphite produced by ball milling under a hydrogen atmosphere. *Carbon* **2006**, *44*, 983–988. [CrossRef]
15. Zhong, C.; Deng, Y.D.; Hu, W.B.; Qiao, J.L.; Zhang, L.; Zhang, J.J. A review of electrolyte materials and compositions for electrochemical supercapacitors. *Chem. Soc. Rev.* **2015**, *44*, 7484–7539. [CrossRef] [PubMed]
16. Peng, H.R.; Yao, B.; Wei, X.J.; Liu, T.Y.; Kou, T.Y.; Xiao, P.; Zhang, Y.H.; Li, Y. Pore and heteroatom engineered carbon foams for supercapacitors. *Adv. Energy Mater.* **2019**, *9*, 9. [CrossRef]
17. Jiang, Y.T.; Li, J.; Jiang, Z.M.; Shi, M.J.; Sheng, R.; Liu, Z.; Zhang, S.; Cao, Y.L.; Wei, T.; Fan, Z.J. Large-surface-area activated carbon with high density by electrostatic densification for supercapacitor electrodes. *Carbon* **2021**, *175*, 281–288. [CrossRef]
18. Pérez, C.R.; Yeon, S.H.; Ségalini, J.; Presser, V.; Taberna, P.L.; Simon, P.; Gogotsi, Y. Structure and electrochemical performance of carbide-derived carbon nanopowders. *Adv. Funct. Mater.* **2013**, *23*, 1081–1089. [CrossRef]
19. Wu, L.Q.; Li, W.W.; Li, P.; Liao, S.T.; Qiu, S.Q.; Chen, M.L.; Guo, Y.F.; Li, Q.; Zhu, C.; Liu, L.W. Powder, paper and foam of few-layer graphene prepared in high yield by electrochemical intercalation exfoliation of expanded graphite. *Small* **2014**, *10*, 1421–1429. [CrossRef]
20. Zheng, Q.W.; Li, X.M.; Yang, Q.Z.; Li, C.M.; Liu, G.Q.; Wang, Y.C.; Sun, P.C.; Tian, H.M.; Wang, C.H.; Chen, X.L.; et al. High performance solid-state supercapacitors based on highly conductive organogel electrolyte at low temperature. *J. Power Sources* **2022**, *524*, 10. [CrossRef]
21. Wang, Q.; Liu, F.Y.; Jin, Z.Y.; Qiao, X.R.; Huang, H.C.; Chu, X.; Xiong, D.; Zhang, H.T.; Liu, Y.; Yang, W.Q. Hierarchically divacancy defect building dual-activated porous carbon fibers for high-performance energy-storage devices. *Adv. Funct. Mater.* **2020**, *30*, 8. [CrossRef]

22. Vonlanthen, D.; Lazarev, P.; See, K.A.; Wudl, F.; Heeger, A.J. A stable polyaniline-benzoquinone-hydroquinone supercapacitor. *Adv. Mater.* **2014**, *26*, 5095–5100. [CrossRef]
23. Heon, M.; Lofland, S.; Applegate, J.; Nolte, R.; Cortes, E.; Hettinger, J.D.; Taberna, P.L.; Simon, P.; Huang, P.H.; Brunet, M.; et al. Continuous carbide-derived carbon films with high volumetric capacitance. *Energy Environ. Sci.* **2011**, *4*, 135–138. [CrossRef]
24. Wu, J.Y.; Zhang, X.; Ju, Z.Y.; Wang, L.; Hui, Z.Y.; Mayilvahanan, K.; Takeuchi, K.J.; Marschilok, A.C.; West, A.C.; Takeuchi, E.S.; et al. From fundamental understanding to engineering design of high-performance thick electrodes for scalable energy-storage systems. *Adv. Mater.* **2021**, *33*, 16. [CrossRef]
25. Liu, H.Y.; Xu, T.; Cai, C.Y.; Liu, K.; Liu, W.; Zhang, M.; Du, H.S.; Si, C.L.; Zhang, K. Multifunctional superelastic, superhydrophilic, and ultralight nanocellulose-based composite carbon aerogels for compressive supercapacitor and strain sensor. *Adv. Funct. Mater.* **2022**, *32*, 12. [CrossRef]
26. Zornitta, R.L.; Barcelos, K.M.; Nogueira, F.G.E.; Ruotolo, L.A.M. Understanding the mechanism of carbonization and KOH activation of polyaniline leading to enhanced electrosorption performance. *Carbon* **2020**, *156*, 346–358. [CrossRef]
27. Chmiola, J.; Yushin, G.; Gogotsi, Y.; Portet, C.; Simon, P.; Taberna, P.L. Anomalous increase in carbon capacitance at pore sizes less than 1 nanometer. *Science* **2006**, *313*, 1760–1763. [CrossRef]
28. Huang, C.; Tang, Q.L.; Feng, Q.S.; Li, Y.H.; Xu, Y.L.; Zhang, Y.; Hu, A.; Zhang, S.Y.; Deng, W.N.; Chen, X.H. Achieving ultrahigh volumetric performance of graphene composite films by an outer-inner dual space utilizing strategy. *J. Mater. Chem. A* **2020**, *8*, 9661–9669. [CrossRef]
29. Shao, H.; Wu, Y.C.; Lin, Z.F.; Taberna, P.L.; Simon, P. Nanoporous carbon for electrochemical capacitive energy storage. *Chem. Soc. Rev.* **2020**, *49*, 3005–3039. [CrossRef]
30. Wang, Y.G.; Song, Y.F.; Xia, Y.Y. Electrochemical capacitors: Mechanism, materials, systems, characterization and applications. *Chem. Soc. Rev.* **2016**, *45*, 5925–5950. [CrossRef]
31. Pan, B.Y.G.; Bai, L.; Hu, C.M.; Wang, X.P.; Li, W.S.; Zhao, F.G. Graphene-indanthrone donor-π-acceptor heterojunctions for high-performance flexible supercapacitors. *Adv. Energy Mater.* **2020**, *10*, 8. [CrossRef]
32. Sevilla, M.; Ferrero, G.A.; Diez, N.; Fuertes, A.B. One-step synthesis of ultra-high surface area nanoporous carbons and their application for electrochemical energy storage. *Carbon* **2018**, *131*, 193–200. [CrossRef]
33. Salanne, M.; Rotenberg, B.; Naoi, K.; Kaneko, K.; Taberna, P.L.; Grey, C.P.; Dunn, B.; Simon, P. Efficient storage mechanisms for building better supercapacitors. *Nat. Energy* **2016**, *1*, 10. [CrossRef]
34. Xu, L.; Shi, R.Y.; Li, H.F.; Han, C.P.; Wu, M.Y.; Wong, C.P.; Kang, F.Y.; Li, B.H. Pseudocapacitive anthraquinone modified with reduced graphene oxide for flexible symmetric all-solid-state supercapacitors. *Carbon* **2018**, *127*, 459–468. [CrossRef]
35. Yang, X.W.; Cheng, C.; Wang, Y.F.; Qiu, L.; Li, D. Liquid-mediated dense integration of graphene materials for compact capacitive energy storage. *Science* **2013**, *341*, 534–537. [CrossRef]
36. Wang, C.J.; Liu, F.; Chen, J.S.; Yuan, Z.W.; Liu, C.; Zhang, X.S.; Xu, M.Y.; Wei, L.; Chen, Y. A graphene-covalent organic framework hybrid for high-performance supercapacitors. *Energy Storage Mater.* **2020**, *32*, 448–457. [CrossRef]
37. Yang, B.J.; Chen, J.T.; Lei, S.L.; Guo, R.S.; Li, H.X.; Shi, S.Q.; Yan, X.B. Spontaneous growth of 3D framework carbon from sodium citrate for high energy- and power-density and long-life sodium-ion hybrid capacitors. *Adv. Energy Mater.* **2018**, *8*, 11. [CrossRef]
38. Peng, L.; Hung, C.T.; Wang, S.W.; Zhang, X.M.; Zhu, X.H.; Zhao, Z.W.; Wang, C.Y.; Tang, Y.; Li, W.; Zhao, D.Y. Versatile nanoemulsion assembly approach to synthesize functional mesoporous carbon nanospheres with tunable pore sizes and architectures. *J. Am. Chem. Soc.* **2019**, *141*, 7073–7080. [CrossRef]
39. Sutarsis; Patra, J.; Su, C.Y.; Li, J.; Bresser, D.; Passerini, S.; Chang, J.K. Manipulation of nitrogen-heteroatom configuration for enhanced charge-storage performance and reliability of nanoporous carbon electrodes. ACS Appl. *Mater. Interfaces* **2020**, *12*, 32797–32805. [CrossRef]
40. Liang, Y.R.; Liang, F.X.; Zhong, H.; Li, Z.H.; Fu, R.W.; Wu, D.C. An advanced carbonaceous porous network for high-performance organic electrolyte supercapacitors. *J. Mater. Chem. A* **2013**, *1*, 7000–7005. [CrossRef]
41. Yang, I.; Yoo, J.; Kwon, D.; Choi, D.; Kim, M.S.; Jung, J.C. Improvement of a commercial activated carbon for organic electric double-layer capacitors using a consecutive doping method. *Carbon* **2020**, *160*, 45–53. [CrossRef]
42. Wu, X.; Liu, R.; Zhao, J.; Fan, Z. Advanced carbon materials with different spatial dimensions for supercapacitors. *Nano Mater. Sci.* **2021**, *3*, 241–247. [CrossRef]
43. Fan, Z.J.; Wang, K.; Wei, T.; Yan, J.; Song, L.P.; Shao, B. An environmentally friendly and efficient route for the reduction of graphene oxide by aluminum powder. *Carbon* **2010**, *48*, 1686–1689. [CrossRef]
44. Hsu, H.L.; Miah, M.; Saha, S.K.; Chen, J.H.; Chen, L.C.; Hsu, S.Y. Three-dimensional bundle-like multiwalled carbon nanotubes composite for supercapacitor electrode application. *Mater. Today Chem.* **2021**, *22*, 10. [CrossRef]
45. Guo, W.; Yu, C.; Li, S.F.; Qiu, J.S. Toward commercial-level mass-loading electrodes for supercapacitors: Opportunities, challenges and perspectives. *Energy Environ. Sci.* **2021**, *14*, 576–601. [CrossRef]
46. Guo, J.B.; Li, L.; Luo, J.; Gong, W.N.; Pan, R.; He, B.; Xu, S.H.; Liu, M.N.; Wang, Y.J.; Zhang, B.H.; et al. Polypyrrole-assisted nitrogen doping strategy to boost vanadium dioxide performance for wearable nonpolarity supercapacitor and aqueous Zinc-Ion battery. *Adv. Energy Mater.* **2022**, *12*, 12. [CrossRef]
47. Raj, C.J.; Manikandan, R.; Thondaiman, P.; Sivakumar, P.; Savariraj, A.D.; Cho, W.J.; Kim, B.C.; Jung, H. Sonoelectrochemical exfoliation of graphene in various electrolytic environments and their structural and electrochemical properties. *Carbon* **2021**, *184*, 266–276. [CrossRef]

48. Li, F.F.; Wang, X.L.; Sun, R.C. A metal-free and flexible supercapacitor based on redox-active lignosulfonate functionalized graphene hydrogels. *J. Mater. Chem. A* **2017**, *5*, 20643–20650. [CrossRef]
49. Long, S.S.; Feng, Y.C.; He, F.L.; Zhao, J.Z.; Bai, T.; Lin, H.B.; Cai, W.L.; Mao, C.W.; Chen, Y.H.; Gan, L.H.; et al. Biomass-derived, multifunctional and wave-layered carbon aerogels toward wearable pressure sensors, supercapacitors and triboelectric nanogenerators. *Nano Energy* **2021**, *85*, 11. [CrossRef]
50. Xie, K.; Qin, X.T.; Wang, X.Z.; Wang, Y.N.; Tao, H.S.; Wu, Q.; Yang, L.J.; Hu, Z. Carbon nanocages as supercapacitor electrode materials. *Adv. Mater.* **2012**, *24*, 347–352. [CrossRef]
51. Yang, Y.Y.; Ma, W.X.; Zhu, H.; Meng, H.X.; Wang, C.J.; Ma, F.Q.; Hu, Z.A. Graphene covalently functionalized with 2,6-diaminoanthraquinone (DQ) as a high performance electrode material for supercapacitors. *New J. Chem.* **2020**, *44*, 16821–16830. [CrossRef]
52. Tian, M.; Wu, J.W.; Li, R.H.; Chen, Y.L.; Long, D.H. Fabricating a high-energy-density supercapacitor with asymmetric aqueous redox additive electrolytes and free-standing activated-carbon-felt electrodes. *Chem. Eng. J.* **2019**, *363*, 183–191. [CrossRef]
53. Isikli, S.; Díaz, R. Substrate-dependent performance of supercapacitors based on an organic redox couple impregnated on carbon. *J. Power Sources* **2012**, *206*, 53–58. [CrossRef]
54. Xiong, C.L.; Zou, Y.B.; Peng, Z.Y.; Zhong, W.B. Synthesis of morphology-tunable electroactive biomass/graphene composites using metal ions for supercapacitors. *Nanoscale* **2019**, *11*, 7304–7316. [CrossRef]
55. An, N.; An, Y.F.; Hu, Z.G.; Guo, B.S.; Yang, Y.Y.; Lei, Z.Q. Graphene hydrogels non-covalently functionalized with alizarin: An ideal electrode material for symmetric supercapacitors. *J. Mater. Chem. A* **2015**, *3*, 22239–22246. [CrossRef]
56. Jaidev; Ramaprabhu, S. Poly(p-phenylenediamine)/graphene nanocomposites for supercapacitor applications. *J. Mater. Chem.* **2012**, *22*, 18775–18783. [CrossRef]
57. Tian, W.Q.; Gao, Q.M.; Tan, Y.L.; Yang, K.; Zhu, L.H.; Yang, C.X.; Zhang, H. Bio-inspired beehive-like hierarchical nanoporous carbon derived from bamboo-based industrial by-product as a high performance supercapacitor electrode material. *J. Mater. Chem. A* **2015**, *3*, 5656–5664. [CrossRef]
58. Chao, Y.Z.; Chen, S.B.; Xiao, Y.C.; Hu, X.J.; Lu, Y.; Chen, H.Q.; Xin, S.X.; Bai, Y.X. Ordinary filter paper-derived hierarchical pore structure carbon materials for supercapacitor. *J. Energy Storage* **2021**, *35*, 7. [CrossRef]
59. Sankari, M.S.; Vivekanandhan, S. Jatropha oil cake based activated carbon for symmetric supercapacitor application: A comparative study on conventional and hydrothermal carbonization processes. *ChemistrySelect* **2020**, *5*, 1375–1384. [CrossRef]
60. Li, H.; Zhang, X.F.; Wang, X.T.; Zhang, J.Y.; Yang, Y.K. One-pot solvothermal incorporation of graphene into chain-engineered polyquinones for metal-free supercapacitors. *Chem. Commun.* **2020**, *56*, 11191–11194. [CrossRef]
61. Xue, Q.; Gan, H.B.; Huang, Y.; Zhu, M.S.; Pei, Z.X.; Li, H.F.; Deng, S.Z.; Liu, F.; Zhi, C.Y. Boron element nanowires electrode for supercapacitors. *Adv. Energy Mater.* **2018**, *8*, 8. [CrossRef]
62. Xu, Y.X.; Lin, Z.Y.; Zhong, X.; Huang, X.Q.; Weiss, N.O.; Huang, Y.; Duan, X.F. Holey graphene frameworks for highly efficient capacitive energy storage. *Nat. Commun.* **2014**, *5*, 8. [CrossRef] [PubMed]

batteries MDPI

Article

Electrochemical Performance of Pre-Modified Birch Biochar Monolith Supercapacitors by Ferric Chloride and Ferric Citrate

Ziyue Song, Tianjie Feng, Donald W. Kirk and Charles Q. Jia *

The Green Technology Lab, Department of Chemical Engineering and Applied Chemistry, University of Toronto, Toronto, ON M5S 3E5, Canada; ziyue.song@alumni.utoronto.ca (Z.S.); tianjie.feng@mail.utoronto.ca (T.F.); don.kirk@utoronto.ca (D.W.K.)

* Correspondence: cq.jia@utoronto.ca

Abstract

This study investigated the electrochemical properties of supercapacitors by pre-modifying thick birch biochar monoliths with $FeCl_3$ or $C_6H_5FeO_7$ solutions prior to wood pyrolysis. The pre-modification introduced iron species to the surface, promoting the specific surface area, charge-stored species, and surface functionalities, which enhanced the gravimetric capacitance. X-ray diffraction confirmed the successful loading of Fe_3O_4 and Fe. SEM implied the wider distribution of iron-rich particulates and porous carbon via self-pyrolysis on the biochar surface modified with 1.0 M $C_6H_5FeO_7$. Contact angle measurements demonstrated the enhanced wettability of the biochar surfaces following pre-modification, with the $C_6H_5FeO_7$-modified samples exhibiting superior wettability compared to the other groups. The gravimetric capacitance of the supercapacitor was dramatically promoted and reached 210 F/g and 219 F/g, respectively, when modified with 1.0M $C_6H_5FeO_7$ and 1.0 M $FeCl_3$ at a 5 mA/g current density. Compared to the birch biochar modified with 1.0 M $FeCl_3$, the 1.0 M $C_6H_5FeO_7$ had a higher current response peak and capacitive behavior in the CV analysis, demonstrated better ion diffusion capacity, and had lower charge-transfer resistance in the EIS results. But, a slight irreversible process on the electrode of the 1.0 M $C_6H_5FeO_7$ group led to a lower level of the supercapacitor capacitance retention. The results using ferric solution pre-impregnation show how iron species doping can improve capacitance behavior, providing a feasible scheme for the modification of thick biochar monolith.

Keywords: biochar monolith; supercapacitors; ferric chloride; ferric citrate; iron species

1. Introduction

Cellulose, abundantly present in natural biomass, is interconnected through hydrogen bonding, giving rise to an intricate tubular bundle structure within living organisms [1]. Biomass that is characterized by a highly developed tubular bundle structure when being pyrolyzed yields a biochar possessing a well-defined pore architecture and a substantial specific surface area. This biochar is a stable carbon-rich product offering a secure and long-term carbon storage solution within the environment [2]. Its porosity minimizes ion diffusion resistance, while the combination of micropores and mesopores provides abundant active sites, facilitating efficient electron and ion transport within the biochar [3]. Consequently, biochar demonstrates favorable physical and electrochemical adsorption properties including valuable attributes encompassing a high surface area, porosity, surface functionality, and sorption capacity, rendering it suitable for utilization as an electrode in supercapacitors for effective energy storage.

The source of the biochar is important. For example, some wood species, such as balsa, are characterized by relatively sparse tubular bundle structures and thus exhibit a lower bulk density before and after pyrolysis [4,5]. This can reduce their charge storage capacity compared to other wood types, such as maple and birch, and have been proven to be effective electrode materials [6]. Recognizing the need to increase charge storage capacity, researchers have developed numerous biochar modification approaches, including physical, chemical, and biological modifications.

The most common physical modification includes the steam gasification to increase the surface areas and improve the surface morphologies through the surface corrosion of the biochar and syngas release [7]. Biological treatment is based on the principle that microbial surfaces are rich in functional groups (amino, carbonyl, carboxyl, and hydroxyl groups), which has improved the sorption properties of biochar through surface grafting [8]. Chemical modification is among the most widely used technique to enhance material properties. It primarily includes acid treatment, alkaline treatment, oxidizing agent modification, and the use of metal salts or oxidizing agents [9]. Compared to relatively mild reagents such as transition metal salt solutions, strong acid and base modification agents such as H_2SO_4 or NaOH lead to secondary pollution and require stringent handling protocols [10].

Transition metals oxides (TMOs) have been widely investigated as electrode materials because of their exceptional properties, such as structural flexibility, conductivity, environmental and cost friendliness, and high specific capacitance [11]. The metal salt solution produces the corresponding oxides during pyrolysis, and the innovative integration of biochar with magnetic transition metals and TMOs to produce magnetic biochar enhances catalytic efficiency by improving porosity, surface area, and active sites [11]. Zhang et al. [12] clearly indicated that the formation of iron oxide in granular-size porous biochar pretreated with impregnated $FeCl_3$ solution generated carbonyl and carboxyl groups after pyrolysis. Meanwhile, magnetic biochar doped with iron and iron oxides exhibited rougher contact surfaces compared to pristine biochar [12]. Therefore, $FeCl_3$ was regarded as the one of modifier agents in this experiment.

Compared to $FeCl_3$, citrate, an organic acid salt with low molecular weight, is naturally abundant, eco-friendly in terms of green chemistry, and exhibits properties similar to those of citric acid [13,14]. Citrate salt can provide the required acid environment, which can introduce acidic functional groups, such as carboxylic groups, onto the surface of carbonized materials. This modified biochar exhibits impressive porosity, with the multiple functional groups increasing the sorption affinity and capacity for electrolyte ions by promoting charge exchange and surface complexation through newly formed active sites [15,16]. Ferric citrate is now widely used as an effective modification agent, with the advantages of being inexpensive and having the capability to improve biochar's physicochemical and surface chemical properties, including specific surface area, pore size, molecular weight, cation exchange capacity, and the quantity and variety of functional groups [17]. In previous studies, citrate was predominantly used as a post-modifier to enhance biochar properties following pyrolysis. For instance, Wang et al. [18] demonstrated the use of ammonium citrate to post-modify ground biochar particles, creating a highly efficient adsorbent for trivalent lanthanum.

In this experimental study, we took a different approach by focusing on the pre-modification of the raw materials before pyrolysis. Meanwhile, we focused on the idea of pre-modification through metal salt solution impregnation of thick biochar monolith compared to general powdery or granular biochar. The enhancement in birch wood biochar monolith's electrochemical performance for supercapacitor applications was pursued through the wood monolith impregnation of a ferric solution including the traditional modification agent $FeCl_3$ and eco-friendly $C_6H_5FeO_7$. The differences in the material and electrochemical characteristics between these two ferric pre-modifiers were observed as well.

2. Materials and Methods

2.1. Sample Preparation

Birch, predominantly found in the eastern part of Canada, widely thrives in cool, moist, and well-drained soil, especially in mixed hardwood forests. Birch turning squares, obtained from KJP Select Hardwood, Ottawa, ON, Canada, were cut into identical rectangular pieces (20 × 10 × 7 mm) and dried at 100 °C in an oven for 3 h until no weight change was measured. Ferric chloride power and ferric citrate powder were obtained from Sigma-Aldrich, St. Louis, MO, USA. The powder was weighed and dissolved in deionized water to prepare sample solutions with three distinct concentrations: 0.5, 0.7, and 1.0 M. The pieces were then submerged in the prepared ferric citrate or ferric chloride solution at 90 °C under stirring for 48 h. After submersion, the samples were allowed to air dry for 3 h before the final drying in the oven at 100 °C for another 3 h. KOH pellets, obtained from Sigma-Aldrich, were dissolved in deionized water to prepare 4 M KOH solutions.

The birch pieces after the pretreatments are shown in Figure 1.

Figure 1. Photographs of wood pieces; from left to right: (1) $FeCl_3$-treated; (2) $C_6H_5FeO_7$-treated; (3) untreated birch.

2.2. Pyrolysis of Biochar Monolith and Cell Assembly

One untreated wood piece and two wood pieces treated with different ferric solutions were held in a metal mesh basket and lowered into the center of a vertical tube quartz furnace (125 cm in length and 10 cm in diameter) for pyrolysis. The heating parameters settings were obtained from Byrne and Nagle's U.S. patent [19] to avoid crack formation in the biochar monolith. Low heating rates were applied in the proposed pyrolysis to avoid the explosive evaporation of the wood's moisture and reach a stable and high carbon yield [19]. Grade 5.0 nitrogen gas was used at a flow rate of 400 mL/min throughout the pyrolysis to achieve anaerobic conditions. The pyrolysis schedule was divided into six continuous stages with holding temperature points of 90, 200, 400, 600, 800, and 1000 °C, with associated predefined heating rates of 8.3, 2.5, 1.3, 3.3, 6.7, and 6.7 °C/min, respectively, as presented in Table 1. After completing the 6th step and reaching a temperature of 1000 °C, the system was cooled to room temperature with nonstop nitrogen gas supply for approximately 18 h.

Table 1. The conditions of the 6 pyrolysis steps.

Step No.	Heating Rate (°C/min)	Final Temp (°C)	Holding Time (min)
1	8.3	90	180
2	2.5	200	6
3	1.3	400	6
4	3.3	600	6
5	6.7	800	6
6	6.7	1000	6

After cooling to room temperature, each thick biochar piece was cut into two pieces, and these pieces were ground to the same weight (±0.001 g), a similar square cross-section, and the same thickness of 1.7 mm (±0.05 mm) using 100-grit and 200-grit sandpaper. The ground biochar pieces were ultrasonically cleaned in 95% ethanol to remove the debris from the material's surface. The pieces were then boiled in 4 M KOH for 1 h to increase their wettability and ion transport ability.

The groups of samples were labeled as 'Untreated Birch' for the unmodified birch biochar monolith; '$FeCl_3$-0.5 M', '$FeCl_3$-0.7 M', and '$FeCl_3$-1.0 M' for the birch biochar monolith pretreated with the corresponding $FeCl_3$ solution; 'FC-0.5 M', 'FC-0.7 M', and 'FC-1.0 M' for the birch biochar monolith pretreated with the corresponding $C_6H_5FeO_7$ solution.

A cell was assembled based on the schematic drawing shown in Figure 2. A pair of birch biochar electrodes was carefully aligned in parallel with each other and separated with a fibrous polysulfone separator 1 mm thick. The middle porous separator and two nickel wire meshes on the sides completely covered the cross -section of the biochar electrode. The upper side of the mesh was stripped of the transverse nickel wire structure and rolled into an extended wire. The extended wire was connected to the poles of the electrochemical measurement station. The entire system was tightly held in place by screwed Lucite plastic plates (5 × 5 × 0.5 cm) submerged in a 4 M KOH solution as the electrolyte. There were four screws located at the corners of the plastic plates that passed through the plates. Nuts held the screws on the other plate.

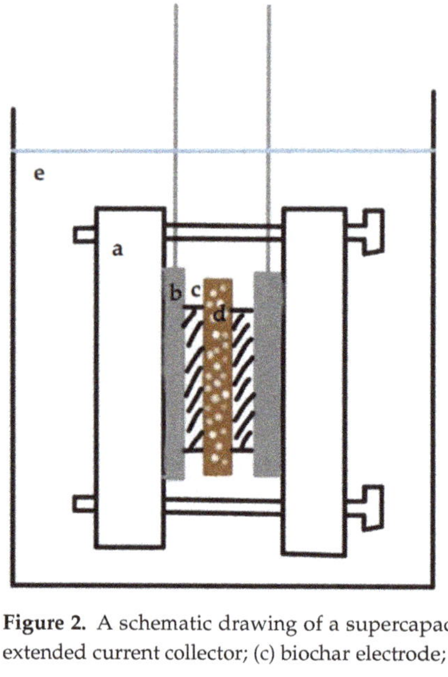

Figure 2. A schematic drawing of a supercapacitor: (a) bolted plastic plate; (b) nickel mesh with extended current collector; (c) biochar electrode; (d) porous separator film; (e) 4 M KOH electrolyte.

2.3. Characterizations of Biochar Material

2.3.1. X-Ray Diffraction

The composition of the birch biochar, both with and without pre-modification, was analyzed using X-ray diffraction (XRD). The untreated and birch biochar pieces treated with 1.0 M $FeCl_3$ and 1.0 M $C_6H_5FeO_7$ were powered for X-ray diffraction using a Rigaku MiniFlex 600 (Rigaku, Tokyo, Japan). The diffraction pattern was analyzed by comparison with the Joint Committee on Powder Diffraction Standards (JCPDS) cards from the International Centre for Diffraction Data.

2.3.2. Scanning Electron Microscopy Analysis

The surface morphology of the biochar samples was analyzed using scanning electron microscopy (Hitachi SU3500 SEM, Hitachi, Tokyo, Japan) at high resolution. Two specific locations chosen on the material surface were investigated by Energy-Dispersive X-ray Spectroscopy (EDX) to determine their composition using AZtec software Version 6.1.

2.3.3. Surface Contact Angle

The contact angle of the liquid on the biochar surface indicated the wettability of the liquid to the biochar modified with ferric solution. The solid–liquid interface interactions of the biochar surface modified with $FeCl_3$ and $C_6H_5FeO_7$ were obtained by measuring the contact angle. Samples of untreated, $FeCl_3$-1.0 M, and FC-1.0 M birch were placed on a horizontal measuring platform, while the capture camera was horizontally placed on the same platform. We made the 10 wt% isopropanol solution (IPA), as the contact angle indicator, by mixing amounts of isopropanol (99.5% purity, obtained from Sigma-Aldrich) in deionized water. The liquid droplets (~5 μL) on the material's horizontal surfaces were captured, and the angles in the captured images were determined using the angle tool of ImageJ Version 1.54 j.

2.4. Electrochemical Characterization of Biochar Supercapacitor

Two nickel mesh current collectors with extended tabs, as shown in Figure 2, were connected to a Solartron Sl 1280B (Solartron Analytical, Leicester, UK) potentiostat to perform galvanostatic charge–discharge (GCD), cyclic voltammetry (CV), and electrochemical impedance spectroscopy (EIS). All the supercapacitor tests included a pre-cycle at 300 mA/g for 100 charge–discharge cycles in the 0–1.0 V potential range prior to conducting any measurement. The pre-cycle tests increased data reproducibility.

2.4.1. Galvanostatic Charge–Discharge

In this study, GCD measurements were conducted by applying a constant current ranging from 5 to 200 mA/g to explore the electrochemical behavior of the biochar supercapacitor. This measurement offered the ability to directly calculate the gravimetric capacitance (F/g), which was normalized to the mass of the active electrode material using Equation (1) [20]:

$$C_g = \frac{I}{m\frac{dV}{dt}} \tag{1}$$

where C_g is the gravimetric capacitance (F/g); I/m is the constant mass current density (A/g); dV/dt (V/s) represents the change rate of the voltage during the discharge process past the region of IR drop.

2.4.2. Cyclic Voltammetry

Cyclic voltammetry (CV) tests were conducted within a potential window of ±0.8 V, with the initial potential set at −0.8 V. The treated samples with the best GCD performance were selected for comparison with the untreated birch biochar. A standard scan rate of 20 mV/s was applied across all groups to establish a consistent baseline for comparison. The choice of this voltage range and scan rate allowed a good evaluation of the material's performance.

2.4.3. Electrochemical Impedance Spectroscopy

In this study, electrochemical impedance spectroscopy (EIS) (by Solartron Sl 1280B) was performed to investigate the electrode–electrolyte interface behavior of the biochar-based supercapacitors produced using untreated birch biochar, $FeCl_3$-1.0 M, and FC-1.0 M over the frequency range of 0.05–20,000 Hz. The Randles equivalent circuit model was used and analyzed based on the Nyquist plot. The Python 3.12 fitting method was used to export the electrolyte resistance (R_s), the charge-transfer resistance (R_{ct}), and Warburg impedance (Z_w) via impedance package analysis.

2.4.4. Supercapacitor Retention

The retention of the supercapacitors was analyzed to reflect the supercapacitor's durability for long-term use and examine the pseudo-capacitance resulting from the potentially irreversible redox reactions that occurred in some groups. Testing was conducted using the galvanostatic charge–discharge method at a current density of 0.5 A/g within a potential window of 0–1.0 V. Gravimetric capacitance values were recorded and calculated every 80 cycles over a total of 3520 cycles.

3. Results and Discussion

3.1. Recorded Mass of Materials

To investigate each material's response to chemical treatment and pyrolysis processing, the precise tracking of the mass analysis was conducted at different stages: (1) of the cut wood pieces before placement in the oven; (2) of the pieces after the first drying for 3 h; (3) of the treated pieces after submerging and drying; (4) of the biochar pieces after pyrolysis at room temperature. The mass data are reported in Tables 2 and 3. The mass analysis revealed distinct trends in the behavior of the wood samples during pre-modification.

Table 2. Mass table of $FeCl_3$ groups at 4 different stages.

Stage	0.5 M		0.7 M		1.0 M	
	Mass (g)	Change %	Mass (g)	Change %	Mass (g)	Change %
1	4.2131		4.8427		4.3845	
2	3.795	9.92%	4.3576	10.02%	3.9535	9.83%
3	3.2406	14.61%	3.6525	16.18%	3.6613	7.39%
4	1.2436	38.38%	1.4234	38.97%	1.4886	40.66%

Table 3. Mass table of $C_6H_5FeO_7$ groups at 4 different stages.

Stage	0.5 M		0.7 M		1.0 M	
	Mass (g)	Change %	Mass (g)	Change %	Mass (g)	Change %
1	4.3928		3.472		2.8594	
2	3.9934	9.09%	3.1242	10.02%	2.5744	9.97%
3	4.2121	5.48%	3.4106	9.17%	2.9142	13.20%
4	1.357	32.22%	0.912	26.74	0.8151	27.97%

For the samples treated with $C_6H_5FeO_7$, the mass increased between the two drying stages and became more pronounced as the concentration of the ferric solution increased, indicating greater mass retention at higher concentrations. The $FeCl_3$-treated samples, conversely, showed a consistent decrease in mass following modification and drying, reflecting an interaction between the $FeCl_3$ solution and the wood material. Based on the findings of Gao et al. [21], where 25 mM $FeCl_3$ hydrolysis at 160 °C for 50 min caused high lignin removal (75.5%) from switchgrass, it is likely that $FeCl_3$ plays a similar role in lignin removal from birch wood. This aligns with the observed mass reductions in the $FeCl_3$-treated samples, suggesting that the chemical modification facilitated lignin extraction from the wood. Figure 3 illustrates the color transformation of the 1.0 M $FeCl_3$ solution before and after immersing the wood pieces for 2 days, indicating large amounts of lignin in the post-modification solution.

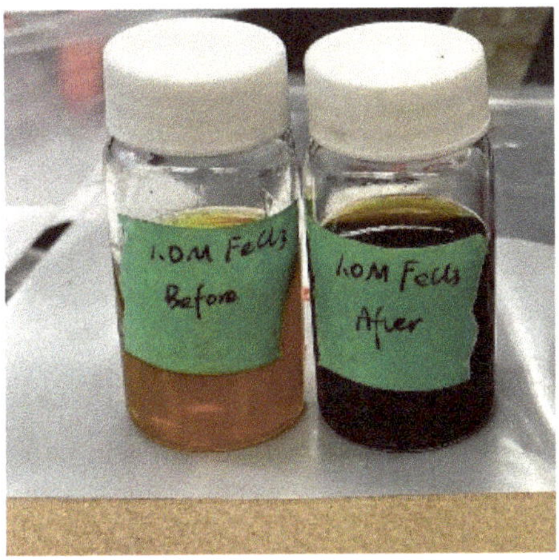

Figure 3. Color of $FeCl_3$ solution before and after pretreatment (from left to right).

3.2. *X-Ray Diffraction Results of Biochar Samples*

The X-ray diffractometer results of the untreated and treated samples are shown in Figure 4.

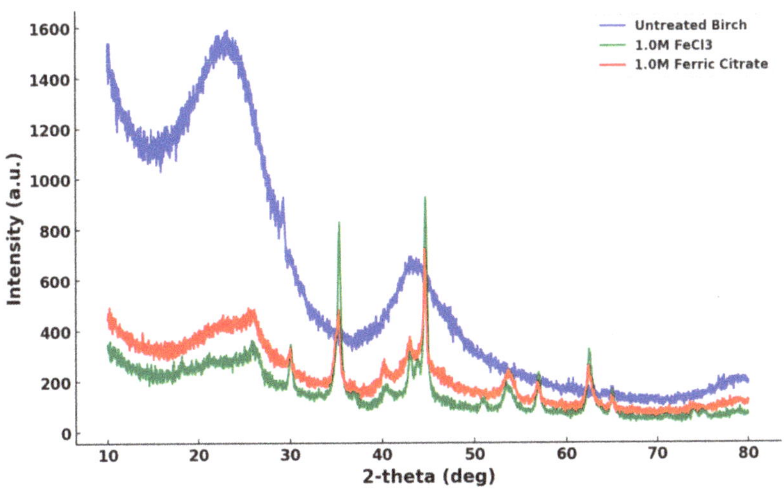

Figure 4. X-ray diffraction patterns of the untreated birch and the treated birch (treated with 1.0 M $FeCl_3$ and 1.0 M $C_6H_5FeO_7$).

The untreated birch showed two broad peaks, while the treated biochar slices subjected to ferric solution pretreatment exhibited multiple major and minor peaks in the 30–50° interval, corresponding to the formation of iron-based compounds. The raw XRD data were compared to the standard diffraction data. Figure 5 illustrates the alignment of the experimental data with the standard XRD patterns.

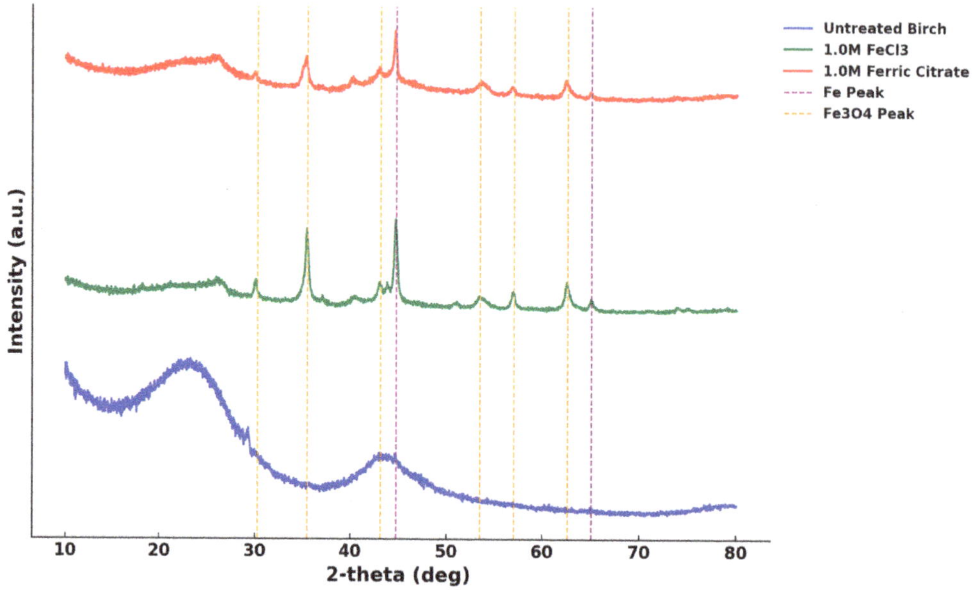

Figure 5. The analyzed XRD of biochar compared to the PDF card with the Fe and Fe_3O_4 peaks highlighted.

Peaks were observed at 30.3°, 35.3°, 43.4°, 53.5°, 56.8°, and 62.6°, well matching the standard diffraction peaks of magnetite (Fe_3O_4) of JCPDS card No. 03-0863, shown as yellow vertical lines in Figure 5, representing the (220), (311), (400), (422), (511), and (440) crystal faces of magnetite, respectively. The two diffraction peaks at 44.68° and 65.05° were consistent with the standard diffraction peaks of iron (Fe) per JCPDS card No. 06-0696, shown as purple vertical lines in Figure 5, representing the (110) and (200) crystal faces of iron. These diffraction peaks confirmed the successful loading of iron and oxides of bivalent and trivalent iron in the modified biochar. Conversely, the untreated birch biochar with broad peaks (approximately $2\theta = 24°$ and 44°), indexed as the (002) and (101) planes, supported the presence of amorphous carbon in the biochar [22].

3.3. Scanning Electron Microscopy Results of Biochar

After X-ray analysis, the untreated birch biochar, $FeCl_3$-1.0 M biochar, and FC-1.0 M biochar samples were separately rinsed using deionized water in an ultrasonic water bath for 15 min to wash off any impurities on the materials' surfaces for SEM analysis. We used 200, 500, 800, and 1000 times magnification for each sample. The 500× backscattered electron images are presented in Figure 6, which were used to evaluate the surface features.

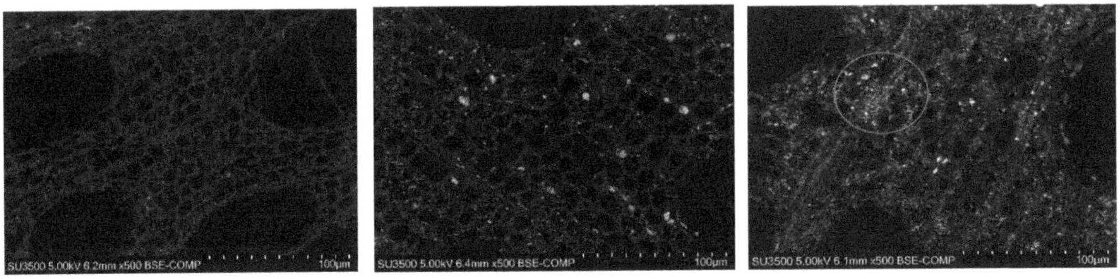

(a) (b) (c)

Figure 6. SEM images of backscattered electrons (BSEs) at 500× of (**a**) untreated birch; (**b**) $FeCl_3$-1.0 M; (**c**) FC-1.0 M.

These results indicated that the birch biochar contained numerous <10 μm micropores and fewer ~100 μm macropores. The pore arrangement in the $FeCl_3$-modified biochar retained the same fixed orientation as the unmodified birch biochar. The SEM captures of the pre-modified samples showed white granular deposits randomly embedded in the micropores for $FeCl_3$-1.0 M and FC-1.0 M. The comparison between the $FeCl_3$-1.0 M and FC-1.0 M micrographs showed that the ferric citrate modification obviously changed the morphology of the surface matrix of the carbons. The white attachments on the surface clearly indicated better adhesion compared to $FeCl_3$-1.0 M. The red-circled region in Figure 6c reveals a distinctive feature: the increased presence of smaller micropores. This structural enhancement facilitated the embedding of white particles, resulting in a brighter biochar surface and the formation of characteristic white clusters in specific areas. The findings are consistent with those in previous work by Cheng et al. [23], who demonstrated the dual functionality of citrate as a precursor and a modifier in the synthesis of porous carbons via the green self-pyrolysis of citrate. And, during pyrolysis, the metal components in citrate are transformed into inorganic carbonates, which act as efficient activators to expand the pore structure [24]. Similarly, our experimental results confirmed that ferric citrate significantly enhanced the development of carbon micropores and promoted particle embedding. This was achieved through the self-pyrolysis process and the formation of carbonates during pyrolysis, further validating the dual role of citrate as an activator and a structural modifier during carbon material synthesis.

EDX was used to analyze the elemental composition of the white inclusions in the treated sample, as shown in Figure 7. The red dots in Figure 7 show the EDX test locations on an $FeCl_3$-1.0 M sample. Red dot (a) centers on a white particle, and the spectrum demonstrates that the deposit was predominantly Fe. While red dot (b) was randomly chosen on the biochar surface beyond the white particles, and the spectrum presents the dominant peak of carbon. And, the comparison of the two EDX spectra shows that there is an oxygen signal associated with the particulate but not the biochar, consistent with the XRD analysis, which showed Fe_3O_4 as well as Fe. The elemental mapping in Figure 8 clearly shows the iron-rich particulates were attached to the carbon substrate, and the 5 μm particles were conglomerates of smaller ~1 μm particles. This indicates that there were significantly more iron-rich particulates deeper than just on the material's surface and within the pores, as shown won in the original SEM image. It is likely that some iron species were much more deeply attached inside the pores, beyond the depth of the XRD analysis.

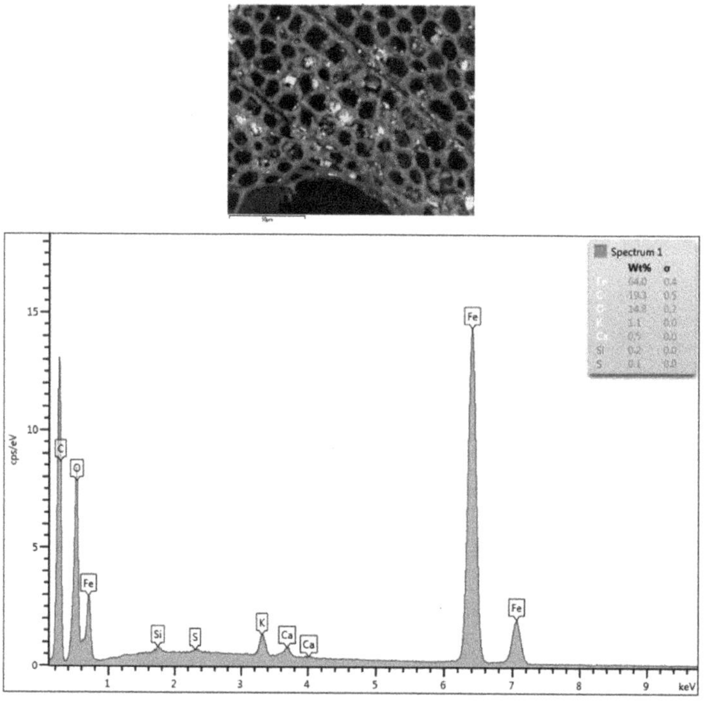

(**a**)

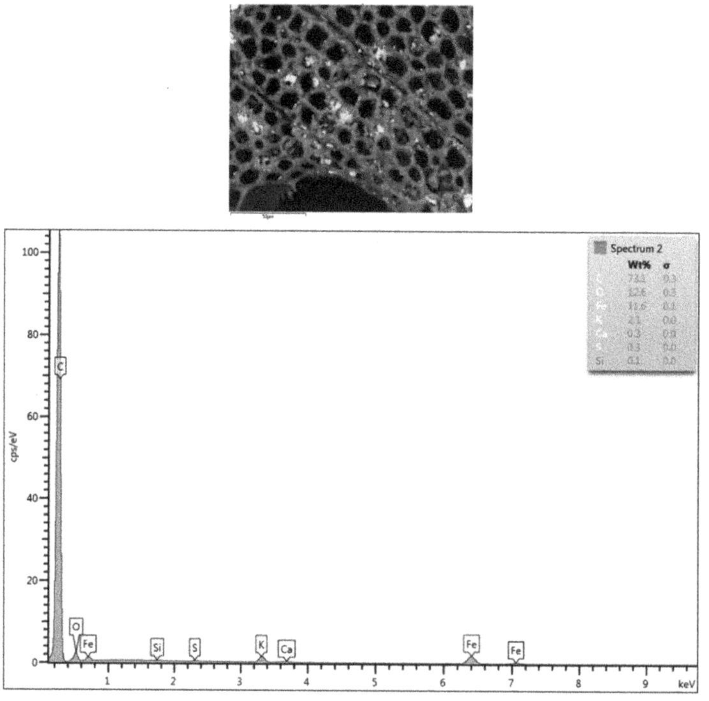

(**b**)

Figure 7. EDX results of $FeCl_3$-1.0 M biochar, with two selected spots highlighted by red dots (**a**,**b**).

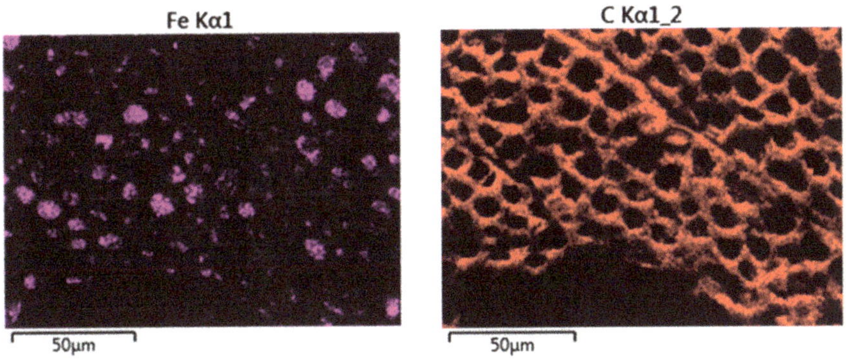

(**a**) (**b**)

Figure 8. EDX spectral imaging of ×800 SEM of $FeCl_3$-1.0 M sample (elements: (**a**) iron; (**b**) carbon).

3.4. Contact Angle Measurements of Biochar Slices

The electrolyte wetting of the treated biochar monolith was an important indicator of capacitance performance; thus, contact angle measurements were undertaken. Three samples, untreated birch biochar, $FeCl_3$-1.0 M biochar, and FC-1.0 M biochar, were boiled in KOH solution in a beaker on a hot plate at 270 °C for 1 h. The addition of the IPA to the deionized water monotonically decreased the contact angle due to the reduction in liquid surface tension. Therefore, it facilitated the visual distinction of the contact angle between the untreated and treated biochar pieces. Figure 9 illustrates horizontal images of a liquid droplet on the biochar surface.

(**a**) (**b**) (**c**)

Figure 9. Images of a drop of solution on the biochar surface: (**a**) untreated birch; (**b**) $FeCl_3$-1.0 M; (**c**) FC-1.0 M.

The captured images, converted to grayscale, were imported to ImageJ to measure the angle between the tangent to the droplet and the surface. The contact angles are presented in Table 4. As illustrated in the table, the introduction of $FeCl_3$ resulted in a 20% reduction in the contact angle of the biochar surface compared to that of the untreated biochar. This is similar to the findings of Li et al. [25], who noted that residue biochar modified with $FeCl_3$ had increased hydrophilic performance, which was attributed to its higher specific surface area, more abundant functional groups, and impregnation with iron during modification. Notably, the most significant decrease in contact angle was observed for the FC-1.0 M biochars, which presented the best wettability among the three groups. The improved wettability, facilitated by the increased hydrophilicity, was reported to enhance the ionic transport kinetics of the electrolyte according to Chai et al. [26].

Table 4. Contact angle of untreated birch biochar, $FeCl_3$-1.0 M biochar, and FC-1.0 M biochar.

Sample	Contact Angle θ
Untreated birch biochar	83°
$FeCl_3$-1.0 M	66°
FC-1.0 M	54°

3.5. Analysis of Electrochemical Properties of Modified Biochar

3.5.1. Galvanostatic Charge–Discharge

The samples were conditioned using a pretest-cycling procedure to stabilize the system. Following pretest cycling, the cell underwent two entire cycles of charge and discharge processes. The charge/discharge results of the untreated and treated biochar cells were reproducible, and the data from the second cycle were utilized for the subsequent calculations and analyses. The supercapacitors were tested under six current densities, including 5, 10, 20, 50, 100, and 200 mA/g. The combined data under the 100 mA/g current density are shown in Figure 10 as an example.

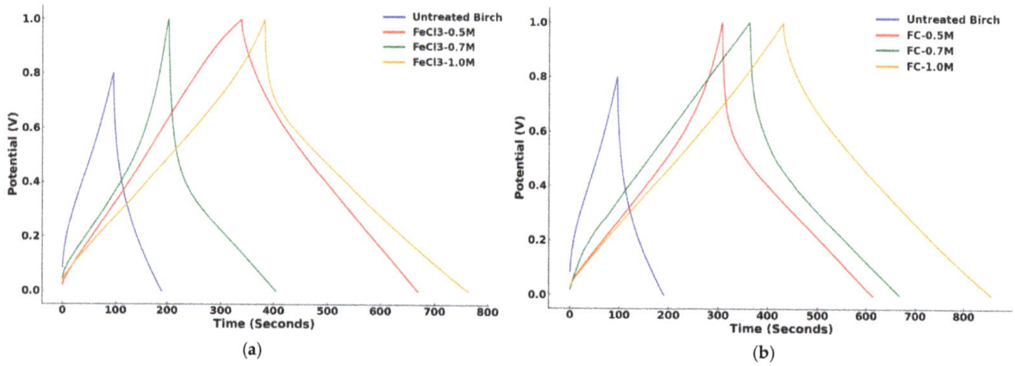

Figure 10. GCD graphs of untreated birch biochar supercapacitors and ones treated with different concentrations of $FeCl_3$ and $C_6H_5FeO_7$ under 100 mA/g: (**a**) untreated birch and $FeCl_3$ groups; (**b**) untreated birch and $C_6H_5FeO_7$ groups.

The figure clearly illustrates that the GCD time of biochar modified with ferric solution was substantially higher in comparison to that of the unmodified biochar. $FeCl_3$-1.0 M and FC-1.0 M demonstrated the longest time for charge and discharge, indicative of their highest specific capacitance. The associated gravimetric capacitance of each sample was calculated under each current density, shown in Table 5 and Figure 11 in detail. The calculations are presented in Appendix A.

Table 5. Summary of calculated gravimetric capacitance (F/g) from GCD.

Sample No.	5 mA/g	10 mA/g	20 mA/g	50 mA/g	100 mA/g	200 mA/g
Untreated Birch	83	78	73	64	53	45
$FeCl_3$-0.5 M	97	88	84	80	79	76
$FeCl_3$-0.7 M	105	97	94	92	90	90
$FeCl_3$-1.0 M	210	185	161	136	125	112
FC-0.5 M	150	130	118	111	107	102
FC-0.7 M	183	160	140	121	116	106
FC-1.0 M	219	187	161	138	127	117

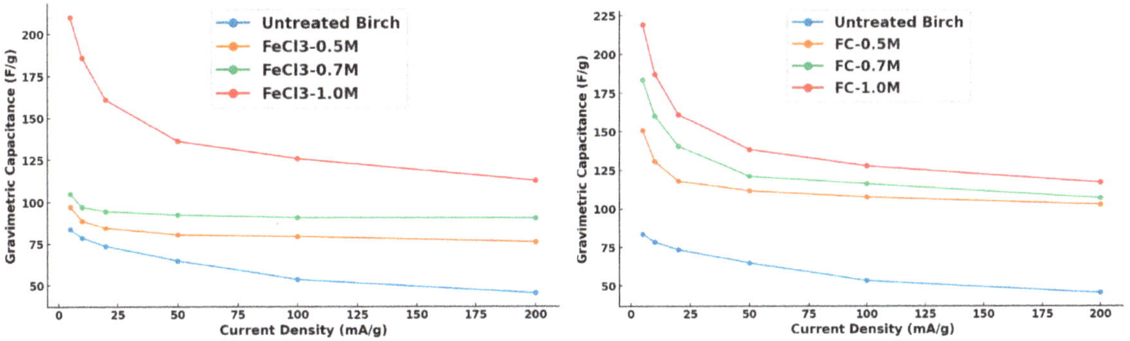

(a) (b)

Figure 11. Gravimetric capacitances of (**a**) untreated birch vs. $FeCl_3$ treated birch and (**b**) untreated birch vs. $C_6H_5FeO_7$ treated birch.

The figure shows that the treated samples were significant improvements over the untreated samples in terms of gravimetric capacitance. The proposed pre-modified supercapacitors had better specific capacitance than the ones of Caguiat et al. [27], with the same battery setup and parameters, and a similar one to microporous hardwood biochar (13–21 F/g in the range of 5–25 mA/g from Caguiat et al.). The capacitance of all biochar samples decreased with the increase in current density. The $FeCl_3$-1.0 M-treated samples (red curve) showed the highest capacitance at 5 mA/g, exceeding 200 F/g. At low current density, the enhanced charge accumulation at the surface facilitated the capacitance development. The capacitance for this modification decreased rapidly as the current density increased, converging around 110 F/g over 200 mA/g. At lower current densities, more of the electrode's surface and active sites contributed to charge storage, but as current density increased, the reduced reaction time limited this, leading to lower capacitance. The treated samples all had better convergence than the untreated birch at higher current densities. Furthermore, at the lower concentrations, the $FeCl_3$-0.5 M and $FeCl_3$-0.7 M samples had similar trends but much lower capacitances than those at the highest concentration ($FeCl_3$-1.0 M), with $FeCl_3$-0.7 M performing slightly better than $FeCl_3$-0.5 M. FC-1.0 M (red curve) showed similar behavior to $FeCl_3$ in terms of having the highest gravimetric capacitance at low current densities, which decreased sharply as the current density increased. The FC-0.7 M and FC-0.5 M curves followed this trend at lower capacitances but correspondingly performed better than the $FeCl_3$ groups. Ferric citrate modification had an even greater effect at all concentrations than $FeCl_3$. It could be inferred from the SEM results that the capacitance performance was better because of the iron-rich particulate distribution on the biochar surface of the samples in the modified groups. The SEM images of the ferric citrate group further revealed denser and smaller surface micropores as well as a wider distribution of iron-rich particles, along with the presence of magnetite and iron clusters in specific regions. These findings again show that the biochar monolith, pre-modified with ferric citrate, had an increased number of active sites, which contributed to its enhanced charge storage capacity. Meanwhile, its better wettability contributed to the capacitance improvement by promoting the ionic transport kinetics of the electrolyte in the supercapacitors. Any reversible or irreversible reactions and the pseudo-capacitance of the treated groups were further revealed by cyclic voltammetry.

3.5.2. Cyclic Voltammetry

To elucidate the rationale behind the increase in the capacitance following the modification, CV tests were conducted. The potential window for the experiment was ±0.8 V, with the initial potential established at −0.8 V. A standard scanning rate of 20 mV/s was employed as a baseline. Figure 12 illustrates the combined CV graph of an untreated biochar sample and two samples treated with 1.0 M $FeCl_3$ and $C_6H_5FeO_7$ within the ±0.8 V potential window at a scanning rate of 20 mV/s. The CV for the untreated birch group did not exhibit any significant redox characteristic peaks throughout any of the cycle. This result indicated a lack of significant pseudo-capacitance in the untreated birch biochar electrode; the behavior was dominated by electrical double-layer capacitance. The small peaks could have been the result of the carboxyl or phenol groups on the biochar surface, which were reported in Oh et al.'s finding [28]. The modified samples exhibited significantly enhance electrochemical performance compared to the untreated birch in the CV graph. The CV peaks demonstrated redox reactions on the electrode, increasing the supercapacitor performance via Faradaic pseudo-capacitance. The enclosed areas of the three curves were calculated, shown in Table 6, which proved the differences in the materials' capacitance, agreeing with the results of the GCD test, where FC-1.0 M had the best energy storage, while the untreated birch had the worst performance. The micropores of the porous carbons generated from citrate self-pyrolysis accommodated the most charges as well, probably contributing to its largest enclosed CV curve.

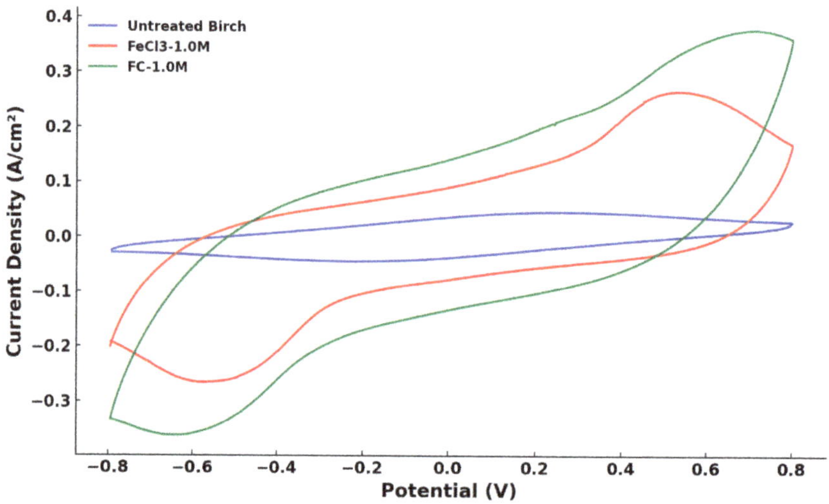

Figure 12. Cyclic voltammetry results of $FeCl_3$-1.0 M and FC-1.0 M birch groups and untreated birch at ±0.8V potential window and 20 mV/s scanning rate.

Table 6. Calculated enclosed areas of CV curves (untreated birch, $FeCl_3$-1.0 M, and FC-1.0 M).

	Area (A·V/cm^2)
Untreated Birch	0.0754
$FeCl_3$-1.0 M	0.3211
FC-1.0 M	0.4319

In addition to the comparisons with untreated birch, the reversible and irreversible redox reactions, as revealed by the curves' peaks, were analyzed between the 1.0 M $FeCl_3$ and 1.0 M $C_6H_5FeO_7$ groups. The calculated potential peaks and absolute current peaks for both the oxidation and reduction directions are presented in Figure 12 and shown in Table 7. The detailed calculation method is provided in Appendix B.

Table 7. Potential and absolute current peaks of $FeCl_3$-1.0 M and FC-1.0 M.

	Potential Peak (Volts)		Absolute Current Peak (A/cm^2)	
	Oxidation	**Reduction**	**Oxidation**	**Reduction**
$FeCl_3$-1.0 M	0.53	−0.58	0.079	−0.075
FC-1.0 M	0.73	−0.66	0.065	−0.064

From Table 7, for $FeCl_3$-1.0 M and FC-1.0 M, the ratios of the absolute current peaks were both close to one, while the peak-to-peak separation potential (ΔEp) values were different. The ΔEp of $FeCl_3$-1.0 M was 50 mV (standard <57 mV @ 25° for reversible process), indicating a reversible redox reaction on the electrodes of the $FeCl_3$-1.0 M supercapacitors [29]. The ΔEp of FC-1.0 M (~65 mV) implied that a slight irreversible process occurred on the material's surface [29]. Citrate can introduce more oxygen functional groups such as carboxyl groups to the surface of biochar [30]. He et al. [31] specifically revealed different electrode/electrolyte interface phenomena in the low, middle, and high potential CV region with the participation of oxygen functional groups, and they emphasized that the process of the hydrated K^+ insertion/de-insertion reaction that occurred in the pore was quasi-reversible due to the squeezing from those functional groups. Despite this, the presence of some micropores and oxygen functional groups barely blocked the diffusion path of the hydrated ions to the bulk solution. The following equation indicates this process in an alkaline KOH electrolyte (where > represents the biochar surface) [31].

$$> C_xO + H_3O^+ + e^- \leftrightarrow > C_xOH(H_2O) \tag{2}$$

The observed peaks are related to the redox processes of iron species probably taking place at the triple Fe/Fe_3O_4/electrolyte interface [32]. The existence of iron species in the 4 M KOH electrolyte (pH = 14) might have been the crystalline forms of $Fe(OH)_3$ and $Fe(OH)_2$ sourced from magnetite based on the stability-field diagram in aqueous conditions (Figure 13), as concluded by Back and Barnes [33]. The two peaks might have resulted from the one-electron transfer between $Fe(OH)_3$ and $Fe(OH)_2$ in the cycle in both the $FeCl_3$-1.0 M and FC-1.0 M groups. However, the electrode modified with $C_6H_5FeO_7$ had an enhanced current response and larger area under the CV curve, and the kinetic differences in the internal system between two ferric solution groups were evaluated using electrochemical impedance spectroscopy.

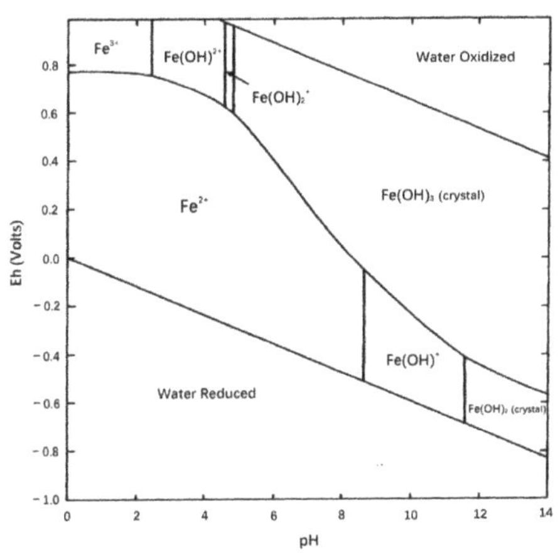

Figure 13. Stability-field diagram for aqueous ferric–ferrous system (Eh is the potential at the hydrogen electrode) [33].

3.5.3. Electrochemical Impedance Spectroscopy (EIS) and Supercapacitor Retention

Electrochemical impedance spectroscopy was conducted to investigate the behavior of the electrolyte resistance (Rs), the charge-transfer resistance (Rct) over the electrode–electrolyte interface, and the Warburg impedance (Zw). Figure 14 shows the complex combined Nyquist plot of the untreated birch, $FeCl_3$-1.0 M, and FC-1.0 M on the same scale of −Z″ and Z′, with the region of −2 Ω −Z″ and Z highlighted′. The results are consistent with the Faradaic EIS based on the Randles equivalent circuit model [34]. Figure 15 presents the data fitting results of the three groups obtained with the Python impedance package. The fitting results are satisfactory and consistent with the model settings. The untreated birch had the best fitting result, while the fitting of FC-1.0 M was the worst. The Rs, Rct, and Zw of the three groups were calculated by Python based on the Randles circuit input, which are presented in Table 8. The Python script is provided in Appendix C.

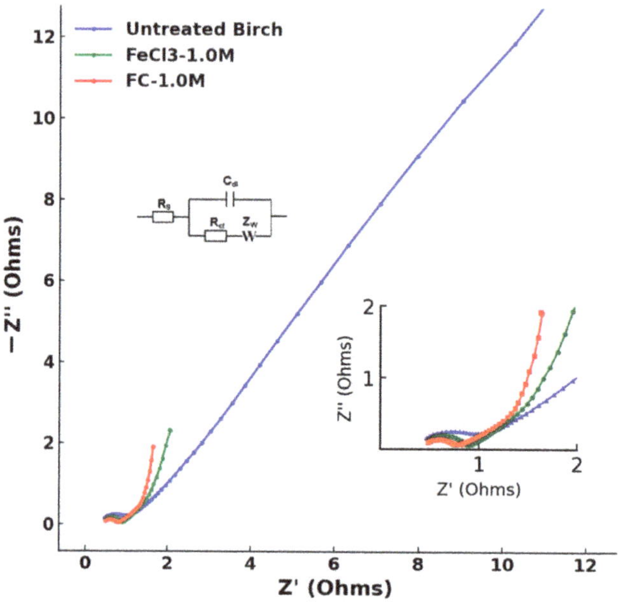

Figure 14. EIS data plotting of untreated birch, $FeCl_3$-1.0 M, and FC-1.0 M.

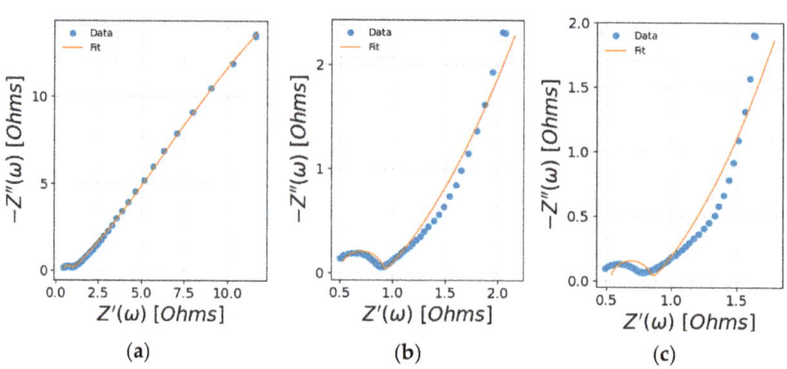

Figure 15. Data fitting results impoarted by Python: (**a**) untreated birch, (**b**) $FeCl_3$-1.0 M, and (**c**) FC-1.0 M.

Table 8. Calculated bulk electrolyte resistance (R_s), charge-transfer resistance (R_{ct}), and Warburg impedance (Z_w) of untreated birch, $FeCl_3$-1.0 M, and FC-1.0 M.

	Untreated Birch	$FeCl_3$-1.0 M	FC-1.0 M
R_s (Ω)	0.47	0.49	0.53
R_{ct} (Ω)	0.52	0.41	0.31
Z_w (Ω)	39.1	17.2	13.2

Bulk electrolyte resistance refers to the resistance encountered by ions as they move through the electrolyte in an electrochemical cell. The R_s values of the three groups intersected within the values of the uncertainty range given by Python, which was consistent with the actual situation for the same 4 M KOH electrolyte. The charge-transfer resistance of the electrode–electrolyte interfacial region correlated with the high-frequency semicircle region. This refers to the resistance to the flow of electrons during the redox reactions at the interface between the electrode and the electrolyte. The observed semicircle diameters followed the trend FC-1.0 M < $FeCl_3$-1.0 M < untreated birch, which corresponded to the increases in the R_{ct} results. A lower R_{ct} implies the kinetics of the heterogeneous electrochemical process are faster [34]. The best redox kinetics of FC-1.0 M correlated with the higher current response peak accelerating the electron transfer in the CV results, while the untreated birch's redox current peak was not significant. The low-frequency region revealed the Warburg impedance of the ions on the electrode. The Warburg impedance expresses the difficulty of the mass transport of redox or electroactive species to an electrode surface [34]. From the values in Table 8, the Warburg impedance was the most significant factor affecting the system's impedance. The highest Warburg impedance of the untreated birch indicated it had the lowest ion diffusion, while the treated samples showed a dramatic decrease in diffusion, with the doped redox iron species being more quickly transported under redox kinetics. The lowest Warburg impedance was observed for FC-1.0 M, indicating its better ion diffusion characteristics compared with $FeCl_3$-1.0 M. The denser distribution of the iron species found in the SEM results and the improved wettability facilitating ion transfer could be the factors that influenced ion diffusion.

The capacitance retention and Coulombic efficiency were investigated to confirm the cycling stability of the biochar supercapacitors treated with $FeCl_3$-1.0 M and FC-1.0 M via GCD at a 0.5 A/g current density under a 0–1.0 V potential window for 3520 cycles.

Figure 16 combines the results of two groups. The gravimetric capacitance values both converged to 90–94 F/g after 3520 cycles, while the decrease in the $C_6H_5FeO_7$ group was larger than in the $FeCl_3$ group. After calculation, the capacitance retention of FC-1.0 M was 81.9%, omitting the first two data points as outliers, after 3520 cycles, while that of $FeCl_3$-1.0 M was 85.0%, when ignoring the outliers. Although the gathering of oxygen functional groups in the pores improved the electrode wettability, the observed irreversible process still prevented hydrated ions from entering the pores during the insertion/de-insertion reaction, which negatively affected the capacitance retention [33]. The effects of the functional groups introduced by the citrate and iron oxides on the surface chemistry, including micropore morphology, functional groups, and their electron transfer, need to be further studied.

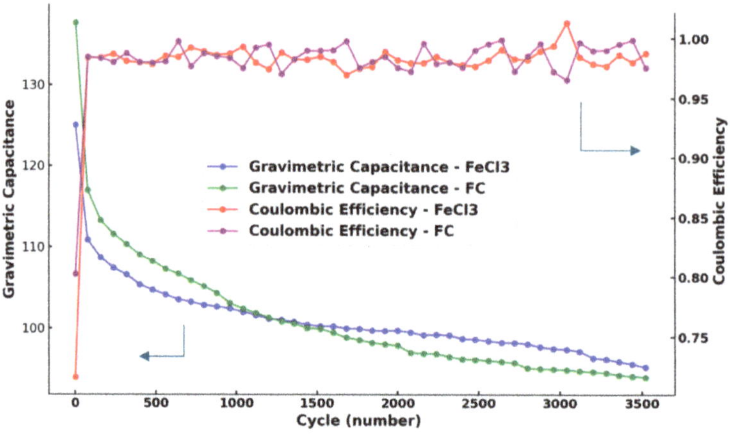

Figure 16. Gravimetric capacitance (left vertical coordinate) and Coulombic efficiency (right vertical coordinate) of $FeCl_3$-1.0 M and FC-1.0 M under 3520 cycles of charge/discharge processes.

4. Conclusions

A thick biochar monolith was pre-modified with $FeCl_3$ and $C_6H_5FeO_7$ solutions, and the surfaces and pores of the modified biochar monolith were successfully loaded with Fe_3O_4 and Fe species. SEM showed that the surface morphology of the biochar changed, leading to denser embedded iron-rich particulates via citrate self-pyrolysis in the 1.0 M $C_6H_5FeO_7$ group. The contact angle measurements of the modified birch biochar demonstrated enhanced wettability and ion mobility, particularly for the $C_6H_5FeO_7$-treated groups. The optimal concentration of the biochar modified with the two solutions was 1.0 M among three gradients of concentrations (0.5 M, 0.7 M, and 1.0 M), and its electrochemical performance was the highest. Its gravimetric capacitance at a current density from 5 mA/g to 200 mA/g was the highest among the concentration gradients. The capacitance of the biochar modified with 1.0 M $FeCl_3$ was 210 F/g at 5 mA/g, while the 1.0 M $C_6H_5FeO_7$ group reached 219 F/g at the same current density, compared to the 83 F/g capacitance of the untreated birch biochar. The CV diagram revealed that the pseudo-capacitance in the capacitor increased due to the redox reaction of $Fe(OH)_2$ and $Fe(OH)_3$ on the surface of the material, and 1.0 M $C_6H_5FeO_7$ had better capacitive performance than the 1.0 M $FeCl_3$ group. EIS revealed that the 1.0 M $C_6H_5FeO_7$ group had the best ion diffusion capacity and least charge-transfer resistance over the electrode–electrolyte interface, which contributed to its having the best kinetics in the electrochemical process. However, after 3520 charge/discharge cycles, the 1.0 M $C_6H_5FeO_7$ group had a lower level of capacitance retention than $FeCl_3$, indicating that the slightly irreversible process of electron transfer on the electrode surface, from the possible side effects of the functional groups, potentially lad to the capacitance drop in the high-potential charge–discharge process.

This overall suggests that the magnetite and iron enhanced the capacitance of the supercapacitor, shedding light on a crucial aspect of the metal salt impregnation of the pre-modified biochar monolith, which warrants further investigation and analysis to obtain a more comprehensive understanding. Meanwhile, the ferric citrate pre-modifier can offer better electrochemical properties or other performance advantages in biochar monolith supercapacitors compared to ferric chloride owing to its specific features, including its porous carbon source and functional group enrichment, which paves the way for further research on the influence of the evolution of citrate during pyrolysis on the surface chemistry of thick biochar monolith.

Author Contributions: Conceptualization, C.Q.J., D.W.K., T.F. and Z.S.; methodology, Z.S. and T.F.; software, Z.S.; validation, Z.S.; formal analysis, Z.S. and T.F.; investigation, Z.S. and T.F.; resources, Z.S., T.F. and C.Q.J.; writing—original draft preparation, Z.S.; writing—review and editing, Z.S. and D.W.K.; visualization, Z.S.; supervision, C.Q.J. and D.W.K.; funding acquisition, C.Q.J. All authors have read and agreed to the published version of the manuscript.

Funding: This research was funded by the Natural Science and Engineering Research Council (NSERC) of Canada, grant number 453859.

Data Availability Statement: The raw data supporting the conclusions of this article will be made available by the authors on request.

Conflicts of Interest: The authors declare no conflicts of interest.

Appendix A. Calculation of Gravimetric Capacitance

Specific calculations methods of gravimetric capacitance were based on the following revised equation and notes.

$$C_g = 2 \times \frac{I}{m \times \frac{\Delta V}{\Delta t}}$$

where 2 represents the two electrodes' materials in the cathode and anode; I is the discharge current recorded with C_{View} software every second; m is the average mass of one electrode; $\Delta V/\Delta t$ is the potential change in the range of every 10 data points in C_{View} (potential change every 10-s interval).

The exported gravimetric capacitance was the last calculated capacitance of the second GCD cycle (the end of the discharge).

Appendix B. Calculation of CV Parameters

Ratio of cathodic and anodic peak current and peak-to-peak separation potential in cyclic voltammetry analysis were both acquired from the following calculation strategy.

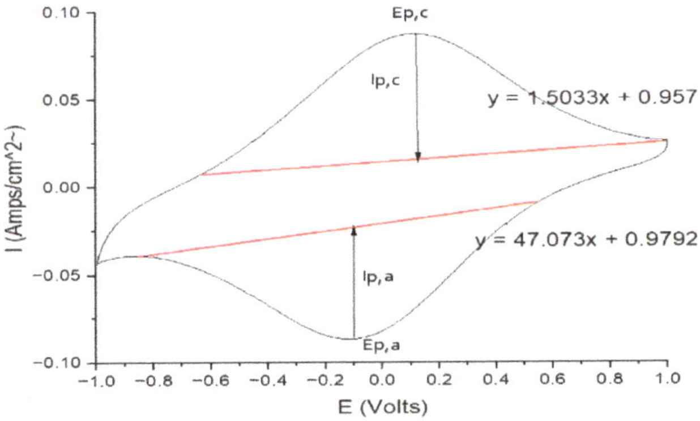

Figure A1. A sample CV with the necessary calculation lines and physical quantities.

The cathodic and anodic peak potentials could be directly acquired from the CV graph datapoints, which were the corresponding current peaks in the two scanning directions.

The cathodic and anodic peak currents were calculated. The red baseline was from the point which was the location that the curve began to increase to the peak, to another point, where the slope of the curve tangent line was equal to 0. The cathodic peak current, Ip,c, was the difference in the red baseline and current peak in the y-axis direction.

Half-wave potential: $E_{1/2}$ = ½ (Ep,a + Ep,c)

Standard potential: E0 = Ep,a + Ep,c

Ratio of cathodic and anodic peak current = | Ip,c/Ip,a |
Peak-to-peak separation potential (ΔEp) = | Ep,a − Ep,c |

Appendix C. Python Script of EIS Data Fitting

```
From impedance import preprocessing:
Frequencies, Z = preprocessing.readCSV('EISdata.csv')
Frequencies, Z = preprocessing.ignoreBelowX(frequencies, Z)
From impedance.models.circuits import CustomCircuit

circuit = 'R0-p(R1-Wo1,C1)'
initial_guess = [.01, .01, .05, 100, 1]
circuit = CustomCircuit(circuit, initial_guess = initial_guess)

circuit.fit(frequencies, Z)
print(circuit)
```

References

1. Somerville, C. Cellulose Synthesis in Higher Plants. *Annu. Rev. Cell Dev. Biol.* **2006**, *22*, 53–78. [CrossRef]
2. Woolf, D.; Amonette, J.E.; Street-Perrott, F.A.; Lehmann, J.; Joseph, S. Sustainable Biochar to Mitigate Global Climate Change. *Nat. Commun.* **2010**, *1*, 56. [CrossRef] [PubMed]
3. Zhai, Y.; Dou, Y.; Zhao, D.; Fulvio, P.F.; Mayes, R.T.; Dai, S. Carbon Materials for Chemical Capacitive Energy Storage. *Adv. Mater.* **2011**, *23*, 4828–4850. [CrossRef]
4. Pan, X.; Zhang, N.; Yuan, Y.; Shao, X.; Zhong, W.; Yang, L. Balsa-Based Porous Carbon Composite Phase Change Material with Photo-Thermal Conversion Performance for Thermal Energy Storage. *Sol. Energy* **2021**, *230*, 269–277. [CrossRef]
5. Gabhi, R.; Tan, K.; Feng, T.; Kirk, D.W.; Giorcelli, M.; Tagliaferro, A.; Jia, C.Q. Intrinsic Electrical Conductivity of Monolithic Biochar. *Biomass Bioenergy* **2024**, *181*, 107051. [CrossRef]
6. Zhang, L.; Jiang, J.; Holm, N.; Chen, F. Mini-Chunk Biochar Supercapacitors. *J. Appl. Electrochem.* **2014**, *44*, 1145–1151. [CrossRef]
7. Agyekum, E.B.; Nutakor, C. Recent Advancement in Biochar Production and Utilization—A Combination of Traditional and Bibliometric Review. *Int. J. Hydrogen Energy* **2024**, *54*, 1137–1153. [CrossRef]
8. Javanbakht, V.; Alavi, S.A.; Zilouei, H. Mechanisms of Heavy Metal Removal Using Microorganisms as Biosorbent. *Water Sci. Technol.* **2014**, *69*, 1775–1787. [CrossRef] [PubMed]
9. Wang, J.; Wang, S. Preparation, Modification and Environmental Application of Biochar: A Review. *J. Clean. Prod.* **2019**, *227*, 1002–1022. [CrossRef]
10. Sevilla, M.; Díez, N.; Fuertes, A.B. More Sustainable Chemical Activation Strategies for the Production of Porous Carbons. *ChemSusChem* **2021**, *14*, 94–117. [CrossRef] [PubMed]
11. Naseem, K.; Tahir, A.; Khan, A.S.; Qin, F.; Usman, M.; Karamat, S.; Al-Salihi, M. Unlocking the Potential of Magnetic Bio-Char as a Low-Cost Cathode Material for the Supercapacitor Application. *Mater. Today Commun.* **2024**, *39*, 108764. [CrossRef]
12. Zhang, Z.; Wang, X.; Wang, Y.; Xia, S.; Chen, L.; Zhang, Y.; Zhao, J. Pb(II) Removal from Water Using Fe-Coated Bamboo Charcoal with the Assistance of Microwaves. *J. Environ. Sci.* **2013**, *25*, 1044–1053. [CrossRef] [PubMed]
13. Ekanayake, A.; Rajapaksha, A.U.; Ahmad, M.; Vithanage, M. Enhanced Adsorption of Hexavalent Chromium from Aqueous Solution by Citric Acid-Modified Biochar from Invasive Plant Biomass. *Water. Air. Soil Pollut.* **2023**, *234*, 461. [CrossRef]
14. Liu, Y.; Zhang, L.; Zhang, Z.; Zhang, Y.; Guan, Y. Citrate-Modified Biochar for Simultaneous and Efficient Plant-Available Silicon Release and Copper Adsorption: Performance and Mechanisms. *J. Environ. Manag.* **2022**, *301*, 113819. [CrossRef] [PubMed]
15. Rajapaksha, A.U.; Chen, S.S.; Tsang, D.C.W.; Zhang, M.; Vithanage, M.; Mandal, S.; Gao, B.; Bolan, N.S.; Ok, Y.S. Engineered/Designer Biochar for Contaminant Removal/Immobilization from Soil and Water: Potential and Implication of Biochar Modification. *Chemosphere* **2016**, *148*, 276–291. [CrossRef] [PubMed]
16. Hama Aziz, K.H.; Fatah, N.M.; Muhammad, K.T. Advancements in Application of Modified Biochar as a Green and Low-Cost Adsorbent for Wastewater Remediation from Organic Dyes. *R. Soc. Open Sci.* **2024**, *11*, 232033. [CrossRef]
17. Mihoub, A.; Amin, A.E.-E.A.Z.; Motaghian, H.R.; Saeed, M.F.; Naeem, A. Citric Acid (CA)–Modified Biochar Improved Available Phosphorus Concentration and Its Half-Life in a P-Fertilized Calcareous Sandy Soil. *J. Soil Sci. Plant Nutr.* **2022**, *22*, 465–474. [CrossRef]
18. Wang, Y.-Y.; Lu, H.-H.; Liu, Y.-X.; Yang, S.-M. Ammonium Citrate-Modified Biochar: An Adsorbent for La(III) Ions from Aqueous Solution. *Colloids Surf. Physicochem. Eng. Asp.* **2016**, *509*, 550–563. [CrossRef]

19. Nagle, D.C.; Byrne, C.E. Carbonized Wood and Materials Formed Therefrom. US6051096A, 26 September 2000.
20. Ge, Y.; Xie, X.; Roscher, J.; Holze, R.; Qu, Q. How to Measure and Report the Capacity of Electrochemical Double Layers, Supercapacitors, and Their Electrode Materials. *J. Solid State Electrochem.* **2020**, *24*, 3215–3230. [CrossRef]
21. Mao, L.; Zhang, L.; Gao, N.; Li, A. $FeCl_3$ and Acetic Acid Co-Catalyzed Hydrolysis of Corncob for Improving Furfural Production and Lignin Removal from Residue. *Bioresour. Technol.* **2012**, *123*, 324–331. [CrossRef] [PubMed]
22. Dell'Era, A.; Pasquali, M.; Tarquini, G.; Scaramuzzo, F.A.; De Gasperis, P.; Prosini, P.P.; Mezzi, A.; Tuffi, R.; Cafiero, L. Carbon Powder Material Obtained from an Innovative High Pressure Water Jet Recycling Process of Tires Used as Anode in Alkali Ion (Li, Na) Batteries. *Solid State Ion.* **2018**, *324*, 20–27. [CrossRef]
23. Cheng, H.; Song, Y.; Bian, Y.; Ji, R.; Wang, F.; Gu, C.; Yang, X.; Ye, M.; Ouyang, G.; Jiang, X. Meso-/Microporous Carbon as an Adsorbent for Enhanced Performance in Solid-Phase Microextraction of Chlorobenzenes. *Sci. Total Environ.* **2019**, *681*, 392–399. [CrossRef] [PubMed]
24. Zhang, Q.; Yan, B.; Feng, L.; Zheng, J.; You, B.; Chen, J.; Zhao, X.; Zhang, C.; Jiang, S.; He, S. Progress in the Use of Organic Potassium Salts for the Synthesis of Porous Carbon Nanomaterials: Microstructure Engineering for Advanced Supercapacitors. *Nanoscale* **2022**, *14*, 8216–8244. [CrossRef] [PubMed]
25. Li, X.; Chu, S.; Wang, P.; Li, K.; Su, Y.; Wu, D.; Xie, B. Potential of Biogas Residue Biochar Modified by Ferric Chloride for the Enhancement of Anaerobic Digestion of Food Waste. *Bioresour. Technol.* **2022**, *360*, 127530. [CrossRef] [PubMed]
26. Chai, W.; Wang, F.; Miao, Z.; Che, N. Hydrophilic Porous Activated Biochar with High Specific Surface Area for Efficient Capacitive Deionization. *Desalination Water Treat.* **2024**, *320*, 100617. [CrossRef]
27. Caguiat, J.N.; Arpino, G.; Krigstin, S.G.; Kirk, D.W.; Jia, C.Q. Dependence of Supercapacitor Performance on Macro-Structure of Monolithic Biochar Electrodes. *Biomass Bioenergy* **2018**, *118*, 126–132. [CrossRef]
28. Oh, Y.J.; Yoo, J.J.; Kim, Y.I.; Yoon, J.K.; Yoon, H.N.; Kim, J.-H.; Park, S.B. Oxygen Functional Groups and Electrochemical Capacitive Behavior of Incompletely Reduced Graphene Oxides as a Thin-Film Electrode of Supercapacitor. *Electrochimica Acta* **2014**, *116*, 118–128. [CrossRef]
29. Elgrishi, N.; Rountree, K.J.; McCarthy, B.D.; Rountree, E.S.; Eisenhart, T.T.; Dempsey, J.L. A Practical Beginner's Guide to Cyclic Voltammetry. *J. Chem. Educ.* **2018**, *95*, 197–206. [CrossRef]
30. Zhang, P.; Duan, W.; Peng, H.; Pan, B.; Xing, B. Functional Biochar and Its Balanced Design. *ACS Environ. Au* **2022**, *2*, 115–127. [CrossRef] [PubMed]
31. He, Y.; Zhang, Y.; Li, X.; Lv, Z.; Wang, X.; Liu, Z.; Huang, X. Capacitive Mechanism of Oxygen Functional Groups on Carbon Surface in Supercapacitors. *Electrochim. Acta* **2018**, *282*, 618–625. [CrossRef]
32. Monteiro, J.F.; Ivanova, Y.A.; Kovalevsky, A.V.; Ivanou, D.K.; Frade, J.R. Reduction of Magnetite to Metallic Iron in Strong Alkaline Medium. *Electrochim. Acta* **2016**, *193*, 284–292. [CrossRef]
33. Back, W.; Barnes, I. *Relation of Electrochemical Potentials and Iron Content to Ground-Water Flow Patterns*; Professional Paper; USGS Numbered Series; U.S. Government Printing Office: Washington, DC, USA, 1965. Available online: https://pubs.usgs.gov/publication/pp498C (accessed on 1 November 2024).
34. Lazanas, A.C.; Prodromidis, M.I. Electrochemical Impedance Spectroscopy—A Tutorial. *ACS Meas. Sci. Au* **2023**, *3*, 162–193. [CrossRef] [PubMed]

batteries

MDPI

Article

Nitrogen Plasma-Assisted Surface Engineering on Multilayer $Ti_3C_2T_x$ Electrodes for Enhanced Interfacial Charge Dynamics and Energy Storage in Ascorbic Acid Electrolyte

Yu-Jie Liu and Chun-Pei Cho *

Department of Applied Materials and Optoelectronic Engineering, National Chi Nan University, Nantou County 54561, Taiwan
* Correspondence: cpcho@ncnu.edu.tw

Abstract

The intrinsic limitations of $Ti_3C_2T_x$ electrodes, specifically low interfacial charge-transfer efficiency and structural degradation in strongly acidic environments, hinder their performance in high-rate aqueous supercapacitors. Herein, we report a synergistic strategy combining nitrogen plasma surface engineering with a redox-active ascorbic acid electrolyte to optimize the electrode/electrolyte interfacial kinetics. By systematic investigation, the $Ti_3C_2T_x$ supercapacitor obtained by a 10-min plasma duration (N10P-AA) achieved the optimal balance between activating surface sites and preserving the conductive Ti–C framework integrity. The ascorbic acid electrolyte broadened the potential window to approximately 0.7 V, and N10P-AA exhibited the lowest charge-transfer impedance and superior rate capability, retaining a relatively high Coulombic efficiency (>72%) even at a high scan rate of 10,000 $mV \cdot s^{-1}$. The EIS results and kinetics analysis (*b* values) confirmed that the moderate plasma activation effectively promoted more surface-dominated charge storage kinetics and mitigated diffusion limitation, consistent with reduced charge-transfer resistance and a smaller Warburg slope. The XPS results revealed that the 10-min treatment suppressed detrimental oxidation during cyclings and facilitated the formation of electrochemically favorable hydroxylated surface functional groups. This work demonstrates a feasible surface electrolyte co-engineering strategy for modulating the interfacial behavior of MXene, which is of great significance for future high-efficiency aqueous electrochemical energy storage and potential biosensing applications.

Keywords: energy storage; transition metal carbide; $Ti_3C_2T_x$; MXene; nitrogen plasma; ascorbic acid; supercapacitor

1. Introduction

MXene represents a novel and rapidly growing family of two-dimensional (2D) transition-metal carbides, nitrides, and carbonitrides [1,2], generally expressed by the formula $M_{n+1}X_nT_x$, which typically originate from etching processes such as acidic or molten salt treatment [2]. Since the initial synthesis of layered $Ti_3C_2T_x$ in 2011 through the selective etching of the A-layer element in MAX phases [1], MXenes have become a research hotspot. The layered structures and the presence of surface terminal groups (T_x, such as –O, –OH, –F, etc.) derived from etching endow these materials with a unique combination of metallic-level conductivity (10^4 $S \cdot cm^{-1}$–10^5 $S \cdot cm^{-1}$), excellent mechanical flexibility, and intrinsic hydrophilicity [3]. Compared to other 2D materials such as highly conductive but surface-inert graphene, or semiconducting transition metal dichalcogenides (TMDs) with low inherent conductivity, MXenes offer the distinct advantage of combining metallic conductivity with tunable surface chemistry [4], making them promising candidates for diverse applications. For example, MXenes have demonstrated high specific capacitance and rapid ion transport in supercapacitors and Li/Na ion batteries [5], served as efficient supports for electrocatalytic hydrogen evolution, oxygen reduction, and CO_2 reduction [6], and acted as effective adsorbents for the removal of heavy metal ions and organic pollutants from water [7]. Furthermore, their exceptional electromagnetic shielding ability and mechanical flexibility render MXenes particularly attractive for applications in wearable electronics, sensors, and electromagnetic interference (EMI) shielding materials [8].

Among MXenes, $Ti_3C_2T_x$ is the most extensively studied material, particularly for high-power energy storage systems such as supercapacitors. Its structure consists of highly conductive Ti–C layers and hydrophilic surfaces [2,3], enabling rapid electron transport and efficient ion diffusion, which leads to a high specific capacitance. The charge storage mechanism of $Ti_3C_2T_x$ often involves intercalation pseudocapacitance, which allows reversible intercalation of various cations (e.g., Li^+, Na^+) accompanied by charge-transfer reactions, significantly enhancing energy storage density [9] compared to conventional electric double-layer capacitors (EDLCs). Aqueous electrolytes (e.g., H_2SO_4, KOH, Na_2SO_4) are widely used due to their high ionic conductivity (>0.1 $S{\cdot}cm^{-1}$) [10] and power output. In an acidic H_2SO_4 electrolyte, the small, highly mobile hydrated protons reversibly intercalate into the $Ti_3C_2T_x$ interlayers [11], enabling high volumetric capacitance and excellent rate performance. For instance, Lukatskaya et al. reported a volumetric capacitance of 900 $F{\cdot}cm^{-3}$ for $Ti_3C_2T_x$ with a high retention rate in 1 M H_2SO_4 [12]. However, acidic environments can lead to surface corrosion and degrade structural integrity after prolonged cycling [13]. More importantly, despite the promising theoretical prospects, maximizing the performance of MXene, especially at high-rate operation, remains fundamentally constrained by interfacial charge-transfer inefficiencies where proton and ion transport become kinetically limited.

While non-aqueous electrolytes offer a broader electrochemical stability window, which is critical for enhancing energy density [12], they often come with trade-offs. Organic electrolytes are limited by their lower ionic conductivity and larger solvent molecules, leading to sluggish intercalation kinetics [14]. Ionic liquids, despite offering outstanding chemical and thermal stability, typically exhibit high viscosity and low ion mobility [15]. To overcome these limitations and enhance energy density without sacrificing power, integrating redox-active electrolytes has become a key strategy. Ascorbic acid (AA), a water-soluble and biocompatible organic molecule, represents a highly promising candidate. It exhibits unique reversible redox activity via its enediol structure, undergoing reversible oxidation to dehydroascorbic acid (DHA) [16]. Its low oxidation potential makes AA an ideal probe molecule for investigating charge-transfer behavior, conductivity, and catalytic activity on the surface of electrode materials [17]. In electrochemical energy storage, AA can act as a functional electrolyte or additive, enhancing pseudocapacitive contribution [18] and increasing the energy density of the storage device. However, the inherently larger molecular size of AA compared to simple protons, along with its susceptibility to surface-induced side reactions, frequently leads to diffusion bottlenecks and unstable reaction pathways when interfacing with MXene surfaces.

Surface modification is thus a critical strategy for precisely tuning the interfacial properties and electronic structure of MXenes [2,3]. Notably, when a molecular redox electrolyte such as AA is employed, the electrolyte is no longer a passive medium; instead, the electrode surface must be engineered to match the interfacial charge-transfer and mass-transport requirements of the redox-active molecules. In this context, nitrogen (N_2) plasma

https://doi.org/10.3390/batteries12010007

surface engineering has emerged as a powerful and low-damage route to reconstruct the outermost atomic layers. This treatment utilizes high-energy nitrogen species to interact with the MXene surface, effectively introducing the desired nitrogen-doped atoms or nitrogen-containing functional groups (e.g., $-NH_2$, O–Ti–N, Ti–O–N) while simultaneously removing detrimental fluorine-containing (–F) groups, thereby enhancing surface polarity and electronic structure [19]. This chemical modification approach promotes interfacial affinity with electrolyte ions and enhances pseudocapacitive contributions. For instance, Zhang et al. reported that nitrogen-doped $Ti_3C_2T_x$ exhibited a high specific capacitance of up to 415 $F{\cdot}g^{-1}$ in 1 M H_2SO_4 and maintained over 90% capacity after 18,000 cycles [20]. However, the key hypothesis of the present study is that, under AA electrolyte conditions, plasma duration functions as a kinetic dial that regulates the MXene–AA interfacial charge-transfer pathway and mitigates diffusion bottlenecks/side reactions that are more pronounced for molecular redox species than for simple ions. Accordingly, an optimized plasma duration is expected to actively modulate the complex MXene-AA interfacial environment, thereby accelerating charge transport and enhancing surface-controlled charge storage, which is necessary to suppress the side reactions and diffusion constraints typical in molecularly active electrolytes.

Based on the above background and hypothesis, this study aims to systematically explore the effects of N_2 plasma treatment duration on the electrochemical performance of multilayer $Ti_3C_2T_x$ electrodes and supercapacitors. The modified $Ti_3C_2T_x$ films were evaluated in two different aqueous systems: a 3 M H_2SO_4 solution (as a reference benchmark), and a 500 μM AA solution (as the target redox-active electrolyte). Through a complementary approach utilizing electrochemical analysis, including cyclic voltammetry (CV), electrochemical impedance spectroscopy (EIS), and kinetics assessment (b value analysis), along with X-ray photoelectron spectroscopy (XPS) for surface composition confirmation, this work establishes a mechanistically explainable strategy. The resulting insights clarify why moderate plasma activation can simultaneously enhance interfacial kinetics in AA while preserving the conductive Ti–C framework, thereby informing the design of high-rate supercapacitors and electrolyte-sensitive electrochemical systems.

2. Experimental

2.1. Preparation of Electrodes and Supercapacitors

Electrodes were fabricated by mixing polytetrafluoroethylene (PTFE, 60 wt% dispersion) with activated carbon powder (Kurary YP-50F) in 95 wt% ethanol to form a mixture, which was stirred until a viscous carbon paste was obtained. The resulting paste was rolled into a uniform thin film and dried overnight at 80 °C to obtain a stable activated carbon film, serving as the counter electrode. Two electrolytes were employed: 3 M sulfuric acid (H_2SO_4) solution and 500 μM AA in 10× PBS. The pH of the 500 μM AA in 10× PBS was verified to be approximately 7 at room temperature using the universal pH paper (pH 0–14). The reference electrode depended on the electrolyte: a saturated calomel electrode (SCE, Hg_2Cl_2/KCl, CHI150) for the AA solution, and a mercury/mercurous sulfate electrode (Hg/Hg_2SO_4, CHI151) for the 3 M H_2SO_4 solution. The working electrode was prepared by attaching a multilayer $Ti_3C_2T_x$ film (Jilin 11 Technology Co., Ltd., Changchun, China) to a glassy carbon electrode (GCE, CHI104) with an inner diameter of 3 mm. The mass loading of the $Ti_3C_2T_x$ film was 0.38 mg per 3-mm-diameter disc (average of $n = 10$), corresponding to ~5.38 $mg{\cdot}cm^{-2}$ based on the geometric area. Celgard 2400 served as the separator, permitting ion transport while preventing direct contact between the two electrodes and thus avoiding internal short circuits. Plasma treatment was performed using a Harrick plasma cleaner (PDC-32G) operated at the medium power setting (10.5 W). Pure N_2 was used as the process gas, and the chamber pressure was maintained at 590 mTorr. Samples

https://doi.org/10.3390/batteries12010007

were placed at the center of the tubular chamber on the standard sample stage with the same orientation for all runs.

2.2. Electrochemical Analysis and Materials Characterization

Electrochemical measurements were performed in a three-electrode configuration using a potentiostat/galvanostat (CHI6273C, CH Instruments, Austin, TX, USA). CV was recorded at scan rates from 2 $mV{\cdot}s^{-1}$ to 10,000 $mV{\cdot}s^{-1}$. EIS was conducted at various open-circuit potentials with a 5 mV sinusoidal perturbation over a frequency range of 0.1 Hz to 100 kHz, to evaluate the impedance at the electrode/electrolyte interface, charge-transfer kinetics, and ion diffusion characteristics. Surface chemical compositions of $Ti_3C_2T_x$ electrodes were analyzed by XPS (PHI 5000 VersaProbe, ULVAC-PHI, Kanagawa, Japan). High-resolution elemental energy spectra including C 1s, F 1s, O 1s, and Ti 2p were acquired, with binding energies calibrated to the C–C/C–H peak of adventitious carbon at 284.8 eV.

3. Results and Discussion

The electrochemical behavior of multilayer $Ti_3C_2T_x$ electrodes with and without nitrogen plasma modification was compared and evaluated by CV in 3 M H_2SO_4 and 500 μM AA electrolytes at different scan rates. As shown in Figure 1a, the unmodified $Ti_3C_2T_x$ supercapacitor in 3 M H_2SO_4 exhibits an extremely narrow stable potential window of approximately 0.14 V (−0.30 V to −0.16 V), which is likely restricted by hydrogen evolution and/or oxidation reactions. In contrast, owing to its neutral to weakly acidic nature and less stringent redox potential constraints, the AA electrolyte significantly extends the stable potential window to approximately 0.7 V. Although AA is redox-active, under our conditions (500 μM AA in 10× PBS on a multilayer $Ti_3C_2T_x$ film), its faradaic contribution may not appear as distinct peak pairs. Instead, it can contribute as a broadened, pseudocapacitive-like current superimposed on electric double-layer capacitance (EDLC), leading to near rectangular CVs despite the enlarged current response and widened potential window relative to H_2SO_4. Although H_2SO_4 provides high ionic conductivity, the excessive proton concentration can accelerate oxidation or hydrogen evolution at the $Ti_3C_2T_x$ surface, particularly when the applied voltage is close to the decomposition potential of water. Figure 1a,b show that at the scan rates of 2 $mV{\cdot}s^{-1}$ and 10 $mV{\cdot}s^{-1}$, the CV curves of various $Ti_3C_2T_x$ supercapacitors treated under different plasma conditions all resemble blunt-rounded rectangles, indicating that charge storage mechanism is primarily EDLC accompanied by some surface pseudocapacitance contributions. At these low scan rates, both the device without plasma treatment (noP-AA) and the device treated with nitrogen plasma for 10 min (N10P-AA) display superior performance; even the unmodified electrode maintains sufficient ion/electron exchange, resulting in a CV area comparable to that of N10P-AA. With increasing scan rate, the CV curves gradually evolve from rectangular to spindle-like shapes, accompanied by larger current responses and more pronounced potential hysteresis, indicating that charge-transfer kinetics and ion diffusion become rate-limiting. As shown in Figure 1e,f, at the high scan rates of 500 $mV{\cdot}s^{-1}$ and 1000 $mV{\cdot}s^{-1}$, the current difference between devices caused by different plasma treatment conditions becomes more pronounced. N10P-AA consistently retains the largest CV area and a faster ion response, demonstrating superior rate capability. In contrast, N15P-AA exhibits a performance decline under certain conditions, which can be attributed to over-etching that induces structural damage and conductivity loss. Meanwhile, noP-H_2SO_4 exhibits the smallest CV curve area with a different shape, reflecting the poorest rate performance.

https://doi.org/10.3390/batteries12010007

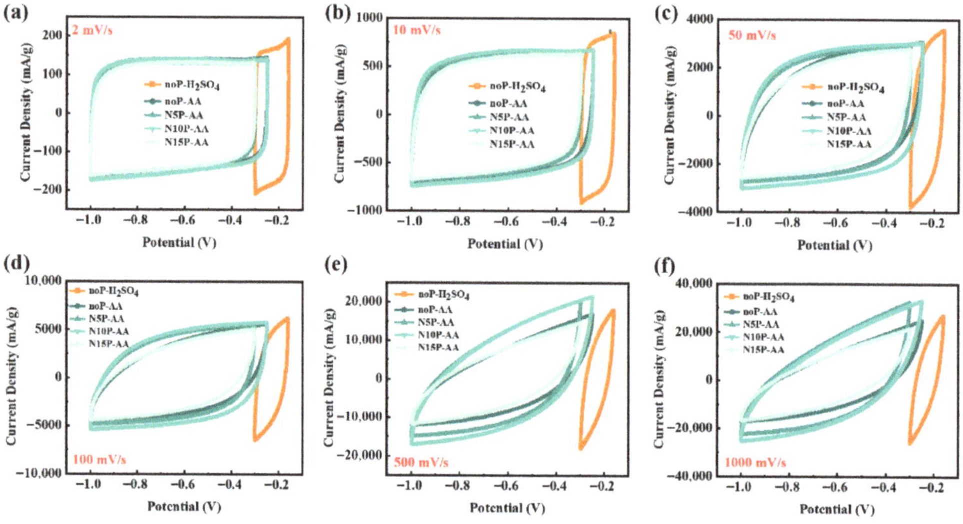

Figure 1. CV curves of $Ti_3C_2T_x$ supercapacitors under various plasma treatment conditions, using 3M H_2SO_4 and 500 μM AA as the electrolytes, at different scan rates: (**a**) 2 $mV{\cdot}s^{-1}$, (**b**) 10 $mV{\cdot}s^{-1}$, (**c**) 50 $mV{\cdot}s^{-1}$, (**d**) 100 $mV{\cdot}s^{-1}$, (**e**) 500 $mV{\cdot}s^{-1}$, and (**f**) 1000 $mV{\cdot}s^{-1}$, in which "noP" indicates no plasma treatment, "N5P" indicates nitrogen plasma treatment duration for 5 min, "N10P" indicates nitrogen plasma treatment duration for 10 min, and so on.

Nitrogen plasma treatment introduces surface functionalities such as pyridinic N, pyrrolic N, and graphitic N, while simultaneously removing surface contaminants and altering oxide states. The modification increases the density of active sites, enhances pseudocapacitance contributions, and improves wettability and conductivity, thereby facilitating charge transport. The optimized nitrogen plasma duration of 10 min strikes a balance between the number of active sites and the integrity of the conductive network, enabling higher current output and a wide potential window even at high scan rates. Additionally, Figure 1d–f reveal that electrolyte composition significantly influences electrochemical behavior. The devices operated in the 500 μM AA solution display markedly larger currents than those in H_2SO_4, presumably attributed to the participation of AA molecules in reversible Faradaic reactions and the enhanced surface pseudocapacitance. However, the chemical activity of AA may also trigger irreversible side reactions, potentially compromising long-term stability and necessitating further testing and verification. Based on the results in Figure 1, it has been confirmed that nitrogen plasma modification is beneficial for improving the potential window and rate capability of $Ti_3C_2T_x$ electrodes and supercapacitors, with N10P identified as the optimal plasma treatment condition. Notably, the performance enhancement does not scale linearly with plasma duration. Prolonged treatment can inhibit performance improvement, suggesting the need to optimize treatment parameters for optimal results in practical applications. An appropriate nitrogen plasma treatment balances introducing active sites and retaining conductivity, thus enabling higher energy storage performance over a wider potential window and at a high scan rate.

As displayed in Figure 2a, all $Ti_3C_2T_x$ supercapacitors exhibit nearly 100% coulombic efficiency at low scan rates (2 $mV{\cdot}s^{-1}$ to 10 $mV{\cdot}s^{-1}$). This indicates a high charge utilization efficiency, as ions have sufficient time to react completely at the electrode surfaces under slow charge–discharge conditions. However, as the scan rate increases to 10,000 $mV{\cdot}s^{-1}$, coulombic efficiency drops significantly. This reduction is most pronounced for the untreated device using the 3 M H_2SO_4 electrolyte (noP), where the efficiency sharply declines

https://doi.org/10.3390/batteries12010007

below 40%. In contrast, the nitrogen plasma-treated supercapacitors using the AA electrolyte (N5P, N10P, and N15P) maintain relatively high coulombic efficiencies (>72%) at high scan rates, demonstrating that nitrogen plasma treatment effectively improves the reversibility of the $Ti_3C_2T_x$ electrodes during rapid charge–discharge cycles. Figure 2b presents the rate capability for all $Ti_3C_2T_x$ supercapacitors. The nitrogen plasma-treated devices consistently outperform untreated ones, particularly in the medium to high scan rate range (100 $mV{\cdot}s^{-1}$ to 1000 $mV{\cdot}s^{-1}$), indicating their better electrochemical kinetics. The phenomenon can be attributed to introducing nitrogen-containing functional groups and surface defects during nitrogen plasma treatment, which enhances interfacial polarization and electronic conductivity, thereby reducing the ion transport impedance under high-rate conditions. Additionally, the rate capability curves of N10P-AA and N15P-AA are flatter than that of N5P-AA, suggesting that extending the nitrogen plasma treatment duration can further optimize surface chemical properties and pore structures, contributing to high-rate capability. Figure 2c shows the corresponding specific capacitance results supporting the above observations. At low scan rates, the noP-H_2SO_4 supercapacitor exhibits the highest initial specific capacitance (approximately 120 $F{\cdot}g^{-1}$), but it decays most rapidly with increasing scan rate. This indicates that while an acidic environment enhances the energy storage at low rates, it limits the cycling stability at high rates. Conversely, while N10P-AA and N15P-AA exhibit slightly lower specific capacitances at lower rates (approximately 100 $F{\cdot}g^{-1}$), their capacitance retention at high rates is better, implying the effectiveness of nitrogen plasma treatment in enhancing both long-term cycling tolerance and high-rate charge and discharge adaptability. The results in Figure 2 confirm that nitrogen plasma modification can improve the electrochemical stability of $Ti_3C_2T_x$ under high rates and the overall performance across different scan rates.

EIS results further elucidate the influences of bias potential and plasma treatment. As seen in Figure 3a,b, the slope of the low-frequency tail in the Nyquist plot gradually flattens and shifts toward 45° as the applied bias becomes more negative. This reflects that the real impedance (Z') and imaginary impedance (Z'') become more equivalent, indicating that the system is gradually controlled by ion diffusion limitation. At more negative voltages (−0.8 to −1.2 V), a reduction reaction occurs on the $Ti_3C_2T_x$ surface, rapidly oxidizing AA to DHA, resulting in a rapid depletion of AA and a steep concentration gradient between the electrode surface and the electrolyte. The system enters a diffusion-controlled regime since the diffusion rate cannot keep up. On the contrary, when a more positive voltage is applied (close to 0 V), the low-frequency tail of the Nyquist plot goes more vertical, signifying a more capacitive behavior with more efficient mass transport to the electrode surface. At this time, the reactants can more effectively diffuse and replenish the electrode surface, preventing the formation of a concentration gradient and thus avoiding a diffusion-limited regime. Thanks to the high conductivity and large specific surface area of $Ti_3C_2T_x$, the electrode exhibits supercapacitor-like behavior, and the interfacial electrochemical reaction is primarily surface-controlled. Figure 3c shows that for the untreated $Ti_3C_2T_x$ supercapacitor noP-AA, the semicircle in the high-frequency region gradually increases as the bias sweeps from 0 V to −1.2 V, indicating a rise in charge-transfer impedance (R_{CT}). Figure 3d demonstrates that N10P-AA obtained by 10-min nitrogen plasma modification exhibits a similar trend. However, its overall R_{CT} is lower than that in noP-AA, indicating that nitrogen plasma treatment can reduce R_{CT} and enhance electron transfer ability. This finding aligns with the aforementioned CV results, where N10P-AA caused a larger current and lower polarization potential, revealing its higher conductivity and more effective interfacial reaction kinetics compared to noP-AA. These also demonstrate that the appropriate nitrogen plasma modification promotes electrode surface activation and accelerates AA redox kinetics. The Nyquist plots in Figure 3 obtained

https://doi.org/10.3390/batteries12010007

from EIS analysis have clearly shown that nitrogen plasma treatment effectively reduces R_{CT} across the entire bias voltage range and mitigates the impact of diffusion limitations at high bias voltages. These findings underscore the potential of nitrogen plasma-modified MXenes for future electrochemical energy storage and biosensing applications.

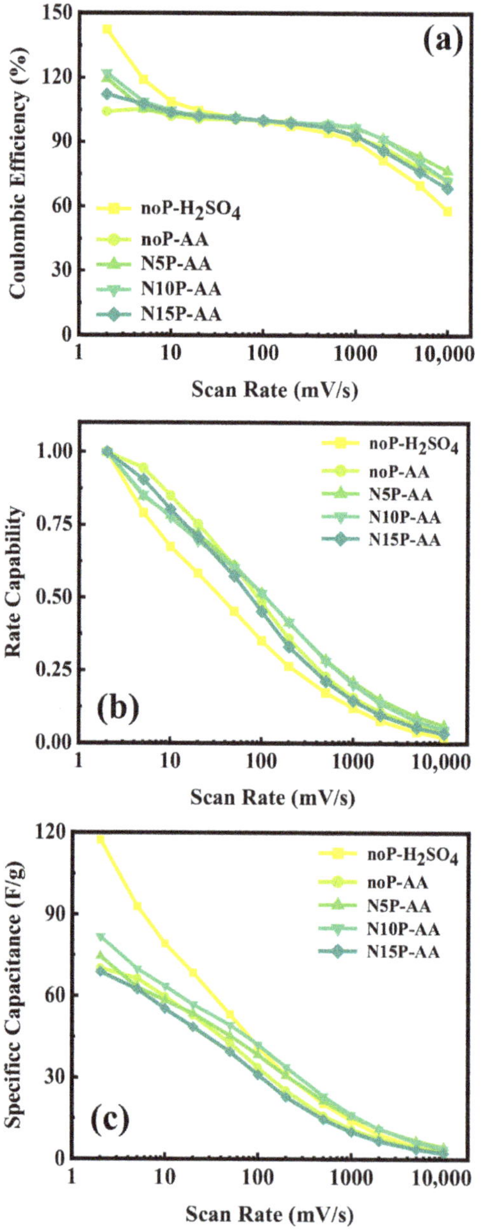

Figure 2. Plots of (**a**) coulombic efficiency, (**b**) rate capability, and (**c**) specific capacitance vs. scan rate, of the supercapacitors with $Ti_3C_2T_x$ electrodes being modified by various plasma treatment conditions and then using 3 M H_2SO_4 and 500 μM AA as electrolytes.

https://doi.org/10.3390/batteries12010007

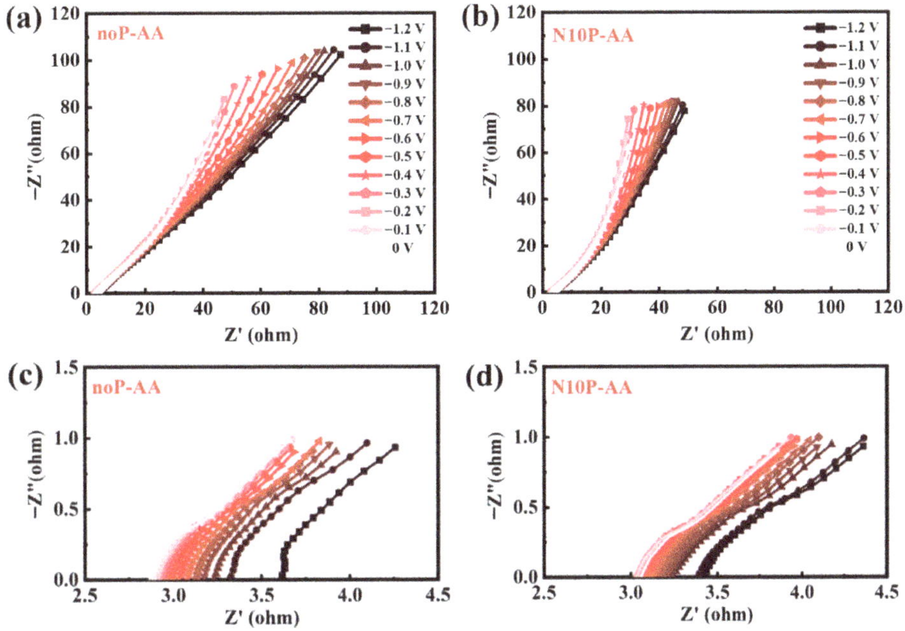

Figure 3. Nyquist plots of $Ti_3C_2T_x$ supercapacitors (**a**) noP-AA and (**b**) N10P-AA under different bias voltages; (**c**) and (**d**) are magnified images of the high-frequency regions of (**a**) and (**b**), respectively.

Figure 4 presents the magnified high-frequency regions of the Nyquist plots for $Ti_3C_2T_x$ supercapacitors under different plasma treatment conditions at the applied bias voltages of −0.3 V and −0.7 V. The intercept on the real axis corresponds to the series resistance (R_s), while the semicircle diameter reflects the magnitude of R_{CT}. As shown in Figure 4a, under −0.3 V, the untreated device in H_2SO_4 (noP-H_2SO_4) exhibits the smallest R_s (approximately 1.8 Ω), primarily due to the high ionic conductivity of the protonic acid electrolyte. In comparison, the noP-AA device displays a larger R_s and a larger semicircle, indicating an increased R_{CT}. Figure 4a,b show that with an increasing nitrogen plasma treatment duration, the Nyquist plot gradually shifts towards a higher real impedance, corresponding to an increase in R_s. As shown in Figure 4b, under −0.7 V, the plots shift further to the right relative to those at −0.3 V in Figure 4a, reflecting a more substantial polarization effect at a more negative potential, leading to increases in both R_s and v. N10P-AA maintains a relatively minor R_{CT}, highlighting its better interfacial charge-transfer capability across different operating bias voltages. By contrast, N15P-AA exhibits the largest R_s and R_{CT}, again confirming that excessive nitrogen plasma treatment can deteriorate charge transport. Collectively, N10P-AA achieves an optimal balance of Rs and a consistently low R_{CT} in the AA electrolyte across different operating bias voltages, thus delivering the most favorable energy-storage characteristics.

https://doi.org/10.3390/batteries12010007

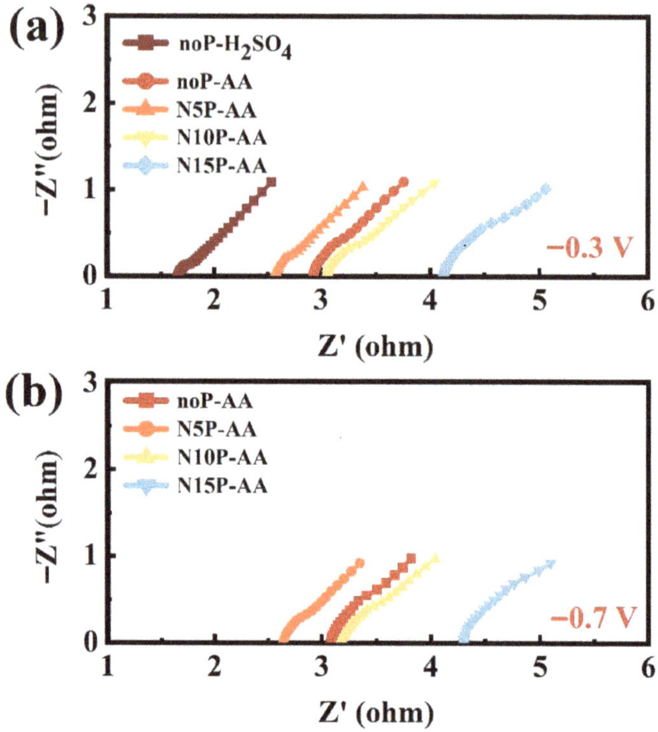

Figure 4. Magnified high-frequency regions of the Nyquist plots for $Ti_3C_2T_x$ supercapacitors at the applied bias voltages of (**a**) −0.3 V and (**b**) −0.7 V under different plasma treatment conditions.

The magnitude of the bias voltage also pronouncedly influences diffusion behavior. As the applied bias shifts from negative to less negative (−1.0 V to −0.3 V), the Warburg slope consistently decreases for all $Ti_3C_2T_x$ supercapacitors. This trend can be attributed to the physicochemical changes at the electrode/electrolyte interface. The variation in bias voltage alters the local electric field intensity and the structure of the electrical double layer near the electrode surface. Under a more negative potential (e.g., −1.0 V), more negative charges are on the electrode surface, exerting a stronger electrostatic interaction on the electrolyte ions and thereby increasing the diffusion resistance. As the potential becomes less negative, the electric field on the electrode surface weakens, reducing electrostatic hindrance, facilitating ion diffusion, and improving transport efficiency. During this process, the electrochemical behavior in the system gradually transits from being diffusion-limited towards being more surface-controlled, exhibiting a more capacitance-dominated behavior. Furthermore, the change in bias potential not only affects the diffusion process but also the activation energy of the electrochemical reactions. At more favorable potentials (e.g., −0.3 V), faster charge-transfer kinetics can more effectively sustain the concentration gradients at the interface and accelerate diffusion, reducing Warburg impedance.

The findings in Figure 5 highly align with the preceding CV and EIS results. The relatively minor Warburg slope observed for N10P-AA indicates that ions can be rapidly replenished at the reaction interface during fast surface reactions, effectively avoiding diffusion bottlenecks. This explains its ability to maintain a larger CV curve area and high current density at high scan rates. In contrast, N15P-AA exhibits larger Warburg slopes across all potentials, in agreement with its larger R_{CT} observed in its Nyquist plots. These suggest that both electron transport and mass transfer are limited, accounting for its inferior

https://doi.org/10.3390/batteries12010007

CV performance. A particularly noteworthy observation is shown in Figure 5c at −0.3 V, where the untreated noP-H_2SO_4 device exhibits the smallest Warburg slope among all $Ti_3C_2T_x$ supercapacitors in this study, indicating that its interfacial behavior is closest to that of an ideal electric double-layer capacitor. This finding contrasts sharply with the large slope observed for noP-AA using the AA electrolyte. It is then inferred that while the pristine $Ti_3C_2T_x$ surface possibly provides a highly efficient ion diffusion pathway in a simple proton (H^+) ion system, it is not conducive to ion transport when larger and more complex active species like AA are introduced. This underscores the critical role of nitrogen functionalities or microstructures, introduced by moderate nitrogen plasma treatment, in enhancing the diffusion efficiency of specific active species. Nevertheless, this optimization effect strongly depends on the electrochemical environment. Taken together, the analysis reconfirms that a nitrogen plasma treatment duration of 5 min to 10 min can be considered the optimal time window for $Ti_3C_2T_x$, as it strikes a balance between generating sufficient active sites and preserving the integrity of the conductive substrate. However, its practical benefits still need further verification for different application systems.

To further clarify the effect of nitrogen plasma treatment on the charge storage mechanism of $Ti_3C_2T_x$ supercapacitors, the b values were calculated by fitting the logarithmic relationship between scan rate and peak current. A b value closer to 1 indicates more surface-controlled behavior, whereas a b value closer to 0.5 reflects more diffusion-controlled processes. Here, "surface-controlled" refers to the dominance of near-surface charge storage kinetics ($i \propto v^b$), rather than requiring rapid long-range diffusion of AA into deep MXene interlayers. Although AA is bulkier than protons, its contribution in this system is mainly interfacial/surface-mediated, and moderate N_2-plasma activation lowers the interfacial electron transfer barrier by introducing polar terminations/active sites, thereby strengthening the surface kinetic contribution reflected by higher b values. The b values of noP-H_2SO_4, noP-AA, N5P-AA, N10P-AA, and N15P-AA were 0.71, 0.74, 0.81, 0.79, and 0.73, respectively. All are within the range of 0.7 to 0.8, indicating that their charge storage involves both surface- and diffusion-controlled contributions, with surface effects being dominant. The lowest b value of noP-H_2SO_4 suggests a greater reliance on proton diffusion at the electrode surface and between layers. The use of AA raises the b value to 0.74 (noP-AA), reflecting the contribution from the surface redox reactions of AA. Nitrogen plasma treatment further increases the b values to 0.81 (N5P-AA) and 0.79 (N10P-AA), demonstrating that nitrogen functional groups enhance surface polarity and electrolyte wettability, and generate more active sites, thereby strengthening surface-controlled behavior. In contrast, prolonged treatment reduces the b value to 0.73 (N15P-AA), likely due to structural degradation and defect accumulation from over-etching, which impede ion/electron transport and shift the mechanism toward diffusion control. Overall, AA combined with moderate nitrogen plasma treatment can effectively reinforce the surface-controlled properties of $Ti_3C_2T_x$ electrodes, while excessive plasma treatment damages their electrochemical activity.

The b-value trend aligns well with the Warburg analysis of mass transport. Therefore, b value mainly reflects kinetic control near the interface, whereas the Warburg slope reflects mass transport limitation. noP-H_2SO_4 exhibits the smallest Warburg slope, indicating faster ion replenishment and minimal diffusion resistance in the protonic electrolyte. By contrast, noP-AA exhibits a larger b value and a markedly larger Warburg slope, reflecting hindered ion transport caused by the bulkier and more complex AA molecules. The large b values of N5P-AA and N10P-AA combine with their lowest Warburg slopes, confirming that nitrogen plasma treatment improves wettability and ion transport efficiency by introducing nitrogen functionalities and tailoring surface structure, thus increasing the contribution from surface-controlled processes. Conversely, N15P-AA exhibits a reduced b value and a larger Warburg slope, indicating

https://doi.org/10.3390/batteries12010007

accumulated defects and transport limitations that deteriorate its electrochemical performance. These results jointly provide quantitative evidence that moderate nitrogen plasma treatment can optimize the kinetic control mechanism of $Ti_3C_2T_x$ electrodes, reaffirming its importance in achieving more efficient, surface-dominated charge storage.

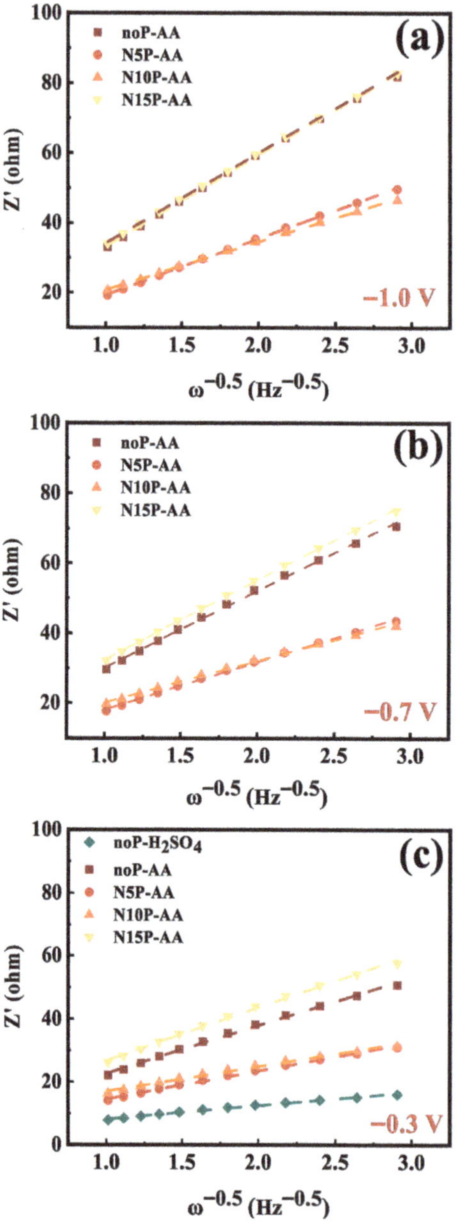

Figure 5. Warburg plots of $Ti_3C_2T_x$ supercapacitors obtained by different plasma treatment conditions under the bias voltages of (**a**) −1.0 V, (**b**) −0.7 V, and (**c**) −0.3 V.

XPS analysis was also conducted to examine the C 1s, O 1s, F 1s, and Ti 2p spectra, aiming to elucidate the evolution of surface chemistry in $Ti_3C_2T_x$ electrodes before and

https://doi.org/10.3390/batteries12010007

after nitrogen plasma treatment, and compare with the aforementioned electrochemical results for verification. As displayed in Figure 6a, the pristine $Ti_3C_2T_x$ exhibits mainly two characteristic peaks at approximately 282.0 eV (C–Ti–T_x) and 284.8 eV (C–C/C–H), indicating structural integrity with negligible contributions from C–O or C=O bonds [21]. As shown in Figure 6b, after electrochemical cyclings in AA, both the electrodes of noP-AA and N10P-AA exhibit CF_x peaks (290.0 eV to 298.0 eV) and intensified C–O (~286.9 eV) and C=O (~288.8 eV) signals, demonstrating surface oxidation and the participation of fluorine species in reactions during the cyclings [21]. A comparison of the C–Ti–T_x peaks reveals the highest intensity for pristine $Ti_3C_2T_x$ and the lowest intensity for N10P-AA, which can be attributed to high-energy ion bombardment during plasma treatment that disrupted Ti–C bonds, generated more adventitious carbon (C–C/C–H), and possibly induced passivation layer formation or altered oxidation kinetics, thereby suppressing excessive oxidation. As shown in Figure 6c, nitrogen plasma treatment also introduces new carbonaceous structures on the electrode surface of N10P-AA. These changes are highly consistent with the aforementioned electrochemical results, that is, the passivation layer in N10P-AA protected its electrode surface. It enabled higher conductivity and rate capability, particularly at high scan rates. As shown in Figure 6d, the F 1s spectra exhibit a dominant peak at approximately 685.0 eV corresponding to C–Ti–F_x bonds and a weak peak at 686.5 eV to 688.7 eV assigned to fluorine impurities [21]. As seen in Figure 6e,f, after electrochemical cyclings in AA, the impurity peak is markedly attenuated, suggesting the reductive effect of AA removes unstable fluorine species. Notably, Figure 6f shows that the 10-min nitrogen plasma treatment effectively eliminates fluorine-containing impurities on the electrode surface of N10P-AA, helps reconstruct the surface, and reduces surface defects.

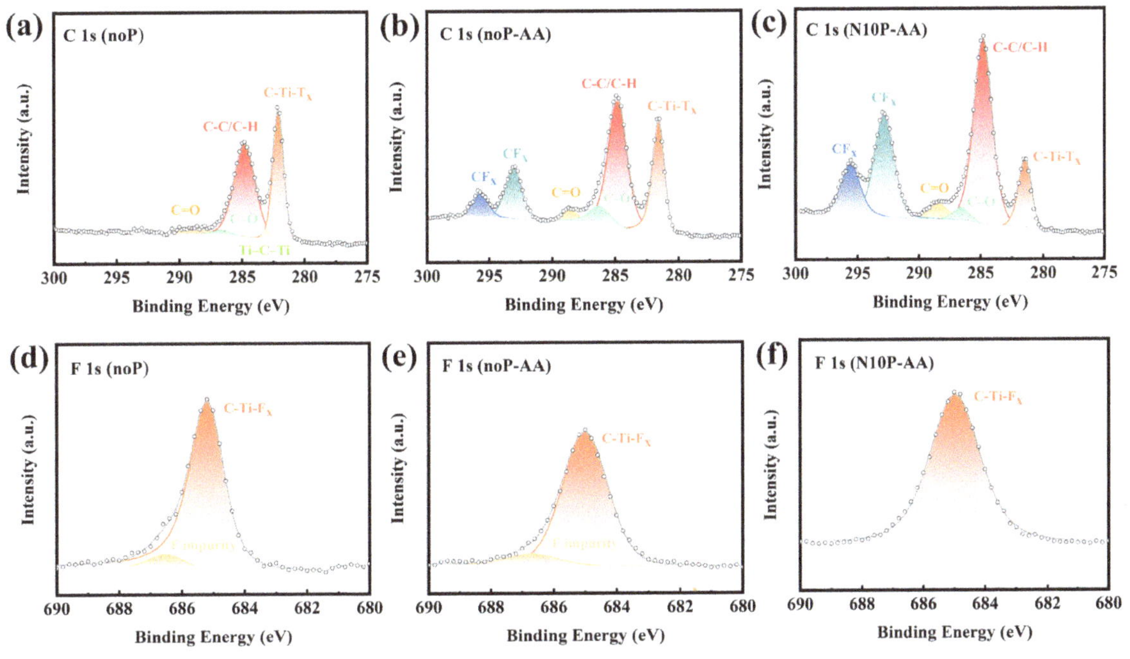

Figure 6. C 1s spectra: (**a**) pristine multilayer $Ti_3C_2T_x$, (**b**) $Ti_3C_2T_x$ electrode of noP-AA, (**c**) $Ti_3C_2T_x$ electrode of N10P-AA; F 1s spectra: (**d**) pristine multilayer $Ti_3C_2T_x$, (**e**) $Ti_3C_2T_x$ electrode of noP-AA, (**f**) $Ti_3C_2T_x$ electrode of N10P-AA.

As displayed in Figure 7a–c, the O 1s spectra comprising the peaks of TiO_2 (529.9 eV), C–Ti–O_x (530.8 eV), C–Ti–OH_x (531.9 eV), and H_2O (533.4 eV), provide more insights into the evolution of surface functional groups [21]. Figure 7a reveals that the pristine $Ti_3C_2T_x$ contains more TiO_2 on its surface. As shown in Figure 7b, after electrochemical cyclings in AA, the peak intensity of TiO_2 on the electrode surface of noP-AA weakens while that of C–Ti–O_x slightly increases, indicating that AA, as a strong reducing agent, removes part of the unstable oxide layer and generates more stable hydroxyl terminal groups (C–Ti–OH_x and –OH). As revealed in Figure 7c, the TiO_2 signal intensity on the electrode surface of N10P-AA is even smaller after 10-min nitrogen plasma treatment, and the C–Ti–OH_x signal intensity becomes relatively larger, strongly indicating that nitrogen plasma promotes surface hydroxylation and mitigates excessive Ti oxidation/passivation during cycling. The hydroxyl terminations impart excellent hydrophilicity and provide more adsorption and proton-buffering sites, contributing to enhanced capacitance and ion diffusion. The findings from Figures 6d–f and 7a–c correlate well with those from CV, EIS, and Warburg analysis, confirming that the $Ti_3C_2T_x$ electrode of N10P-AA possesses more stable and electrochemically favorable surface functional groups.

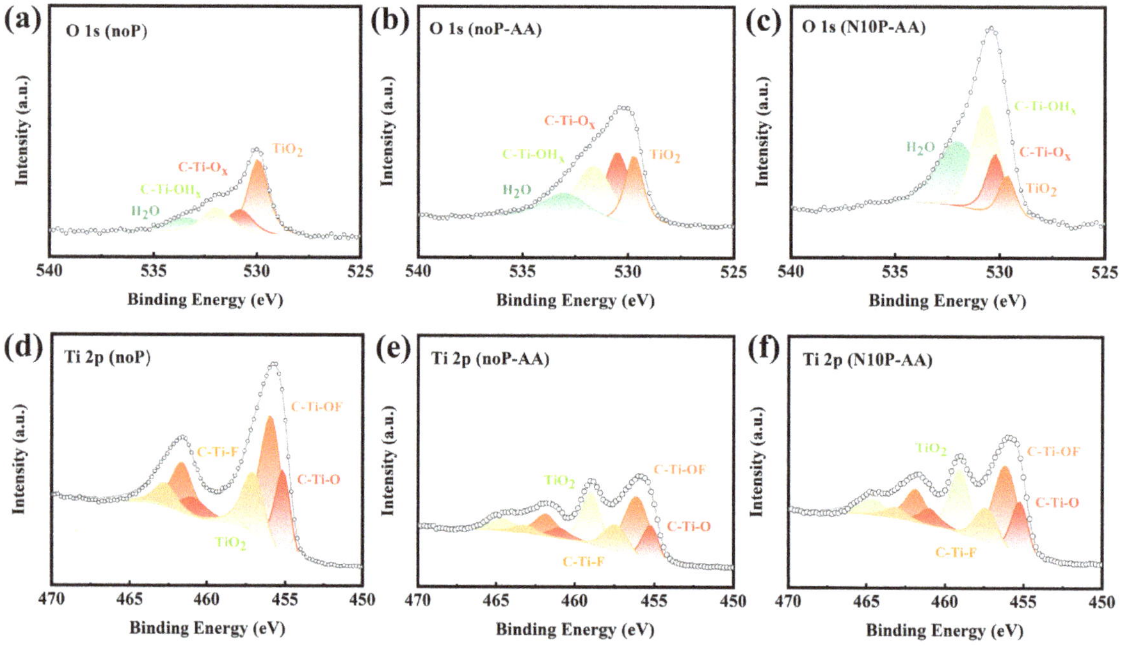

Figure 7. O 1s spectra: (**a**) pristine multilayer $Ti_3C_2T_x$, (**b**) $Ti_3C_2T_x$ electrode of noP-AA, (**c**) $Ti_3C_2T_x$ electrode of N10P-AA; Ti 2p spectra: (**d**) pristine multilayer $Ti_3C_2T_x$, (**e**) $Ti_3C_2T_x$ electrode of noP-AA, (**f**) $Ti_3C_2T_x$ electrode of N10P-AA.

Figure 7d–f presents the Ti 2p spectra of $Ti_3C_2T_x$ under different conditions. As seen in Figure 7d, the pristine material exhibits four pairs of feature peaks corresponding to C–Ti–O, C–Ti–OF, C–Ti–F, and TiO_2, indicating that its surface has been partially oxidized during synthesis and exposure [21]. As displayed in Figure 7e, the peak intensity markedly decreases after electrochemical cyclings in AA, reaffirming the reductive role of AA in suppressing the formation of TiO_2. Meanwhile, the ratios of C–Ti–F and C–Ti–O increase, reflecting the redistribution of surface terminations. As shown in Figure 7f, after 10-min nitrogen plasma treatment to achieve N10P-AA, the peak intensity of TiO_2 increases instead,

https://doi.org/10.3390/batteries12010007

suggesting that the plasma introduces more oxygen-containing groups. Compared with the electrode material in noP-AA, that in N10P-AA exhibits more prominent C–Ti–O and C–Ti–OF peaks, indicating that nitrogen plasma promotes oxygen affinity and also reinforces re-oxidation tendency on the surface. These results highlight the competing effect of AA and plasma: AA tends to suppress the formation of TiO_2, whereas nitrogen plasma inversely promotes the formation of TiO_2 and related oxygen-containing functional groups. The interaction between the two largely determines the chemical composition and stability of the $Ti_3C_2T_x$ surface. To avoid ambiguity, it should be emphasized that XPS peak intensities mainly reflect the relative distribution of surface species within the probing depth, rather than a monotonic change in the absolute amount of a single phase. After electrochemical cycling in AA, the reducing nature of AA can attenuate unstable oxide-rich features, whereas N_2 plasma exposure can simultaneously promote the redistribution of surface terminations and expose additional Ti sites. Therefore, the increased Ti–O/TiO_2-related contribution observed in Ti 2p spectrum after plasma treatment is best interpreted as a reorganization toward oxygen-containing terminations (e.g., Ti–O/Ti–OH/Ti–OF), rather than the growth of a thick, insulating TiO_2-rich passivation layer. In this context, the apparently different Ti–O/TiO_2-related contributions in the O 1s and Ti 2p spectra can be reconciled within a unified framework of termination/oxidation-state reorganization, rather than contradictory oxidation behaviors.

Looking at Figures 6 and 7, the C 1s, F 1s, O 1s, and Ti 2p spectra obtained from different conditions show no significant chemical shifts in their respective peak positions, implying that the bonding environment and chemical states of $Ti_3C_2T_x$ remain relatively stable. The effects of nitrogen plasma and using AA are primarily associated with regulating surface properties by changing the relative proportions of surface functional groups rather than altering the intrinsic bulk structure of $Ti_3C_2T_x$. The enhanced performance of N10P-AA over other devices is attributed to adjusting the number and distribution of surface functional groups rather than the difference in the intrinsic chemical state. The XPS results provide a surface chemistry foundation [21], strongly supporting the aforementioned conclusions from electrochemical analysis. They clearly reveal that nitrogen plasma treatment (particularly 10 min) can effectively inhibit excessive oxidation of the $Ti_3C_2T_x$ electrode during electrochemical cyclings while preserving the conductive framework. This protective effect is consistent with reduced R_{CT} and mitigated diffusion limitation for N10P-AA in the AA electrolyte, resulting in a larger b value and a smaller Warburg slope, and thus the overall best electrochemical performance among the AA-based devices. In contrast, noP-AA, which has not been treated with nitrogen plasma, suffers from severe oxidation and surface passivation, which impede both electron and ion transport, leading to inferior electrochemical performance.

4. Conclusions

This study demonstrated that nitrogen plasma modification effectively optimized the electrochemical performance of multilayer $Ti_3C_2T_x$ electrodes and supercapacitors. Two electrolytes, 3 M H_2SO_4 and 500 μM AA, were compared. The results showed that AA significantly broadened the stable potential window to approximately 0.7 V, while the H_2SO_4 solution only limited it to approximately 0.14 V. The $Ti_3C_2T_x$ electrode treated with nitrogen plasma for 10 min exhibited the best energy storage performance, achieving an ideal balance between introducing active sites and maintaining the integrity of the conductive network. The N10P-AA supercapacitor exhibited consistently small R_{CT} in the AA electrolyte across the investigated bias range and achieved the best balance of R_s and R_{CT} among the AA-based devices, demonstrating better energy storage characteristics. At medium to high scan rates (100 mV·s^{-1} to 1000 mV·s^{-1}), the electrode of N10P-AA outperformed the untreated electrode, maintaining the largest CV curve area and rapid ion response, demonstrating higher rate capability. Furthermore, even at an ultra-high scan rate (10,000 mV·s^{-1}), N10P-AA maintained a relatively higher coulombic efficiency (>72%), demonstrating that this method indeed improved the reversibility during rapid charge and discharge cycles. However, prolonged plasma treatment time (for N15P-AA) could cause structural damage and conductivity loss due to over-etching, resulting in performance degradation.

Mechanism analysis further revealed the reasons for performance improvement: (1) Nitrogen plasma treatment introduced nitrogen-containing functional groups and surface defects, enhancing interface polarization, wettability, and conductivity. The EIS results showed that the R_{CT} in N10P-AA was effectively reduced over the entire bias voltage range, compared to the untreated devices. The b values ranging from 0.7 to 0.8 indicated that the charge storage of all devices involved both surface-controlled and diffusion-controlled contributions, with the surface effect being dominant. N10P-AA exhibited a combination of a higher b value (0.79) and a relatively minor Warburg slope, suggesting that its rapid surface reactions and the effective ion replenishment to the interface could avoid diffusion bottlenecks. (2) Surface chemical analysis (XPS) results confirmed that the 10-min nitrogen plasma treatment effectively suppressed excessive oxidation during electrochemical cyclings, promoted the formation of hydroxylated surface functional groups (such as C–Ti–OH_x) beneficial for electrochemical reactions, and removed surface fluorine impurities, providing a more stable and electrochemically active surface. In summary, moderate nitrogen plasma modification offers a feasible and optimized strategy for future high-efficiency electrochemical energy storage and potential biosensing applications.

Author Contributions: Conceptualization, C.-P.C.; Methodology, C.-P.C.; Validation, Y.-J.L. and C.-P.C.; Formal analysis, Y.-J.L. and C.-P.C.; Investigation, Y.-J.L. and C.-P.C.; Resources, C.-P.C.; Data curation, Y.-J.L. and C.-P.C.; Writing—original draft, C.-P.C.; Writing—review & editing, C.-P.C.; Visualization, Y.-J.L. and C.-P.C.; Supervision, C.-P.C.; Project administration, C.-P.C.; Funding acquisition, C.-P.C. All authors have read and agreed to the published version of the manuscript.

Funding: This research was funded by National Science and Technology Council, Taiwan, grant number NSTC 114-2635-E-260-001.

Data Availability Statement: The data presented in this study are openly available in [OSF] [https://osf.io/w4tks/overview?view_only=fb781d48867b4415ad7a1724496ade4d] (accessed on 16 November 2025).

Acknowledgments: Financial support from the National Science and Technology Council, Taiwan, is greatly appreciated.

Conflicts of Interest: The authors declare no conflict of interest.

References

1. Naguib, M.; Mashtalir, O.; Carle, J.; Presser, V.; Lu, J.; Hultman, L.; Gogotsi, Y.; Barsoum, M.W. Two-dimensional nanocrystals produced by exfoliation of Ti_3AlC_2. *Adv. Mater.* **2011**, *23*, 4248–4253. [CrossRef] [PubMed]
2. Sun, Y.; Chen, D.; Liang, Z. Two-dimensional MXenes for energy storage and conversion applications. *Mater. Today Energy* **2017**, *5*, 22–36. [CrossRef]
3. Anasori, B.; Lukatskaya, M.R.; Gogotsi, Y. 2D metal carbides and nitrides (MXenes) for energy storage. *Nat. Rev. Mater.* **2017**, *2*, 16098. [CrossRef]
4. Gogotsi, Y.; Anasori, B. The rise of MXenes. *ACS Nano* **2019**, *13*, 8491–8494. [CrossRef]
5. Lukatskaya, M.R.; Mashtalir, O.; Ren, C.E.; Dall'Agnese, Y.; Rozier, P.; Taberna, P.L.; Naguib, M.; Simon, P.; Barsoum, M.W.; Gogotsi, Y. Cation intercalation and high volumetric capacitance of two-dimensional titanium carbide. *Science* **2013**, *341*, 1502–1505. [CrossRef]
6. Salmasi, M.Z.; Omidkar, A.; Nguyen, H.M.; Song, H. MXenes as electrocatalysts for hydrogen production through the electrocatalytic water splitting process: A mini review. *Energy Rep.* **2024**, *10*, 100070. [CrossRef]
7. Janjhi, F.A.; Ihsanullah, I.; Bilal, M.; Castro-Muñoz, R.; Boczkaj, G.; Gallucci, F. MXene-based materials for removal of antibiotics and heavy metals from wastewater-a review. *Water Resour. Ind.* **2023**, *29*, 100202. [CrossRef]

https://doi.org/10.3390/batteries12010007

8. Shahzad, F.; Alhabeb, M.; Hatter, C.B.; Anasori, B.; Hong, S.M.; Koo, C.M.; Gogotsi, Y. Electromagnetic interference shielding with 2D transition metal carbides (MXenes). *Science* **2016**, *353*, 1137–1140. [CrossRef]
9. Wang, X.; Bak, S.M.; Han, M.; Shuck, C.E.; McHugh, C.; Li, K.; Li, J.; Tang, J.; Gogotsi, Y. Surface redox pseudocapacitance of partially oxidized titanium carbide MXene in water-in-salt electrolyte. *ACS Energy Lett.* **2022**, *7*, 30–35. [CrossRef]
10. Zainab, S.; Awan, S.U.; Hussain, D.; Rizwan, S.; Iqbal, T.; Fraz, S. Exploring electrolyte specific effects on multisheets 2-dimensional $Ti_3C_2T_x$-$BiFeO_3$ nanocomposites electrodes for high-performance supercapacitors applications. *Sci. Rep.* **2025**, *15*, 22848. [CrossRef]
11. Yadav, S.; Kurra, N. Diffusion kinetics of ionic charge carriers across $Ti_3C_2T_x$ MXene-aqueous electrochemical interfaces. *Energy Storage Mater.* **2024**, *65*, 103094. [CrossRef]
12. Ghidiu, M.; Lukatskaya, M.R.; Zhao, M.Q.; Gogotsi, Y.; Barsoum, M.W. Conductive two-dimensional titanium carbide 'clay' with high volumetric capacitance. *Nature* **2014**, *516*, 78–81. [CrossRef]
13. Doo, S.; Chae, A.; Kim, D.; Oh, T.; Ko, T.Y.; Kim, S.J.; Koh, D.Y.; Koo, C.M. Mechanism and kinetics of oxidation reaction of aqueous $Ti_3C_2T_x$ suspensions at different pHs and temperatures. *ACS Appl. Mater. Interfaces* **2021**, *13*, 22855–22865. [CrossRef]
14. Lin, Z.; Daffos, B.; Taberna, P.L.; Van Aken, K.L.; Anasori, B.; Gogotsi, Y.; Simon, P. Capacitance of $Ti_3C_2T_x$ MXene in ionic liquid electrolyte. *J. Power Sources* **2016**, *326*, 575–579. [CrossRef]
15. Zang, X.; Wang, J.; Qin, Y.; Wang, T.; He, C.; Shao, Q.; Zhu, H.; Cao, N. Enhancing capacitance performance of $Ti_3C_2T_x$ MXene as electrode materials of supercapacitor: From controlled preparation to composite structure construction. *Nanomicro Lett.* **2020**, *12*, 77. [CrossRef] [PubMed]
16. Hu, I.F.; Kuwana, T. Oxidative mechanism of ascorbic acid at glassy carbon electrodes. *Anal. Chem.* **1986**, *58*, 3235–3239. [CrossRef]
17. Kairy, P.; Hossain, M.M.; Khan, M.A.R.; Almahri, A.; Rahman, M.M.; Hasnat, M.A. Electrocatalytic oxidation of ascorbic acid in the basic medium over electrochemically functionalized glassy carbon surface. *Surf. Interfaces* **2022**, *33*, 102200. [CrossRef]
18. Shakil, R.; Shaikh, M.N.; Shah, S.S.; Reaz, A.H.; Roy, C.K.; Chowdhury, A.N.; Aziz, M.A. Development of a novel bio-based redox electrolyte using pivalic acid and ascorbic acid for the activated carbon-based supercapacitor fabrication. *Asian J. Org. Chem.* **2021**, *10*, 2220–2230. [CrossRef]
19. Xu, W.; Li, S.; Zhang, W.; Ouyang, B.; Yu, W.; Zhou, Y. Nitrogen-doped $Ti_3C_2T_x$ MXene induced by plasma treatment with enhanced microwave absorption properties. *ACS Appl. Mater. Interfaces* **2021**, *13*, 49242–49253. [CrossRef] [PubMed]
20. Zhang, T.; Xiao, J.; Li, L.; Zhao, J.; Gao, H. A high-performance supercapacitor electrode based on freestanding N-doped $Ti_3C_2T_x$ film. *Ceram. Int.* **2020**, *46*, 12842–12849. [CrossRef]
21. Natu, V.; Benchakar, M.; Canaff, C.; Habrioux, A.; Célérier, S.; Barsoum, M.W. A critical analysis of the X-ray photoelectron spectra of $Ti_3C_2T_z$ MXenes. *Matter* **2021**, *4*, 1224–1251. [CrossRef]

https://doi.org/10.3390/batteries12010007

batteries

MDPI

Article

MnO_2/AgNPs Composite as Flexible Electrode Material for Solid-State Hybrid Supercapacitor

Borislava Mladenova, Mariela Dimitrova and Antonia Stoyanova *

Institute of Electrochemistry and Energy Systems, Bulgarian Academy of Sciences, G. Bonchev Str. 10, 1113 Sofia, Bulgaria; borislava.mladenova@iees.bas.bg (B.M.); mariela.dimitrova@iees.bas.bg (M.D.)

* Correspondence: antonia.stoyanova@iees.bas.bg

Abstract: A MnO_2/AgNP nanocomposite was synthesized using a sonochemical method and investigated as an electrode material in a solid-state hybrid supercapacitor. Aquivion's sodium and lithium electrolyte membrane serves as an electrolyte and separator. For comparison, MnO_2 was used as the active material. The developed supercapacitor containing a carbon xerogel as a negative electrode, the MnO_2/AgNP composite as a positive electrode and a Na^+-exchange membrane demonstrated the highest performance characteristics. These results indicate that the incorporation of silver nanoparticles into the MnO_2 structure is a prospect for obtaining an active composite electrode material for solid-state supercapacitors.

Keywords: MnO_2/AgNP nanocomposite; solid-state supercapacitor; Aquivion electrolyte membrane

1. Introduction

Supercapacitors (SCs) are energy storage devices that have extremely high power densities but do not reach the energy density of batteries and fuel cells. In recent years, research interest has focused on hybrid supercapacitors, which provide higher specific capacitance and improved energy density characteristics without changing the power density compared to existing electric double-layer capacitors (EDLCs) and pseudocapacitors [1]. There is also a growing interest in SCs with polymer electrolytes and flexible electrodes as they provide highly efficient and safe energy storage solutions for the rapidly developing application areas in consumer electronics and microelectronics [2].

Manganese oxide is a well-known electroactive material for SCs because of its low price, ideal capacitive behavior, theoretical specific capacity around 1370 F g^{-1} and high cyclic stability (over 5000 cycles) [3–5]. However, conventional MnO_x-based supercapacitors suffer from limitations such as low electrical conductivity and insufficient interface surface area. To address these challenges, the hybridization of MnO_x with other highly conductive materials and increasing its porosity have emerged as promising approaches [6].

Ag stands out among the recently explored conductive materials because of its excellent conductivity and affordability. Lately, Ag nanoparticles (AgNPs) have found an application as conductive additives in energy storage systems. Recent research indicates that electrode materials doped with these elements can establish electron transport pathways owing to their superior electrical conductivity and notable surface activity. Consequently, the incorporation of Ag helps sustain the low internal resistance of metal oxide electrodes while enhancing proton diffusion across the electrodes. For example, RuO_2 doped with Ag shows a much higher specific capacitance than pure RuO_2. It has been found that the incorporation of silver nanoparticles into the structure of MnO_2 during discharge leads to the formation of similar channels and thus significantly improves the conductivity of the oxide and shows an increase in specific capacitance [7–10].

Silver-containing hybrid MnOx nanostructures (silver–manganese oxide (Ag-MnOx) and silver–manganese oxide with cetyltrimethylammonium bromide (c-Ag-MnOx)) were synthesized and investigated as a positive electrode material in SCs [6]. Utilizing silver as a conductive additive to MnO_x is motivated not only by its cost-effectiveness compared to other noble metals but also by its comparatively greater operational stability in comparison to transition metals and conducting polymers like polypyrrole [11,12]. Using gas-phase evaporation-induced self-assembly, silver nanoparticles are uniformly dispersed within the hybrid nanostructure, establishing a conductive pathway at the interface of the MnO_x crystallite cluster [6].

Despite the advantages shown, high SC values have not yet been achieved in Ag–metal oxide composites as there are problems related to their reduced surface area, ease of oxidation and tendency to agglomerate [13].

It should be noted that the optimization of any electrode material should always be carried out in combination with the electrolyte used, as the electrode and electrolyte should show good interaction and increased cycling stability. Polymer electrolyte and ion-exchange membranes can be effectively used as separators and electrolytes for the development of supercapacitor devices. The Aquivion membrane exhibits superior proton conductivity and thermal stability in comparison to the Nafion membrane. This is attributed to its shorter side chain, higher glass transition temperature, increased crystallinity and lower equivalent weight. Additionally, the Aquivion membrane demonstrates excellent chemical stability and is capable of operating at elevated temperatures, rendering it highly promising for utilization in energy systems [14]. The advantages of solid-state SCs, such as their light weight, small size, easy operation, excellent reliability and wider temperature range, are also significant factors for their development [15]. Neutral electrolytes are widely used in supercapacitor systems. Various factors such as the electrolyte pH, type of cations and anions, concentration of salts, different additives and solution temperature have been found to affect the device performance [16,17]. Their advantages, such as wider operating potential windows and increased protection, make them applicable in asymmetric and hybrid supercapacitors. Na_2SO_4 is the most widely used neutral electrolyte, promising for many pseudocapacitive materials (especially MnO_2-based materials). Li (e.g., LiCl, Li_2SO_4 and $LiClO_4$), K (e.g., KCl, K_2SO_4 and KNO_3), Ca ($Ca(NO_3)_2$) and others are also common conducting salts in neutral electrolytes [18]. Alkali or alkaline earth metal cations, despite their similarities, exhibit variations in ionic sizes and hydrated ion sizes. These differences lead to distinct diffusion coefficients and ionic conductivity, consequently affecting the specific capacitance of the supercapacitor in diverse ways. For sulfate electrolytes of alkali metals, such as Li_2SO_4, Na_2SO_4 and K_2SO_4, some studies have shown that the values of the asymmetric specific capacitance of SCs follow the order $Li_2SO_4 < Na_2SO_4 < K_2SO_4$ [19–21]. MnO_2 stands as one of the most extensively investigated pseudocapacitive compounds in neutral electrolytes to date. Its charge–discharge cycle involves the alteration of Mn oxidation states from III to IV, accompanied by the surface adsorption–desorption or intercalation–deintercalation of electrolyte cations such as K^+, Na^+ and Li^+ as well as protons (H^+). Given the direct involvement of electrolyte ions in the charge storage mechanism, the nature of neutral electrolytes is anticipated to exert a notable impact on the pseudocapacitive properties [18].

The aim of this research is to develop a new type of composite material containing MnO_2 and AgNPs to improve the performance of the positive electrode on solid-state supercapacitors and increase their energy density. The synthesized MnO_2 and MnO_2/AgNP samples were physicochemically characterized and electrochemically tested as flexible electrode materials in a hybrid supercapacitor cell. Na^+ and Li^+ forms of the commercial membrane Aquivion® E87-05S (FullCellStore, Texas/USA) were used as the electrolyte and separator. The relationship between the nature of the electrolyte, the presence of AgNPs in the composite electrode material and the supercapacitor performance was investigated.

2. Materials and Methods

2.1. Synthesis of Materials

The MnO_2 and MnO_2/AgNP composite was synthesized in order to study its electrochemical behavior as an electrode material in supercapacitor cells. MnO_2 was obtained by the ultrasound technique using manganese (II) chloride (Sigma-Aldrich, St. Louis, MO, USA) and potassium permanganate (≥99.0% Sigma-Aldrich) as reagents. These compounds were previously dissolved in distilled water and mixed, then the resulting suspension was treated for 1 h at room temperature in an ultrasonic bath (Siel, UST 5.7-150, Siel Ltd.Sofia, Sofia, Bulgaria). Precipitate formation was observed during synthesis.

To obtain the MnO_2/AgNP composite, AgNPs were initially synthesized by the chemical reduction of aqueous solution using ultrasonic field. The resulting nanoparticles were added to the suspension of $MnCl_2$ and $KMnO_4$, during the synthesis of MnO_2. A precipitate was formed which was washed repeatedly using a centrifuge (NEYA-16) at 5000 rpm, dried and subjected to further heat treatment at 200 °C to enhance the characteristics of the resulting material.

The carbon xerogel was prepared by the sol–gel methodology described in [22]. Here, resorcinol $C_6H_6O_2$ (Indspec, Petrolia, PA, USA, 99.6% purity) and formaldehyde CH_2O (Merck, Rahway, NJ, USA, 37% aqueous solution) were used as precursors. The pH of the precursor mixture was controlled, being 3 at the beginning of the process and increased to 6.5 using NaOH. This facilitates the polymerization reaction, and it is possible to have good control over the physicochemical properties of the resulting material.

2.2. *The Preparation of the Electrodes and the Activation of the Polymer Electrolyte Membrane*

The carbon electrodes were obtained by preparing an ink containing 80 wt.% xerogel, 10 wt.% ABG 1005 EG1 graphite and 10 wt.% polyvinylidene difluoride (PVDF) dissolved in 1-methyl-2-pyrrolidone [22]. The prepared ink was thoroughly mixed to ensure homogeneity before being coated onto a glass plate and subsequently dried at 70 °C for 12 h. Following this, each layer was meticulously removed from the glass plate and subjected to further drying at 120 °C for 1 h, followed by a final drying step at 160 °C for 20 min.

The synthesized MnO_2 and MnO_2/AgNP powders were used to fabricate a positive composite electrode. The ink consists of 75 wt% MnO_2 or MnO_2/AgNP, 10 wt% graphite ABG 1005 EG1, 5 wt% carbon fibers and 10 wt% PVDF binder in 1-Methyl-2-pyrrolidone solvent. Subsequently, the composite electrodes were manufactured in a process similar to that of carbon electrodes. The mass ratio between the carbon xerogel and composite electrodes is 1:1, the mass loading being 0.002 g.

A perfluorosulfonic acid (PFSA) membrane, specifically the Aquivion® E87-05S model, supplied by Sigma-Aldrich in Italy, with a thickness of 50 μm and an equivalent weight of 870 g mol^{-1}, was used both as an electrolyte and as a separator. Prior to use, the polymer membrane was activated to Na^+ and Li^+ forms by immersion in solutions of 1 M Na_2SO_4 and 1 M Li_2SO_4. H_2O, respectively, for three hours at room temperature.

2.3. *The Physicochemical Characterization of the Prepared Materials*

The synthesized materials were analyzed using an X-ray powder diffractometer (XRD), in particular a PANalytical Aeris, operating with CuKα radiation (λ = 1.5406Å) and θ-θ using Bragg–Brentano geometry. The 2θ scanning range starts from 20 to 80° (scan step 0.02°, 60 s). The analysis of diffractograms was conducted utilizing the PDF-2 2022 database from the ICDD. The determination of the elemental composition was accomplished through energy-dispersive X-ray fluorescence spectrometry (EDXRF). The spectrometer Epsilon 1 (Panalytical, Malvern, UK) was employed to gather the spectra, equipped with a Ag target X-ray tube capable of operating at a voltage of 50 kV and a maximum power of 10 W. The elements that can be measured by this instrument unit range from sodium (Na) to americium (Am).

The sample morphology was examined using a scanning electron microscope, SEM (Zeiss Evo 10), following the careful application of a carbon coating to avoid surface charge accumulation. Energy-dispersive analysis (EDX) was conducted using the Zeiss Smart EDX system. Additionally, transmission electron microscope (TEM) analyses were performed using the JEOL JEM 2100, operating at 80–200 kV (Jeol Ltd. Akishima, Japan).

The porous texture of the obtained materials was investigated with a Quantachrome (Boynton Beach, FL, USA) AutoSorb iQ apparatus. The pore size distribution was determined using the BJH method. The obtained silver nanoparticles were analyzed using a scanning spectrophotometer (Shimadzu UV-1900i, Kyoto, Japan) in the range from 300 to 700 nm.

The measurement of the wetting angle consisted of dropping 10 μL of the electrolyte onto the electrode and observing the drop shape. Each determination was performed 5 times to obtain an average value for each sample using a KRÜSS K100 Force Tensiometer K100.

2.4. Electrochemical Characterization

The solid-state supercapacitors were electrochemically investigated to determine their performance. Cyclic galvanostatic charge–discharge (GCD) and long-term durability tests were performed using an Arbin BT-2000 instrumented system. Galvanostatic charge–discharge measurements were conducted from 0.05 to 1.6 V at a current load of 60, 90, 120, 240, 480, 600, 900 and 1200 mAg^{-1}, involving 30 cycles per step. Long-term cell tests were carried out at a current load of 240 mAg^{-1} for up to 1000 cycles. CV experiments were performed using the Multi PalmSens system (model 4, Houten, The Netherlands) at a constant voltage window and different scan rates of 1, 10, 20 and 40 mVs^{-1}. Electrochemical impedance spectroscopy (EIS) measurements were performed on the same setup, with frequency ranges from 10 MHz to 1 mHz.

The specific capacitance, energy density E and power density P of the developed supercapacitors were calculated using Equations (1)–(4) [23].

The specific capacitance Cs (Fg^{-1}) was determined based on cyclic voltammetric curves using the equation as follows:

$$Cs = [4(I/(dV/dt))/m], \quad (1)$$

where I is the current (A), dV/dt is the voltage scan rate (Vs^{-1}) and m is the mass of the active material (g).

The specific capacitance C (Fg^{-1}) derived from the galvanostatic charge–discharge tests was calculated from Equation (2):

$$C = [2(I \times \Delta t)/(m \times \Delta V)], \quad (2)$$

where I (A), Δt (s), m (g) and ΔV (V) are the discharge current, discharge time, mass of the active composite material and voltage window, respectively.

The energy density E ($Whkg^{-1}$) and the power density P (Wkg^{-1}) were calculated using Equations (3) and (4):

$$E = C\,\Delta V^2/7.2 \quad (3)$$

$$P = 3600\,E/t \quad (4)$$

3. Results and Discussion

In Figure 1, the UV-Vis spectrum of a 5 mM $AgNO_3$ solution containing silver nanoparticles obtained by treatment in an ultrasonic bath at 50 °C is shown. The figure shows an absorption peak with a maximum at a wavelength around 413 nm, characteristic of spherical particles. The appearance of this peak in the spectrum is due to the Surface Plasmon Resonance (SPR) and is associated with the formation of silver particles of colloidal size. As a consequence of the successfully synthesized nanoparticles, we continued with their further integration into our materials.

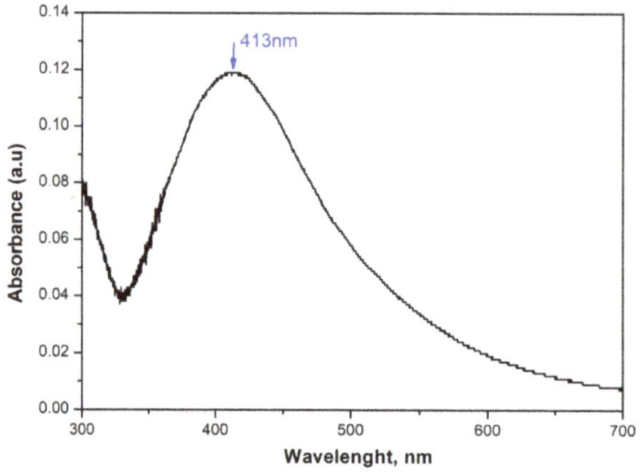

Figure 1. The UV-Vis spectrum of a solution synthesized in an ultrasonic field with 5 mM $AgNO_3$.

Figure 2 presents the X-ray diffraction (XRD) results of the obtained materials without being normalized. The appearance of broad peaks in the XRD patterns of both samples indicates the fine dispersity of the materials. The diffractogram of the synthesized MnO_2 shows a pronounced peak at 37.52° 2θ. The appearance of broader peaks with lower intensity at 25.71° and 66.69° 2θ corresponds to the (211), (220) and (112) planes characteristic of the α-MnO_2 structure. In the composite material MnO_2/AgNP, the increase in the intensity of the MnO_2 main peak can be associated with the presence of the introduced Ag nanoparticles and their partial oxidation as a result of the additional thermal treatment of the material at a temperature of about 200 °C. The main peak of MnO_2 overlaps with those of the cubic structure of Ag in the (111) direction at 37.71° 2θ, also of monoclinic AgO at 37.23 2θ. The availability of AgO was proven by the appearance of its main peak at 32.33° and others at 39.43 and 56.76 2θ, corresponding to the (−111), (−202) and (−113) crystallographic planes of the monoclinic structure of AgO.

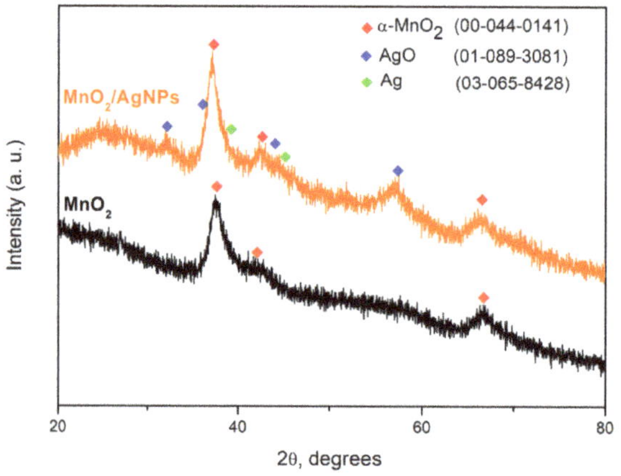

Figure 2. X-ray powder diffraction patterns of ultrasonically synthesized MnO_2 and MnO_2/AgNP and comparison with ICSD database.

The crystallite sizes (t) were calculated using Scherer's formula: $t = \kappa\lambda/B\cos\theta$ (5), where κ represents the shape factor (with a value of 0.9 for spherical crystals), λ (in Å) denotes the wavelength, θ is the diffraction angle of the peak and B (in radians) corresponds to the line broadening at the full width at half maximum (FWHM) values of the peaks.

The crystallite size at 37.5° 2θ (the primary orientation for the three crystallized phases) is roughly 5 nanometers, affirming the fine-grained characteristic of the materials.

Table 1 presents the results of the Rietveld analysis. It can be seen that structural modifications occur after the incorporation of AgNPs into the material. A change in the unit cell parameters of MnO_2 was observed, leading to a decrease in its volume.

Table 1. Unit cell parameters of synthesized MnO_2 and MnO_2/AgNP materials.

Sample	Phases	Quantity [wt %]	Unit Cell Parameters							t [nm]
			a [Å]	b [Å]	c [Å]	α [°]	β [°]	γ [°]	V [Å³]	
MnO_2	MnO_2	100	9.9340	9.9340	2.8382	90	90	90	280.08	4.76
MnO_2/AgNP	MnO_2	93.51	9.8038	9.8038	2.8385	90	90	90	272.82	5.36
	AgO	4.04	6.0884	3.4175	5.4073	90	109.107	90	106.31	
	Ag	2.45	4.0855	4.0855	4.0855	90	90	90	68.19	

The qualitative and quantitative elemental analysis of the materials was determined by the XRF technique and was consistent with the XRD study. The quantitative analysis is shown in Table 2. The registration of K in the samples can be explained by the presence of unreacted potassium permanganate which, due to its low percentage content, was not registered as an available phase in the X-ray structural analysis.

Table 2. Qualitative and quantitative elemental analysis of synthesized nanocomposites.

Sample	Element (%)		
	Mn	Ag	K
MnO_2	96.54	---	3.46
MnO_2/AgNP	91.75	7.09	1.16

The pore size distribution of MnO_2/AgNP and MnO_2 is shown in Figure 3, and Table 3 presents the main characteristics and active surface area of both materials.

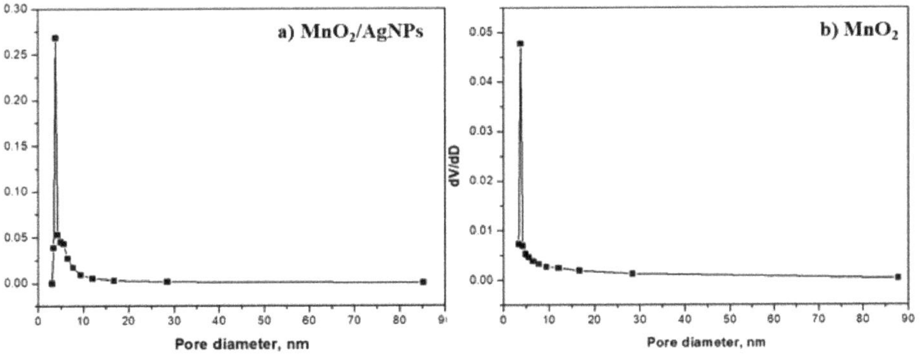

Figure 3. Pore size distribution for (**a**) MnO_2/AgNP and (**b**) MnO_2.

Table 3. Main porous parameters for the MnO_2/AgNP and MnO_2 obtained from N_2 adsorption: specific surface area (SBET), total pore volume (V_t) and average pore diameter (nm).

Samples	SBET (Multi-Point BET), m^2g^{-1}	V_t, cm^3g^{-1}	Dav (4V/S), nm
MnO_2	47	0.124	11
MnO_2/AgNP	162	0.346	9

The active surface area of the composite material containing MnO_2 and AgNPs is 162.3 m^2g^{-1}, and it is about four times smaller at 46.6 m^2g^{-1} for MnO_2, which is a prerequisite for its better properties as an electrode material. The pore size distribution (Barrett–Joyner–Halenda, BJH) of the MnO_2/AgNP sample exhibited a sharp and narrow peak at 3.8 nm with a relatively narrow range from 3 to 17 nm. For the MnO_2 sample, the results were similar, with the main peak centered again at 3.8 nm and the pore size distribution ranging from 3.5 to 16.4 nm. The average pore size for the two materials was almost the same, 9 nm (for MnO_2/AgNP) and 11 nm (for MnO_2), and mesopores exist with the presence of macropores.

The SEM images of both samples (Figure 4) show multiple irregularly shaped particles as well as individual larger agglomerates. The particle size ranges from approximately 300 to 900 nm. In the MnO_2/AgNP composite, lighter particles (under 100 nm) were observed located on the MnO_2 surface (Figure 4, highlighted in red). It is assumed that these are AgNPs, in agreement with the results of UV-VIS and XRD analyses. The morphological observation confirms the conclusions already drawn regarding the nanostructural nature of the materials. EDX analysis also shows a clear high-intensity peak with energy corresponding to Ag in the MnO_2/AgNP sample, while this peak is absent in pure MnO_2. The presence of K recorded in the spectra of both samples is related to the unreacted $KMnO_4$, as previously discussed.

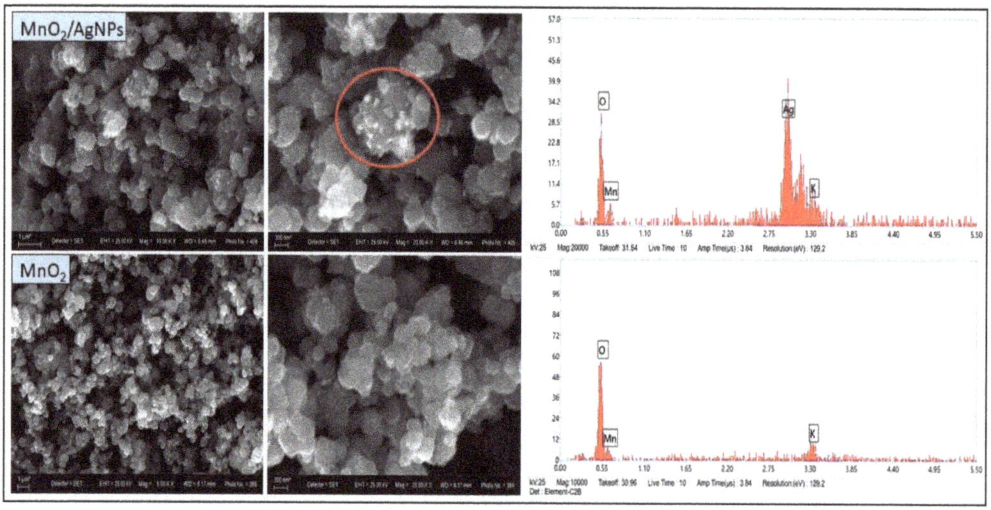

Figure 4. SEM images (**left**) and EDX spectra (**right**) of MnO_2 obtained by chemical reduction in aqueous medium.

The morphological characteristics of the synthesized materials (MnO_2 and MnO_2/AgNP) were further investigated with a transmission electron microscope, at a magnification of 40,000 times, and the results are presented in Figure 5. The TEM image of MnO_2 shows the formation of thread-like particles interconnected and thus forming a network. In the TEM image of the MnO_2/AgNP composite, the same network was observed, but in this case, it is denser due to the silver nanoparticles integrated into the manganese oxide structure. Single silver nanoparticles located on the periphery of the material with a size of about 20 nm are clearly visible, and in some places, the formation of individual clusters is also noticeable.

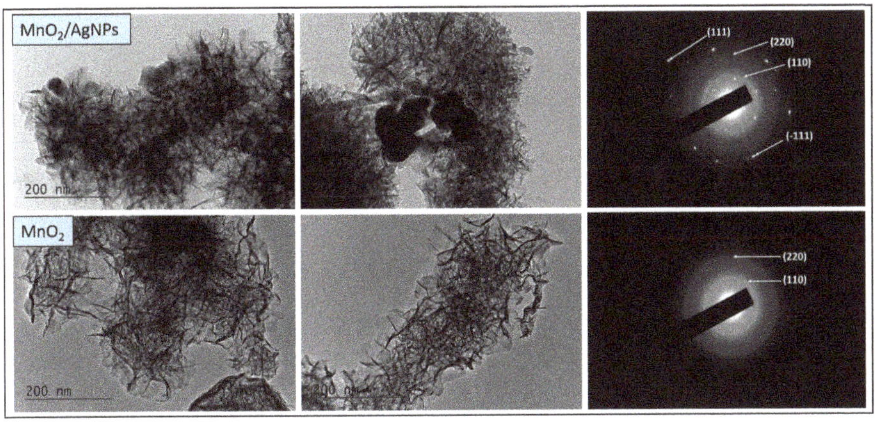

Figure 5. TEM images (**left**) and corresponding SAED patterns (**right**) of MnO_2/AgNP and MnO_2.

The clearly defined concentric rings in the SAED images indicate the interplanar spacing in the crystal structure corresponding to the (110) and (220) orientations of α-MnO_2. In the composite material, the appearance of additional rings was established, in accordance with the (111) direction of the face-centered cubic crystal lattice of silver nanoparticles, as well as (-111) of the monoclinic structure of silver oxide. These results fully confirm the XRD and SEM-EDX data.

The effect of the presence of Ag nanoparticles in the composite material on the performance of the supercapacitors was evaluated in two asymmetric configurations: AX(−)//MnO_2(+) and AX(−)//MnO_2/AgNP(+). The cells were electrochemically tested under identical experimental conditions using Na^+- and Li^+-exchange Aquivion membranes as the electrolyte and separator (Figure 6).

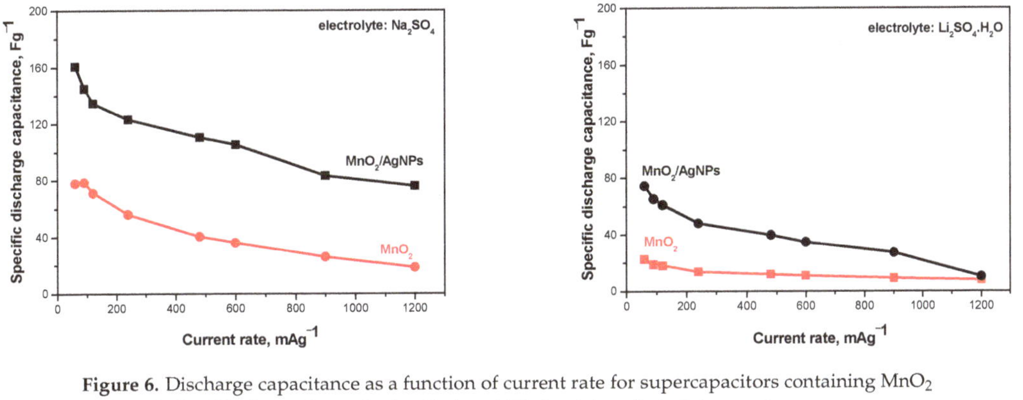

Figure 6. Discharge capacitance as a function of current rate for supercapacitors containing MnO_2 and MnO_2/AgNP electrodes and using Na^+ and Li^+ Aquivion electrolyte membranes.

It can be seen that the supercapacitors containing the MnO_2/AgNP composite as the positive electrode show a higher discharge capacitance compared to SCs with MnO_2 in both electrolytes. The result obtained can be associated with the different pore structure and surface morphology of the two materials mentioned above, which determine their different capacitive behaviors. It is understood that the distribution of pore sizes can influence the accessibility of electrolyte ions to the electrode surface, thus influencing the system's capacitance [24]. MnO_2/AgNP has a much larger active surface area, which affects the diffusion of ions. As is known, the presence of mesopores in the material supports ion diffusion to the macropores, thus transporting the ions from the electrolyte and forming an electric double layer [25]. It should also be kept in mind that the electrical conductivity of silver is very high, which is a prerequisite for Ag-doped metal oxides (in this case MnO_2) to exhibit high efficiency in the studied systems [13,26].

Figure 7 compares the cycle voltammograms of MnO_2/AgNP in lithium- and sodium-based polymer electrolyte membranes. The voltammograms with the Na^+-exchange membrane show a typical profile for supercapacitor systems with an almost rectangular shape. The curves overlap when the scan rate is increased from 1 to 40 mVs^{-1}, suggesting the excellent capacitive behavior of the devices [27]. For the Li^+ electrolyte, the rectangular shape of the CV profiles is not as well defined, and the curves at 1 mV/s differ significantly from those at higher rates. The oval profile of the CV curves using the Li^+-exchange membrane as an electrolyte may be due to either slow electron transfer or slow/insufficient ion diffusion, i.e., limited charge–discharge kinetics [28].

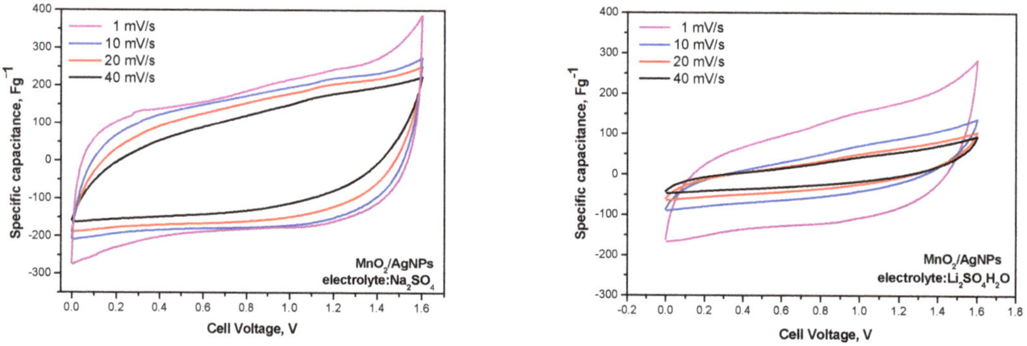

Figure 7. Cyclic voltammograms of asymmetric supercapacitors with MnO_2/AgNP electrodes and Na^+ and Li^+ Aquivion membranes performed at scan rates ranging from 1 to 40 mVs^{-1} and a voltage window from 0.0 to 1.6 V.

The results of galvanostatic charge–discharge tests carried out at different current loads (0.3–1.2 Ag^{-1}) of MnO_2/AgNP composite electrode supercapacitors in different electrolytes (respectively 1 M Na_2SO_4 and 1M $Li_2SO_4{\cdot}H_2O$) are presented in Figure 8. It can be seen that the solid-state supercapacitor using the Na^+-form Aquivion membrane shows the highest and most stable specific discharge capacitance (161 Fg^{-1}–75 Fg^{-1}), followed by $Li_2SO_4{\cdot}H_2O$ (75 Fg^{-1}–10 Fg^{-1}). These data are consistent with the literature data showing that the neutral electrolyte is suitable for pseudocapacitive materials, especially those based on MnO_2 [29]. Furthermore, the capacitive characteristics are due to the different conductivity of the electrolyte and its interaction with the functional groups of the electrode material [30].

The different wettability of the electrode materials from the electrolyte is also a possible reason for the obtained result. It is well known that surface wettability effects affect the thermodynamic and dynamic properties of electrochemical energy storage systems [31] and can contribute to increasing the effective surface area, improving the specific capacitance and also the energy density of supercapacitors [32]. To this end, in the present study, the wetting angle of MnO_2/AgNP electrodes versus Na_2SO_4 and Li_2SO_4 electrolytes was measured (Figure 9).

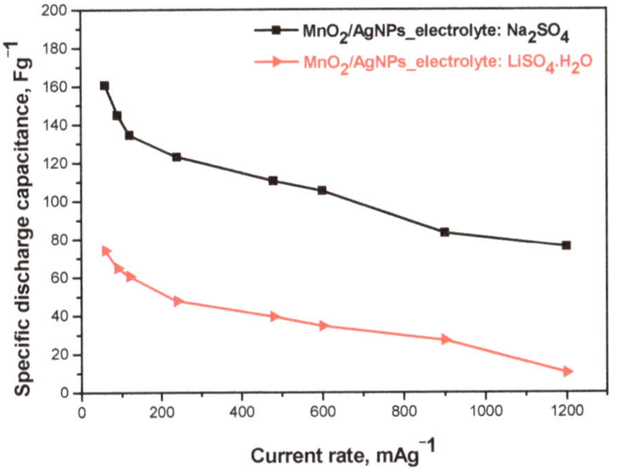

Figure 8. Comparison: discharge capacitance as function of current rate of MnO_2/AgNP SC with different electrolytes.

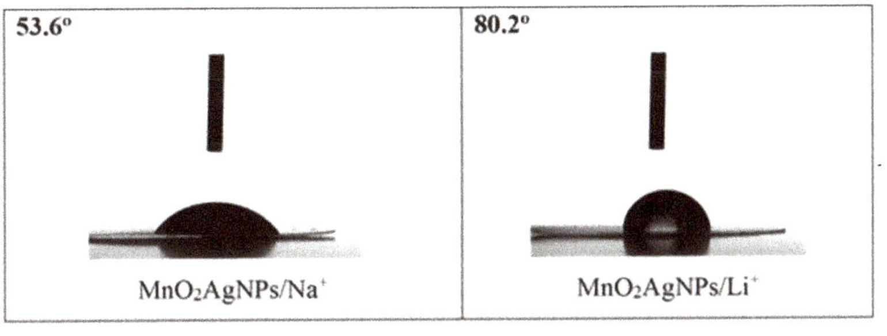

Figure 9. Wetting angles of MnO_2/AgNP electrode versus 1 M Na_2SO_4 and 1 M Li_2SO_4.

The results show that the composite electrode shows better wettability with Na_2SO_4 than with Li_2SO_4 (the wetting angle is almost two times lower). Furthermore, in our previous work, it was shown that the Aquivion membrane exhibits lower wettability, with no difference in response to the two electrolytes [22]. This may be largely related to their further electrochemical behaviors in the cell.

The good supercapacitive properties of hybrid devices with a MnO_2/AgNPs positive electrode and Na^+ electrolyte can be seen from the Ragone plot. The comparative curves of cells using the Li^+-exchange Aquivion membrane show lower energy values compared to Na^+ (Figure 10).

The features of composite MnO_2/AgNP electrodes with Na^+-form Aquivion membranes were additionally investigated focusing on their cycling stability. Long-term testing covered 1000 charge and discharge cycles under a current load of 240 mAg^{-1} (Figure 11).

From the long-term cycling, it is seen that SCs in Na_2SO_4 and $Li_2SO_4 \cdot H_2O$ have a 16 and 25% capacitance loss, respectively, with pronounced fluctuations in $Li_2SO_4.H_2O$. This probably indicates that the performance of the supercapacitor is affected by factors other than the ionic conductivity of the electrolyte and the ionic sizes of Li^+ and Na^+ [33]. The GCD curves at a discharge current rate of 240 mAg^{-1} for SCs with a Na^+-form membrane (Figure 10) show a typical shape for the behavior of a hybrid supercapacitor [34]. The iR drop is very low (0.083 V) and negligible considering the solid-state configuration of the cell [35].

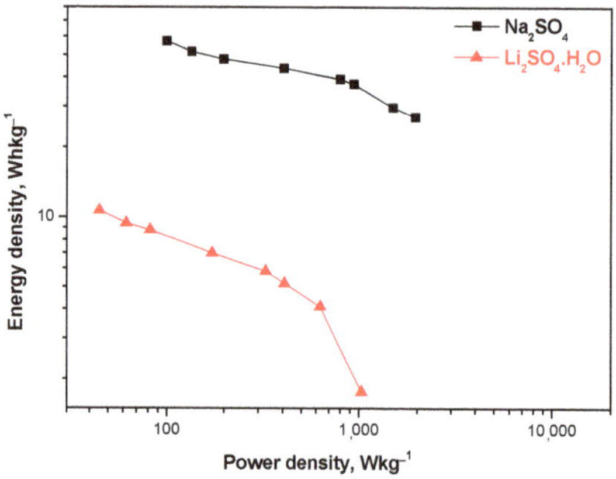

Figure 10. Energy density versus power density (Ragone plot) for SCs with MnO_2/AgNP and Na+- and Li+-exchange Aquivion membrane.

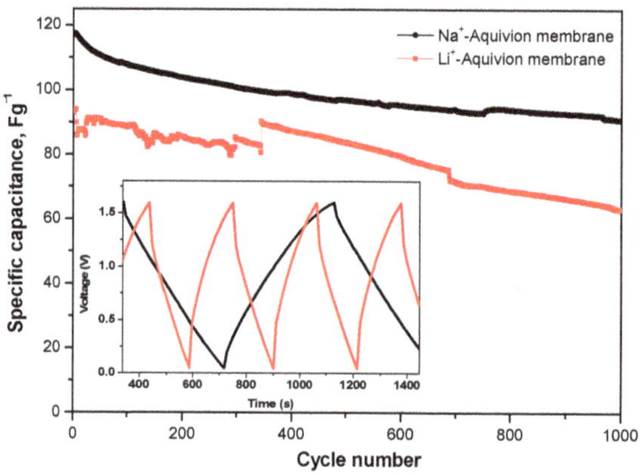

Figure 11. Long-term GCD tests at 240 mAg^{-1} for supercapacitors with MnO_2/AgNP using Na^+- and Li^+-form of Aquivion electrolyte membrane. Inset figure—charge–discharge profile.

The EIS measurement of the cell with MnO_2/AgNP using the Na^+ form of the Aquivion electrolyte membrane was conducted additionally. The Nyquist plot (Figure 12) is typical for an electrochemical supercapacitor. The inset figure represents the zoomed high-frequency region. The hinted depressed semicircle at high frequencies is a superposition of the semicircles due to the pore resistance R_p and charge transfer resistance R_{ct} of the two electrodes with a total value of 0.45 Ω. The high frequency (left) intersection of the semicircle with the Z′ axis gives the value of the series resistance (Rs = 2 Ω) and is usually due to the resistance of the electrolyte as well as the electrical contacts. In the middle and low frequencies, the Nyquist diagram appears as straight lines with low (<45°) and high (>80°) slopes that describe the diffusion process and capacitive behavior, respectively. The intersection of the continuation of the capacitive line with the abscissa gives the internal resistance of the supercapacitor Rint = $R_S + R_P + R_{ct} + R_D$ = 3.9 ohms (R_D is the diffusion resistance). This resistance is comparable to that of symmetrical carbon supercapacitors.

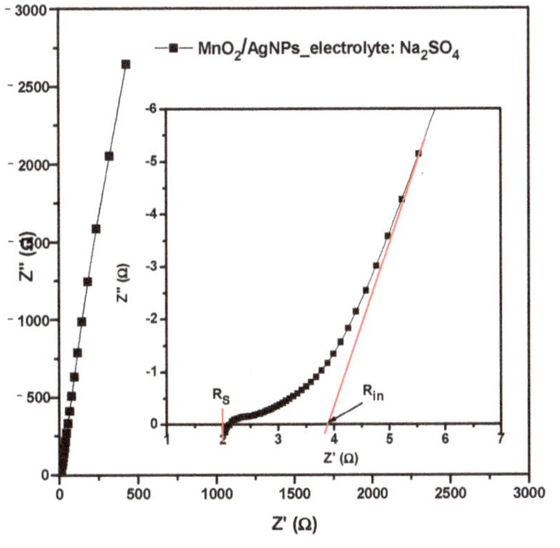

Figure 12. Nyquist plot for supercapacitor with MnO_2/AgNP using Na^+ form of Aquivion electrolyte membrane. Inset figure is zoomed high-frequency region.

The shapes of the CV curves (Figure 7) show that both the Faraday reaction and the electric double layer contribute to the charge storage. Together with the Nyquist plots (Figure 12), this suggests an ideal capacitive behavior due to the fact that Ag nanoparticles can reduce the charge transfer resistance and improve electron transfer [36].

4. Conclusions

The MnO_2/AgNP composite material was successfully synthesized and characterized structurally and morphologically using various analyses. The resulting material has a four-times-higher active surface area (162 m^2g^{-1}) than that of MnO_2 synthesized and investigated as a reference material. Electrochemical tests were carried out with the Na^+ and Li^+ form of the commercial membrane Aquivion® E87-05S used as an electrolyte. The results show that MnO_2/AgNP is a suitable flexible electrode material for solid-state supercapacitors. The introduction of AgNPs into the manganese oxide nanostructure increases cell capacitance. The highest performance is achieved when a Na^+-exchange membrane is used as the electrolyte (supercapacitors provide sufficiently high specific capacitances of 110−115 F g^{-1} at 0.2 A g^{-1}, along with an energy density of 45 Wh kg^{-1} at 405 Wkg^{-1} and in a voltage window of 0–1.6 V).

Author Contributions: Conceptualization, A.S., methodology, A.S. and. B.M; validation, B.M. and M.D.; investigation, B.M.; data curation, B.M. and M.D.; writing—original draft preparation, B.M.; writing—review and editing, A.S.; visualization, B.M; project administration, B.M.; funding acquisition, B.M. All authors have read and agreed to the published version of the manuscript.

Funding: This research was funded by the Bulgarian Ministry of Education and Science under the National Program "Young Scientists and Postdoctoral Students-2".

Data Availability Statement: Data is contained within the article.

Acknowledgments: The authors gratefully acknowledge the support of the Bulgarian Ministry of Education and Science under the National Program "Young Scientists and Postdoctoral Students-2". The authors thank Ana Arenillas for the carbon xerogel provided.

Conflicts of Interest: The authors declare no conflicts of interest. The funders had no role in the design of the study; in the collection, analyses, or interpretation of the data; in the writing of the manuscript; or in the decision to publish the results.

References

1. Muzaffar, A.; Ahamed, M.B.; Deshmukh, K.; Thirumalai, J. A review on recent advances in hybrid supercapacitors: Design, fabrication and applications. *Renew. Sustain. Energy Rev.* **2018**, *101*, 123–145. [CrossRef]
2. Gao, H.; Lian, K. Proton-Conducting Polymer Electrolytes and Their Applications in Solid Supercapacitors: A Review. *RSC Adv.* **2014**, *4*, 33091–33113. [CrossRef]
3. Huang, M.; Li, F.; Dong, F.; Zhang, Y.X.; Zhang, L.L. MnO_2-based nanostructures for high-performance supercapacitors. *J. Mater. Chem. A* **2015**, *3*, 21380–21423. [CrossRef]
4. Obeidat, A.M. Solid-state supercapacitors based on poly (3,4-ethylendioxythiophene) (PEDOT)—Manganese oxide (MnO_2) composite electrodes synthesized by single-step Co-Deposition for electrical energy storage. *Mater. Today Energy* **2018**, *10*, 81–88. [CrossRef]
5. Li, C.; Dong, X.; Zhang, Y.; Hu, J.; Liu, W.; Cui, X.; Hao, A. MnOx nanosheets anchored on a bio-derived porous carbon framework for high-performance asymmetric supercapacitors. *Appl. Surf. Sci.* **2020**, *527*, 146842. [CrossRef]
6. Sun, Y.-A.; Chen, L.-T.; Hsu, S.-Y.; Hu, C.-C.; Tsai, D.-H. Silver Nanoparticles-Decorating Manganese Oxide Hybrid Nanostructures for Supercapacitor Applications. *Langmuir* **2019**, *35*, 14203–14212. [CrossRef] [PubMed]
7. Xia, H.; Hong, C.; Shi, X.; Li, B.; Yuan, G.; Yao, Q.; Xie, J. Hierarchical hetero structures of Ag nanoparticles decorated MnO_2 nanowires as promising electrodes for supercapacitors. *J. Mater. Chem. A* **2014**, *3*, 1216–1221. [CrossRef]
8. Yuksel, R.; Coskun, S.; Unalan, H.E. Coaxial silver nanowire network core molybdenum oxide shell supercapacitor electrodes. *Electrochimica Acta* **2016**, *193*, 39–44. [CrossRef]
9. Nagamuthu, S.; Vijayakumar, S.; Muralidharan, G. Ag incorporated Mn_3O_4/AC nanocomposites based supercapacitor devices with high energy density and power density. *Dalton Trans.* **2014**, *43*, 17528–17538. [CrossRef]
10. Ma, L.; Shen, X.; Ji, Z.; Zhu, G.; Zhou, H. Ag nanoparticles decorated MnO_2/reduced graphene oxide as advanced electrode materials for supercapacitors. *Chem. Eng. J.* **2014**, *252*, 95–103. [CrossRef]
11. Chang, W.-C.; Tai, J.-T.; Wang, H.-F.; Ho, R.-M.; Hsiao, T.-C.; Tsai, D.-H. Surface PEGylation of silver nanoparticles: Kinetics of simultaneous surface dissolution and molecular desorption. *Langmuir* **2016**, *32*, 9807–9815. [CrossRef] [PubMed]
12. Yang, J.; Lan, T.; Liu, J.; Song, Y.; Wei, M. Supercapacitor electrode of hollow spherical V_2O_5 with a high pseudocapacitance in aqueous solution. *Electrochimica Acta* **2013**, *105*, 489–495. [CrossRef]
13. Guan, Y.; Guo, Z.; Che, H.; Mu, J.; Zhang, X.; Zhang, Z.; Wang, G.; Bai, Y.; Xie, H. Core/shell nanorods of MnO_2/carbon embedded with Ag nanoparticles as high-performance electrode materials for supercapacitors. *Chem. Eng. J.* **2018**, *331*, 23–30. [CrossRef]
14. Aricò, A.S.; Di Blasi, A.; Brunaccini, G.; Sergi, F.; Dispenza, G.; Andaloro, L.; Ferraro, M.; Antonucci, V.; Asher, P.; Buche, S.; et al. High temperature operation of a solid polymer electrolyte fuel cell stack based on a new ionomer membrane. *Fuel Cells* **2010**, *10*, 1013–1023. [CrossRef]
15. Lu, X.; Yu, M.; Wang, G.; Tong, Y.; Li, Y. Flexible solid-state supercapacitors: Design, fabrication and applications. *Energy Environ. Sci.* **2014**, *7*, 2160–2181. [CrossRef]
16. Xu, C.; Wei, C.; Li, B.; Kang, F.; Guan, Z. Charge storage mechanism of manganese dioxide for capacitor application: Effect of the mild electrolytes containing alkaline and alkaline-earth metal cations. *J. Power Sources* **2011**, *196*, 7854–7859. [CrossRef]
17. Wen, S.; Lee, J.-W.; Yeo, I.-H.; Park, J.; Mho, S.-I. The role of cations of the electrolyte for the pseudocapacitive behavior of metal oxide electrodes, MnO_2 and RuO_2. *Electrochimica Acta* **2004**, *50*, 849–855. [CrossRef]
18. Sajjad, M.; Khan, M.I.; Cheng, F.; Lu, W. A review on selection criteria of aqueous electrolytes performance evaluation for advanced asymmetric supercapacitors. *J. Energy Storage* **2021**, *40*, 102729. [CrossRef]
19. Fic, K.; Lota, G.; Meller, M.; Frackowiak, E. Novel insight into neutral medium as electrolyte for high-voltage supercapacitors. *Energy Environ. Sci.* **2011**, *5*, 5842–5850. [CrossRef]
20. Qu, Q.; Wang, B.; Yang, L.; Shi, Y.; Tian, S.; Wu, Y. Study on electrochemical performance of activated carbon in aqueous Li_2SO_4, Na_2SO_4 and K_2SO_4 electrolytes. *Electrochem. Commun.* **2008**, *10*, 1652–1655. [CrossRef]
21. Shao, J.; Li, X.; Qu, Q.; Wu, Y. Study on different power and cycling performance of crystalline $KxMnO_2 \cdot nH_2O$ as cathode material for supercapacitors in Li_2SO_4, Na_2SO_4, and K_2SO_4 aqueous electrolytes. *J. Power Sources* **2013**, *223*, 56–61. [CrossRef]
22. Karamanova, B.; Mladenova, E.; Thomas, M.; Rey-Raap, N.; Arenillas, A.; Lufrano, F.; Stoyanova, A. Electrochemical Performance of Symmetric Solid-State Supercapacitors Based on Carbon Xerogel Electrodes and Solid Polymer Electrolytes. *Gels* **2023**, *9*, 983. [CrossRef] [PubMed]
23. Soserov, L.; Mladenova, B.; Karamanova, B.; Lefterova, E.; Arenillas, A.; Stoyanova, A. Comparative study of electrochemical performance of symmetric supercapacitors between aqueous electrolyte and polymer exchange membrane. *Monatshefte Für Chem. Chem. Mon.* **2024**. [CrossRef]
24. Yan, R.; Antonietti, M.; Oschatz, M. Toward the Experimental Understanding of the Energy Storage Mechanism and Ion Dynamics in Ionic Liquid Based Supercapacitors. *Adv. Energy Mater.* **2018**, *8*, 1800026. [CrossRef]
25. Supiyeva, Z.; Pan, X.; Abbas, Q. The critical role of nanostructured carbon pores in supercapacitors. *Curr. Opin. Electrochem.* **2023**, *39*, 101249. [CrossRef]
26. Karami, Z.; Youssefi, M.; Raeissi, K.; Zhiani, M. Effect of the morphology of silver layer on electrical conductivity and electrochemical performance of silver/reduced graphene oxide/cotton fabric composite as a flexible supercapacitor electrode. *J. Energy Storage* **2021**, *42*, 103042. [CrossRef]

27. Rey-Raap, N.; Menéndez, J.A.; Arenillas, A. RF xerogels with tailored porosity over the entire nanoscale. *Microporous Mesoporous Mater.* **2014**, *195*, 266–275. [CrossRef]
28. Huang, C.; Zhang, J.; Young, N.P.; Snaith, H.J.; Grant, P.S. Solid-state supercapacitors with rationally designed heterogeneous electrodes fabricated by large area spray processing for wearable energy storage applications. *Sci. Rep.* **2016**, *6*, 25684. [CrossRef] [PubMed]
29. Rodriguez-Romero, J.; de Larramendi, I.R.; Goikolea, E. Nanostructured Manganese Dioxide for Hybrid Supercapacitor Electrodes. *Batteries* **2022**, *8*, 263. [CrossRef]
30. Sahoo, M.K.; Rao, G.R. A high energy flexible symmetric supercapacitor fabricated using N-doped activated carbon derived from palm flowers. *Nanoscale Adv.* **2021**, *3*, 5417–5429. [CrossRef]
31. Zhao, S.; Song, Z.; Qing, L.; Zhou, J.; Qiao, C. Surface Wettability Effect on Energy Density and Power Density of Supercapacitors. *J. Phys. Chem. C* **2022**, *126*, 9248–9256. [CrossRef]
32. Liu, T.; Wang, K.; Chen, Y.; Zhao, S.; Han, Y. Dominant role of wettability in improving the specific capacitance. *Green Energy Environ.* **2019**, *4*, 171–179. [CrossRef]
33. Li, W.; Lu, H.; Zhang, N.; Ma, M. Enhancing the Properties of Conductive Polymer Hydrogels by Freeze–Thaw Cycles for High-Performance Flexible Supercapacitors. *ACS Appl. Mater. Interfaces* **2017**, *9*, 20142–20149. [CrossRef] [PubMed]
34. Guo, J.; Zhao, Y.; Jiang, N.; Liu, A.; Gao, L.; Li, Y.; Wang, H.; Ma, T. In-situ grown $Ni(OH)_2$ nanosheets on Ni foam for hybrid supercapacitors with high electrochemical performance. *J. Electrochem. Soc.* **2018**, *165*, A882–A890. [CrossRef]
35. Thomas, M.; Veleva, S.; Karamanova, B.; Brigandì, A.; Rey-Raap, N.; Arenillas, A.; Stoyanova, A.; Lufrano, F. Highly stable and reliable asymmetric solid-state supercapacitors with low self-discharge rates. *Sustain. Mater. Technol.* **2023**, *38*, e00770. [CrossRef]
36. Li, Z.; Wang, X.; Yin, Z.; Zhao, J.; Song, M.; Wu, Z.; Li, H.; Wang, X. Ag nanoparticles decorated N/S dual-doped graphene nanohybrids for high-performance asymmetric supercapacitors. *Carbon* **2020**, *161*, 726–735. [CrossRef]

batteries

MDPI

Article

Two-Step Synthesis of $ZnS-NiS_2$ Composite with Rough Nanosphere Morphology for High-Performance Asymmetric Supercapacitors

Meng Jiang, Muhammad Abdullah, Xin Chen *, Yi E, Liyi Tan, Wei Yan, Yang Liu and Wenrui Jiang

Key Laboratory for Ultrafine Materials of Ministry of Education, Shanghai Key Laboratory of Advanced Polymeric Materials, School of Materials Science and Engineering, East China University of Science and Technology, Shanghai 200237, China

* Correspondence: xinchen73@ecust.edu.cn

Abstract: Transition metal sulfides have excellent electrochemical performance and show great potential for improving the energy density of asymmetric supercapacitors. This study demonstrates a two-step synthesis technique and highlights the enhanced energy storage efficiency of $ZnS-NiS_2$ composite materials for asymmetric supercapacitors. The composite materials of ZnS nanosheets and NiS_2 nanocrystals are characterized by a rough surface and spherical shape. The sample with the optimal ratio ($ZnS-NiS_2$-1:7) exhibits a maximum specific capacitance of 1467.9 F g^{-1} (550.5 C g^{-1}) at 1 A g^{-1}. The specific capacitance of the $ZnS-NiS_2$-1:7 sample is 26.1% higher compared to the pure NiS_2 sample. Furthermore, the assembled $ZnS-NiS_2$-1:7//AC device shows a high specific capacitance of 127.8 F g^{-1} (217.3 C g^{-1}) at 1 A g^{-1} and an energy density of 51.3 Wh kg^{-1} at a power density of 820.8 W kg^{-1}. The $ZnS-NiS_2$-1:7 sample has exceptional energy storage capability on its own, but it can also be composited with graphene to further increase the specific capacitance (1681.0 F g^{-1} at 1 A g^{-1}), suggesting promising prospects for the $ZnS-NiS_2$-based composite material in the future.

Keywords: nickel disulfide; zinc sulfide; nanosphere; electrochemical characterization; asymmetric supercapacitor

1. Introduction

Environmental issues resulting from fossil fuel consumption have prompted researchers to concentrate on energy conversion and storage devices as practical alternatives to the ever-increasing demand for fossil fuels. Therefore, the evolution of energy storage devices, like batteries, fuel cells, and supercapacitors (SCs), has been the subject of numerous studies [1,2]. SCs are among the most promising electrochemical energy storage devices, with advantages such as high energy density, quick charge/discharge, and high cycling stability [3]. However, SCs have a disadvantage in terms of their relatively low energy density in comparison to batteries. As a result, searching for stable electrode materials designed for providing high energy density is never ending [4]. According to the energy storage mechanism, SCs can be divided into "pseudo-capacitors" and electric double-layer capacitors (EDLC) [5]. In order to address the limitations associated with comparatively low energy density, asymmetric supercapacitor (ASC) devices were assembled with a pseudo-capacitor electrode and an EDLC electrode to show high energy density [6]. The energy density of ASC devices can be improved by effectively employing the potential gap between two kinds of electrodes and expanding the entire voltage window.

Transition metal compounds such as transition metal oxide [7,8], transition metal hydroxide [9,10], and transition metal sulfide (TMS) are considered appropriate pseudo-capacitor materials. Among them, TMS has the advantages of high energy density, high conductivity, and good cycle stability. According to the literature, as the electronegativity of sulfur is lower than that of oxygen, the TMS has more flexibility in structure design and higher ion diffusivity [11–13]. In the field of TMS, nickel sulfide is very attractive because of its high theoretical capacity, good cyclic stability, simple synthesis, and low cost. Xie [14] et al. synthesized a composite material of α-NiS, β-NiS, Ni_3S_4, and rGO by reducing thiourea. The composite material exhibited a high specific capacity of 609.4 C g^{-1} at 1 A g^{-1} and a high coulombic efficiency of 99.5%. Wang [15] et al. developed MoS_2/NiS composite electrode materials with electrochemical deposition, ionic layer adsorption, and reaction methods. The MoS_2/NiS composite showed a specific capacitance of 721.4 mF cm^{-2} at a current density of 1 mA cm^{-2}. Dar [16] et al. synthesized manganese-doped tin sulfide (SnS) NPS with varying manganese doping ratios (XMn = 0–9%) using the solvothermal method. The resulting single-phase orthorhombic crystal structure exhibited an increase in specific capacitance values from 9 to 30 (F g^{-1}) with Mn dopant, suggesting that external impurities have enhanced the potential of pure SnS NPs as an alternative source for energy storage devices. Zhang [17] et al. synthesized nanoparticles of NiS and NiS_2 through a solid-phase synthesis method. The morphology of NiS and NiS_2 is controlled by the molar ratio of Ni and S. The obtained nanoparticles showed a specific capacitance of 1072.6 F g^{-1} at 2 A g^{-1}. Furthermore, synthesizing binary metal sulfides and controlling the morphology of electrode materials has been proven to be an effective way to promote the electrochemical properties of the electrode materials [18].

ZnS is a proficient metal conductor that facilitates the transport of ions and electrons. It has been widely used in different applications, such as inorganic–organic hybrid photovoltaics [19], solar cells [20], and photocatalysts [13]. Furthermore, ZnS also has a high theoretical capacitance, a wide bandgap of 3.5 to 3.8 eV, and rich redox activity, making it attractive for electrode applications [21]. Palanisamy [22] et al. reported composite electrode materials of WO_3-ZnS prepared through a microwave assisted method. The morphology of the WO_3-ZnS composites was irregular nanoparticles, which can be controlled through selective absorption of the microwave. The WO_3-ZnS composites showed a specific capacitance of 215.0 F g^{-1}, which is 171.0 F g^{-1} higher than that of WO_3. This notable enhancement in specific capacitance can be attributed to the high charge transfer rate of ZnS. Yu [23] et al. prepared $Cu_5Sn_2S_7$-ZnS composite materials using a solvothermal method. The surface of the $Cu_5Sn_2S_7$-ZnS composite showed a flower-like structure. According to electrochemical tests, the addition of ZnS may significantly improve the specific capacitance of $Cu_5Sn_2S_7$, which can be attributed to the synergistic effect between $Cu_5Sn_2S_7$ and ZnS. The assembled supercapacitor device showed an energy density of 11.1 Wh kg^{-1} at a power density of 461.0 W kg^{-1}. In order to construct a supercapacitor with excellent energy density, composite electrode materials must be designed properly.

According to our review of the relevant literature, no research has been conducted on the facile synthesis and characterization of ZnS-NiS_2 composite electrode materials. Our research suggests that composite electrode materials made of ZnS and NiS_2 will have improved performance due to the synergistic effect of the high specific capacitance provided by NiS_2 and the higher ion and charge transfer rates delivered through ZnS. The unique combination of flake ZnS and spherical NiS_2 provides a new idea for high performance composite materials. Additionally, ZnS-NiS_2 composite materials have the advantages of simple composition, cost-effectiveness, and environmental friendliness. This paper suggests an interesting research direction for synthesizing optimized composite materials for supercapacitors.

Herein, we have synthesized the composite electrode materials of ZnS and NiS_2 by using the colloidal chemical method and the solvothermal method, respectively. The obtained ZnS-NiS_2 composites have the morphology of nanospheres with rough surfaces. The electrochemical performances of the NiS_2-based composites were improved by introducing a low portion of ZnS. The optimized ZnS-NiS_2-1:7 sample showed a high specific capacitance of 1467.9 F g^{-1} at 1 A g^{-1} and outstanding energy storage properties. After assembly with commercial activated carbon (AC), the ZnS-NiS_2-1:7//AC device showed a high energy density of 51.3 Wh kg^{-1} at a power density of 820.8 W kg^{-1} and good cyclic stability of 94.9% retention after 5000 cycles. Furthermore, $ZnS-NiS_2$-1:7 can be composited with graphene nanosheets to achieve an even higher specific capacitance of 1681.0 F g^{-1} at 1 A g^{-1}.

2. Materials and Methods

2.1. Synthesis of ZnS Nanosheets

ZnS nanosheets were prepared through a colloidal chemical method. First, 0.5 mmol of zinc acetylacetonate ($Zn(C_5H_7O_2)_2$) was dissolved in 20 mL of oleylamine (OLA). The mixed liquid was magnetically stirred for 30 min to obtain a uniform solution. The solution was transferred into a four-necked flask and heated to 120 °C under a N_2 atmosphere for 30 min. Subsequently, 3.6 mmol of sulfur powder, dissolved in 5 mL of OLA, was gradually added to the four-necked flask. After that, the temperature was quickly increased to 330 °C and maintained for 6 h. After cooling down to room temperature naturally, the solution was washed several times with a mixture of cyclohexane and ethanol and dried at 60 °C for 12 h. Eventually, ZnS samples were annealed under the N_2 atmosphere at 500 °C for 2 h before sealed storage.

2.2. Synthesis of $ZnS-NiS_2$ Composites

$ZnS-NiS_2$ composites were synthesized by a solvothermal method. A certain amount of ZnS nanosheets prepared in the previous step was added to 50 mL of N,N-dimethylformamide (DMF) and ultrasonicated for an hour. After that, 1.5 mmol of nickel acetate ($Ni(CH_3COO)_2 \cdot 4H_2O$) and 9.3 mmol of thioacetamide (CH_3CSNH_2) were added to 25 mL of ethanol and magnetically stirred for 0.5 h. The above two solutions were mixed and transferred into a 100 mL Teflon-lined stainless-steel autoclave. The autoclave was sealed and heated to 150 °C for 1.5 h. After cooling down to room temperature, the $ZnS-NiS_2$ samples were collected by centrifugation and washed several times with deionized water and ethanol. Finally, the $ZnS-NiS_2$ composites were dried at 60 °C for 12 h. Different composite ratios of ZnS and NiS_2 (0:1, 1:9, 1:7, 1:5) were achieved by controlling the amount of ZnS nanosheets and were named $ZnS-NiS_2$-0:1, 1:9, 1:7, and 1:5, respectively.

2.3. Synthesis of $ZnS-NiS_2$-G Composites

$ZnS-NiS_2$-G composites were synthesized by a facile physical method. An amount of 100 mg of the prepared $ZnS-NiS_2$-1:7 sample was added to 20 mL of ethanol. A certain amount of graphene powder was weighed and added to the above-mentioned mixed liquid. The mixed liquid was further processed by ultrasonic treatment for 30 min to form a uniform dispersion. The dispersion was magnetically stirred for 24 h, and the container was sealed to avoid ethanol volatilization. Finally, the black powders were collected through centrifugation. $ZnS-NiS_2$-G composites were washed several times with deionized water and ethanol and dried at 60 °C for 12 h. The $ZnS-NiS_2$-G composites with different graphene mass ratios were synthesized by controlling the weight of graphene powder and were named $ZnS-NiS_2$-G (3%, 5%, 7%, 9%).

2.4. Materials Characterization

The crystal structures of the samples were characterized by X-ray diffraction (XRD) in a D/max2550VB/PC diffractometer with Cu Kα radiation (λ = 1.5418 Å), and 2θ ranged from 10° to 80°. The X-ray photoelectron spectrometer (XPS, Thermo Fisher ESCALAB 250Xi, Waltham, MA, USA) was used to examine the elemental composition and their chemical states on the shallow layer of the sample surface. The surface morphology of samples was studied by field emission scanning electron microscopy (S-4800, Hitachi, Tokyo, Japan). Transmission electron microscopy (TEM) was performed on a JEL-2100 electron microscope to take higher-resolution images. Selected area electron diffraction (SAED) was performed to analyze the crystal properties of samples.

2.5. Electrochemical Measurements

The electrochemical properties of the samples were characterized by a three-electrode system. A 3 M KOH solution was used as the electrolyte. The nickel foam coated with synthesized electrode material (1 cm × 3 cm) was used as the working electrode. A piece of platinum foil (1 cm × 1 cm) was utilized as the counter electrode, and a saturated calomel electrode (SCE) was used as the reference electrode.

Cyclic voltammetry (CV) and galvanostatic charge/discharge (GCD) tests were performed on an electrochemical workstation (Bio-Logic, Seyssinet-Pariset, France). The electrochemical impedance spectra (EIS) test was operated on a CHI 660E electrochemical workstation. The amplitude of the sinusoidal voltage was set at 5 mV, with a frequency range of 0.01 Hz to 100 kHz.

The working electrode was prepared using the following method: The obtained samples, carbon black, and polyvinylidene fluoride (PVDF), were mixed with a mass ratio of 80:10:10. A few drops of N-methylpyrrolidone (NMP) solutions were added to the mixture. The mixture was stirred by hand for 25 min and brushed onto a piece of nickel foam. The prepared nickel foam was dried at 60 °C for 12 h. The loading mass of electrode materials on each nickel foam was about 2–3 mg.

A supercapacitor device was assembled with prepared samples and commercial AC. A 3 M KOH solution was used as the electrolyte, and a circular polytetrafluoroethylene (PTFE) film was utilized as the separator to separate the two electrodes. The assembled HSC device was passed through the manual tablet press, and the thickness of the active material electrodes was negligible at the micron scale. The values' specific capacitance (C), energy density (E), and power density (P) can be calculated with the following equations [24,25]:

$$C = It/mV \quad (1)$$

$$E = 1/2\, CV^2 \quad (2)$$

$$P = E/t \quad (3)$$

where C (F g^{-1}) is the mass specific capacitance of the active material, I (A g^{-1}) is the discharging current, t (s) is the discharging time, m (g) is the mass of the active materials, and V (V) is the working potential in the charging–discharging process. The loading mass of the positive electrode and the negative electrode in supercapacitor devices can be calculated by the following equation [24]:

$$m^+/m^- = C^-\ V^-/C^+V^+ \quad (4)$$

where m^+ and m^- (g) are loading masses of active materials, C^+ and C^- (F g^{-1}) are the specific capacitances, and V^+, V^- (V) are the working potential windows of the positive and negative electrodes.

3. Results and Discussion

The synthesis process of the ZnS-NiS_2 composite sample is shown in Figure 1. First, zinc acetylacetonate powder and sulfur powder were dispersed in oleylamine, respectively. After that, the two dispersions were heated in a four-necked flask at 330 °C under the N_2 atmosphere. The products were annealed at 500 °C to remove organic impurities and the ZnS nanosheets were collected for further use. The composites of ZnS-NiS_2 were synthesized by a solvothermal method, which was achieved at a relatively low temperature and in a short time.

The crystal structure of the ZnS-NiS_2 samples was studied by XRD analysis. Figure 2 shows the XRD patterns of ZnS-NiS_2 samples with different composite ratios. For the pure NiS_2 sample (ZnS-NiS_2-0:1), the peaks at 31.5°, 35.4°, and 53.5° can be matched with the (200), (210), and (311) crystal planes of NiS_2 (ICDD 65-3325), respectively. For the ZnS-NiS_2 composite samples, diffraction peaks of both ZnS and NiS_2 crystals can be found. The diffraction peaks at about 31.0°, 37.8°, and 53.5° can be matched with the (200), (211), and (311) crystal planes of NiS_2 (ICDD 65-3325). According to our search results, ICDD 65-3325 is the most matched stand card, where the diffraction peak of the (211) crystal plane in the standard card is located at 38.8°, but the peak in the XRD pattern was 1.0° lower than the standard card value. According to the Bragg equation, the expansion of crystal plane spacing was calculated to be 0.04 Å. Such a peak shift happens when ZnS and NiS_2 are composited together, which could be related to the stress at the ZnS/NiS_2 interfaces and/or Zn doping into the NiS_2 nanocrystal lattices. The size of the crystalline domains in a material can also influence the XRD pattern. In ZnS-NiS_2 composites, the presence of NiS_2 nanoparticles dispersed within the ZnS matrix can lead to size effects. The size of the nanoparticles can affect the peak positions due to changes in the crystallite size, strain, and lattice parameters [26].

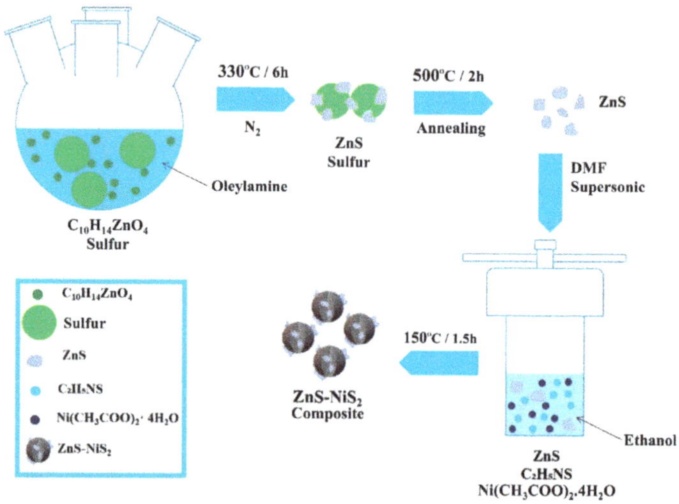

Figure 1. Synthesis process of ZnS-NiS_2 composite sample.

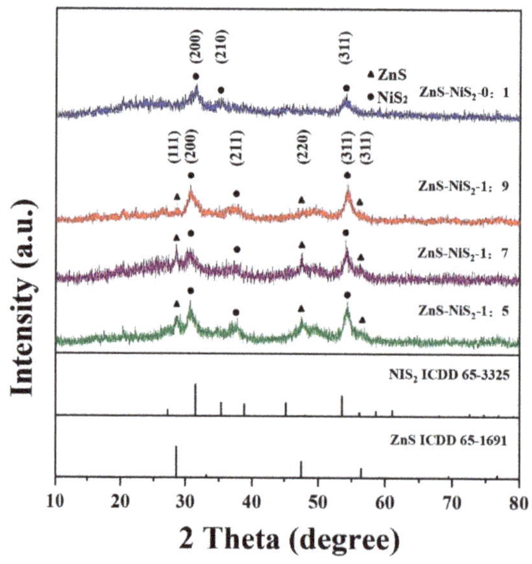

Figure 2. XRD pattern of ZnS-NiS_2 samples with different composite ratios.

The peaks at 28.6°, 47.5°, and 56.4° can be matched with the (111), (220), and (311) crystal planes of the ZnS (ICDD 65-1691), confirming the formation of the ZnS-NiS_2 composite. Furthermore, the peaks of NiS_2 in the composites are relatively wide and weak, indicating that the low crystallinity of NiS_2 is due to the short heating time and low heating temperature during the solvothermal process. The Debye-Scherrer equation was used to calculate the average size of NiS_2 crystals in the directions perpendicular to the (200) and (311) crystal planes as 7.7 and 9.5 nm. As the concentrations of ZnS in the composites are low, the ZnS signals are generally weaker than those of NiS_2. However, in the ZnS-NiS_2-1:7 sample, the ZnS signals are narrower and have higher intensities, suggesting that the ZnS nanosheets are larger in size than in the other composites, which will be discussed further below.

Figure 3 displays the FESEM images of the ZnS-NiS_2 samples. The morphology of the composite samples is quite similar; they all appear to be rough spheres. As illustrated in Figure 3a, when the ZnS composite ratio is 0, roundish protrusions can be observed on the surface of NiS_2 nanospheres. However, the average particle size is about 265 nm, and they gather into a cluster. The ZnS-NiS_2 composites in Figure 3b–d exhibit sharper protrusions on the surfaces of the spheres, indicating the presence of ZnS nanosheets. Among these, the nanosheet morphology on the surface is most clearly observed in Figure 3c. This observation is consistent with the XRD analysis, showing significant growth in the size of the nanosheets in the ZnS-NiS_2-1:7 sample. In Figure 3b, the particle size is reduced to some extent compared with Figure 3a. On the other hand, the ZnS-NiS_2-1:7 and ZnS-NiS_2-1:5 samples show similarity to Figure 3a. As shown in Figure 3c, the ZnS-NiS_2-1:7 sample exhibits a greater amplitude of surface fluctuation. This characteristic facilitates the enhancement of a larger specific surface area, hence exposing a greater number of active sites for redox reactions. When the ZnS concentration is further increased, as shown in Figure 3d, there are more surface height fluctuations at a certain length. These fluctuations may be attributed to the higher ZnS concentration. However, the surface fluctuation amplitude is not as high as that of the ZnS-NiS_2-1:7 sample. This observation suggests that the ZnS nanosheets are now relatively closely attached to the sphere surfaces. As a result, the growth of ZnS into larger nanosheets during the synthesis process is slowed down.

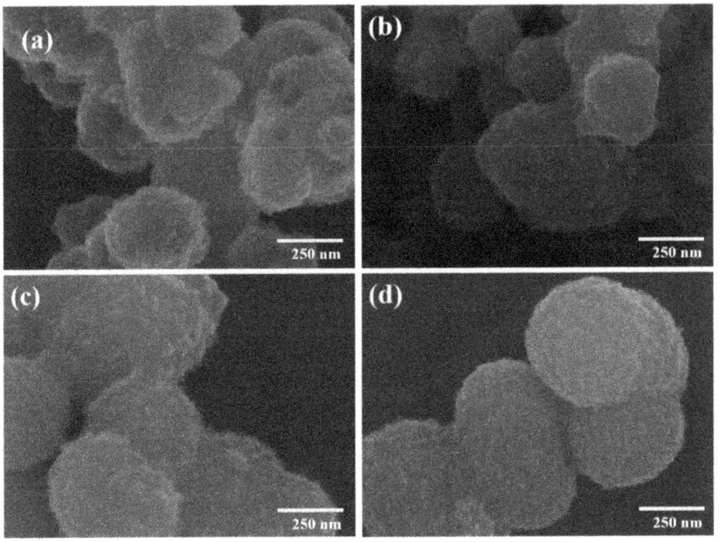

Figure 3. FESEM images of. (**a**) ZnS-NiS_2-0:1, (**b**) ZnS-NiS_2-1:9, (**c**) ZnS-NiS_2-1:7, (**d**) ZnS-NiS_2-1:5.

Furthermore, the morphology of the ZnS-NiS_2-G (5%) sample was also studied. As illustrated in Figure S1 in the Supplementary Materials, the sample exhibited both graphene nanosheets and ZnS-NiS_2-1:7 nanoparticles.

EDS mapping tests were performed on the ZnS-NiS_2-1:7 sample to identify the distribution of S, Ni, and Zn elements in the sample, as shown in Figure 4. The FESEM image of the ZnS-NiS_2-1:7 sample is shown in Figure 4a, while Figure 4b–d show the distribution of S, Ni, and Zn elements, respectively. It can be seen that the distribution areas of Ni and S elements are consistent with the areas of the ZnS-NiS_2-1:7 nanospherical composite. The signals of the Zn elements are characterized by a comparatively low level and scattered distribution. However, there is a slightly higher density of signal spots observed in the composite sample area, which aligns with the sample region depicted in Figure 4a. The low Zn signal level is caused by the low concentration of Zn in the composite sample.

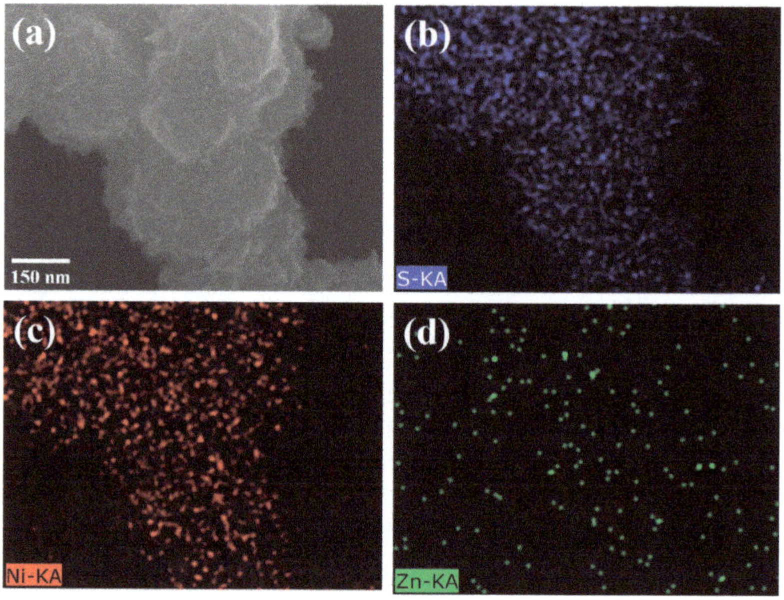

Figure 4. (**a**) FESEM image of ZnS-NiS_2-1:7sample, (**b–d**) EDS mapping of S, Ni, and Zn elements.

The detailed morphology and internal structure of the ZnS-NiS_2-1:7 sample were further studied through TEM, as shown in Figure 5. Figure 5a shows the TEM image of the ZnS-NiS_2-1:7 sample, which clearly shows that the sample is composed of nanospheres with nanosheets visible on the sphere surfaces. This observation should not simply be interpreted as evidence of core-shell structures with a NiS_2 core and ZnS nanosheet shell for the rough spheres. In the synthesis procedures, the nanosheets are initially prepared, followed by the synthesis of NiS_2 nanocrystals from the surrounding solutions. Subsequently, the limited signal penetration depth of the TEM technology hinders the clarity of the inner regions of the spheres. However, the existence of nanosheets within the spheres cannot be dismissed. Conversely, the TEM image provides substantial evidence of nanosheets within the central region of numerous spherical particles. Figure 5b,c show HRTEM images of the sample. Upon closer zooming of the region enclosed by the yellow dashed box in Figure 5b, a set of lattice stripes with a spacing of 0.125 nm is found. This observation suggests a correlation between these lattice stripes and the (331) crystal plane of ZnS. Similarly, upon zooming in on the region enclosed by the dashed box in Figure 5c, a set of lattice stripes with a spacing of 0.235 nm is observed, which corresponds to the (211) crystal plane of NiS_2. The HRTEM images not only demonstrate the presence of both ZnS and NiS_2, but also eliminate the proposed core-shell structure by observing both ZnS and NiS_2 on the sphere surfaces. Figure 5d shows the SAED image of the ZnS-NiS_2-1:7 sample, which shows diffraction spots distributed on several concentric diffraction rings. These rings can be matched with the (311) and (220) crystal planes of ZnS and (211), (210), and (200) crystal planes of NiS_2, respectively. The SAED image confirms the existence of ZnS and NiS_2 crystalline phases in the composite material. XPS was used to examine the surface composition and chemical states of the ZnS-NiS_2-1:7 sample, as shown in Figure S3. Furthermore, TEM and SAED images of the ZnS-NiS_2-G (5%) sample can be seen in Figure S2.

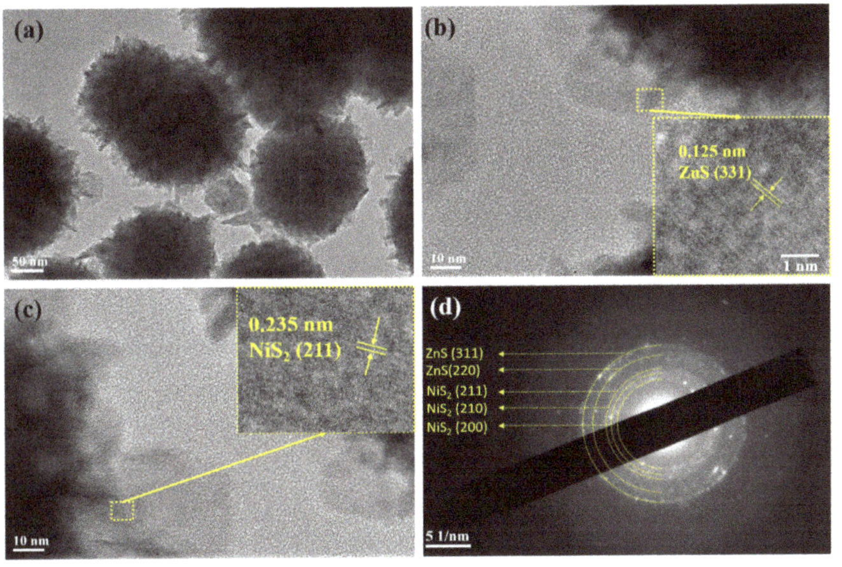

Figure 5. (**a**) TEM image of ZnS-NiS_2-1:7 sample, (**b**,**c**) HRTEM image of ZnS-NiS_2-1:7 sample, (**d**) Selected area electron diffraction image of ZnS-NiS_2-1:7 sample.

The electrochemical performance of the ZnS-NiS_2 samples was tested in a three-electrode system. A 3 M KOH solution was used as the electrolyte. However, a CV test was performed to study the redox behavior of the ZnS-NiS_2 samples during the charge and discharge processes. Figure 6a shows the CV curves of the ZnS-NiS_2-1:7 sample at different scan rates. The range of the potential window was set to be −0.25–0.65 V. Notably, redox peaks can be observed clearly in Figure 6a. This observation confirms that faradaic reactions occurred during the charge and discharge processes, and the ZnS-NiS_2-1:7 sample exhibits pseudo-capacitive behavior. When the scan rate increases, the positions of the redox peaks shift to higher and lower potentials, respectively. This may be caused by polarization at high scanning rates. Therefore, the potential window of the CV test was set to be −0.25–0.65 V for the complete redox peaks. The chemical reaction related to redox peaks may be described with the following equations [27,28]:

$$ZnS + OH^- \leftrightarrow ZnSOH + e^- \quad (5)$$

$$NiS_2 + OH^- \leftrightarrow NiS_2OH + e^- \quad (6)$$

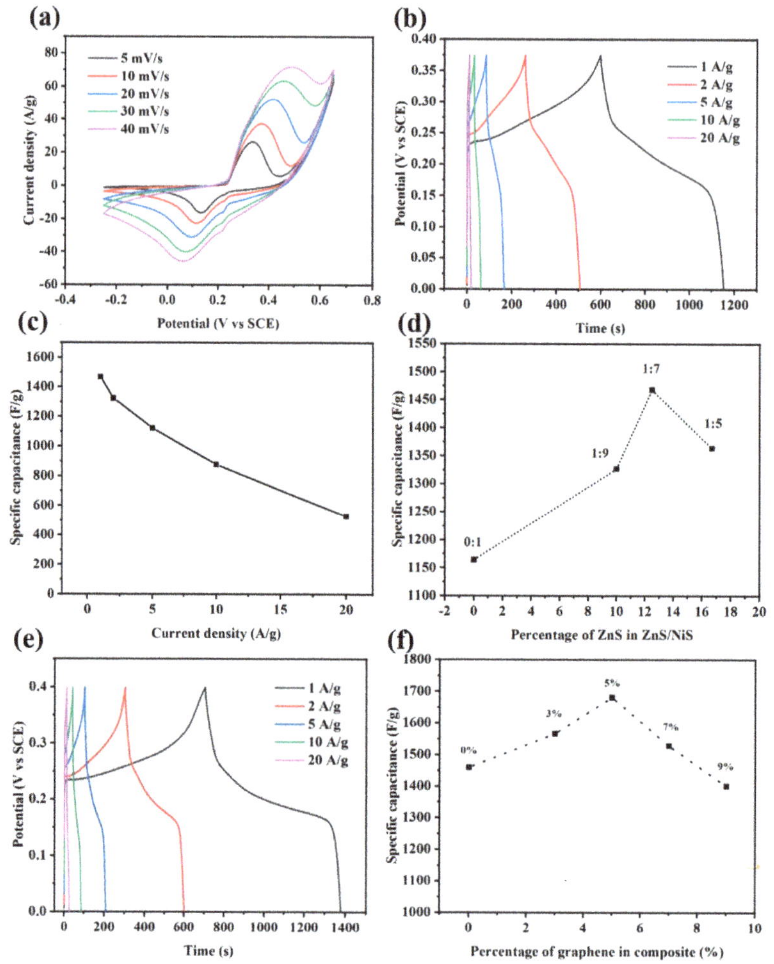

Figure 6. (**a**) CV diagram of ZnS-NiS_2-1:7, (**b**) GCD diagram of ZnS-NiS_2-1:7, (**c**) rate performance of ZnS-NiS_2-1:7, (**d**) specific capacitance of different ZnS-NiS_2 at 1 A g^{-1}, (**e**) GCD curves of ZnS-NiS_2-G(5%), (**f**) specific capacitance of ZnS-NiS_2-G samples with different graphene mass ratios at 1 A g^{-1}.

The observed dip and rise in the CV curve at about 0.2 V might be due to potential phase transitions or structural changes in the ZnS or NiS_2 material within that particular potential range, consequently causing variations in the CV data [29]. Figure 6b shows the GCD curves of the ZnS-NiS_2-1:7 sample. The potential range was set to be 0–0.375 V. Additionally, distinct shoulders can be observed in both the charge and discharge curves. It implies that the capacitance is mainly attributed to the faradic reaction and that the sample is a pseudo-capacitive material [30]. Furthermore, the specific capacitance of the ZnS-NiS_2-1:7 sample was calculated using the Formula (1). The ZnS-NiS_2-1:7 sample has specific capacitances of 1467.9, 1323.4, 1122.3, 877.7, and 527.8 F g^{-1} at 1, 2, 5, 10, and 20 A g^{-1}, respectively. In GCD curves, the platform area appears to be around 0.25 V. Notably, there is some difference between redox peaks in CV curves and platform area in GCD curves. This is because CV tests were performed under constant scanning speed, while GCD tests were performed under constant current density conditions. So the position of redox peaks and platform area can be different [31]. Figure 6c shows the rate performance of the ZnS-

NiS_2-1:7 sample, wherein the specific capacitance remains 35.9% at 20 A g^{-1}. Furthermore, the CV and GCD curves of ZnS-NiS_2-0:1, 1:9, and 1:5 samples are shown in Figure S3. The electrochemical performance of the ZnS-NiS_2 composite increases as the composite ratio increases. This is due to the increase in specific surface area and the synergistic effect between ZnS and NiS_2. When the composite ratio reaches 1:5, ZnS nanosheets are relatively closely attached to the sphere surfaces, resulting in a decrease in the specific capacitance. Figure 6d shows the specific capacitance of the ZnS-NiS_2 samples with different composite ratios. The ZnS-NiS_2-1:7 sample has the highest specific capacitance of 1467.9 F g^{-1}, while the ZnS-NiS_2-0:1 sample has a specific capacitance of 1164.3 F g^{-1}. Additionally, the ZnS-NiS_2-1:7 sample exhibits a notable enhancement in specific capacitance, with a value of 26.1% increase compared to the ZnS-NiS_2-0:1 sample. The high specific capacitance of the ZnS-NiS_2-1:7 sample may be attributed to the synergistic effect between NiS_2 and ZnS. Furthermore, the surface morphology also plays a role in the observed increase in specific capacitance. The ZnS-NiS_2-1:7 sample has a larger surface fluctuation amplitude, which may help get a larger specific surface area and more active sites. Table 1 summarizes the comparison of our research outcomes with those of the other relevant literature.

Table 1. The supercapacitor performance in the current study is compared with previous reports based on the same material composition.

Type of Material	Electrolyte	Specific Capacity (F g^{-1})	Current Density	Cycle Stability	Ref.
PANI/ZnS QDs	1 M H_2SO_4	893.75	0.5 A g^{-1}	59.7%@1000	[32]
ZnS/SOM-C	6 M KOH	1158	1 A g^{-1}	71%@8000	[33]
ZnS/RGO/PANI	6 M KOH	1045.3	1 A g^{-1}	160%@1000	[34]
NiS_2/Ti_3C2T_x	1 M KOH	518.4	1 A g^{-1}	77.27%@9000	[35]
Co-NiS_2/C	3 M KOH	1080	1 A g^{-1}	89.2%@8000	[36]
NiS_2@C-rGO	1 M H_2SO_4	1297	1 A g^{-1}	95.3%@10,000	[37]
Pure pyrite NiS_2	6 M KOH	1072.6	2 A g^{-1}	78.1%@1000	[17]
ZnS-NiS_2	3 M KOH	1467.9	1 A g^{-1}	94.9%@5000	This work

The electrochemical performance of the ZnS-NiS_2-G samples was also tested. Among these, Figure 6e shows the GCD diagrams of the optimized ZnS-NiS_2-G (5%) sample. The GCD curves of the ZnS-NiS_2-G (5%) sample are similar to those of the ZnS-NiS_2-1:7 sample. It implies that the addition of a small amount of graphene would not change the overall pseudo-capacitive behavior of the electrode material. The specific capacitance of the ZnS-NiS_2-G (5%) sample was calculated to be 1681.0, 1532.4, 1391.2, 1261.6, and 937.5 F g^{-1} at 1, 2, 5, 10, and 20 A g^{-1}, respectively. The specific capacitance remained 55.8% at 20 A g^{-1}. After adding graphene, the specific capacitance of ZnS-NiS_2-G (5%) is higher than that of the ZnS-NiS2-1:7 sample. Furthermore, the CV and GCD curves of the ZnS-NiS_2-G (3%, 7%, and 9%) samples are shown in Figure S4. Figure 6f shows the effect of the graphene mass ratio on the specific capacitance. The ZnS-NiS_2-G (0%, 3%, 5%, 7%, and 9%) samples showed specific capacitances of 1460.0, 1566.7, 1681.0, 1528.8, and 1401.1 F g^{-1} at 1 A g^{-1}, respectively. Additionally, the ZnS-NiS_2-G (5%) sample has the highest specific capacitance among all of the electrode materials. Furthermore, adding an appropriate amount of graphene has been demonstrated to increase the specific capacitance of the ZnS-NiS_2-based composite material.

The comprehensive rate performance of the ZnS-NiS_2-G samples is shown in Figure 7a. It is observed that ZnS-NiS_2-G (3%, 5%, 7%) samples have a higher specific capacitance compared to ZnS-NiS_2-G (0%). Notably, this improvement is more pronounced at higher current densities. This enhancement is due to the excellent conductivity of the graphene nanosheets. The improvement in conductivity of the composite material facilitates the attainment of sufficient redox reactions, especially at high current densities. On the other hand, the specific capacitance of the ZnS-NiS_2-G (9%) sample significantly decreases when the mass ratio of graphene is too high. The relatively low specific capacitance of graphene leads to negative effects on overall performance. In addition, the ultra-high mass ratio of graphene may cause stacking of the graphene sheets and reduce the practical specific surface area of the sample.

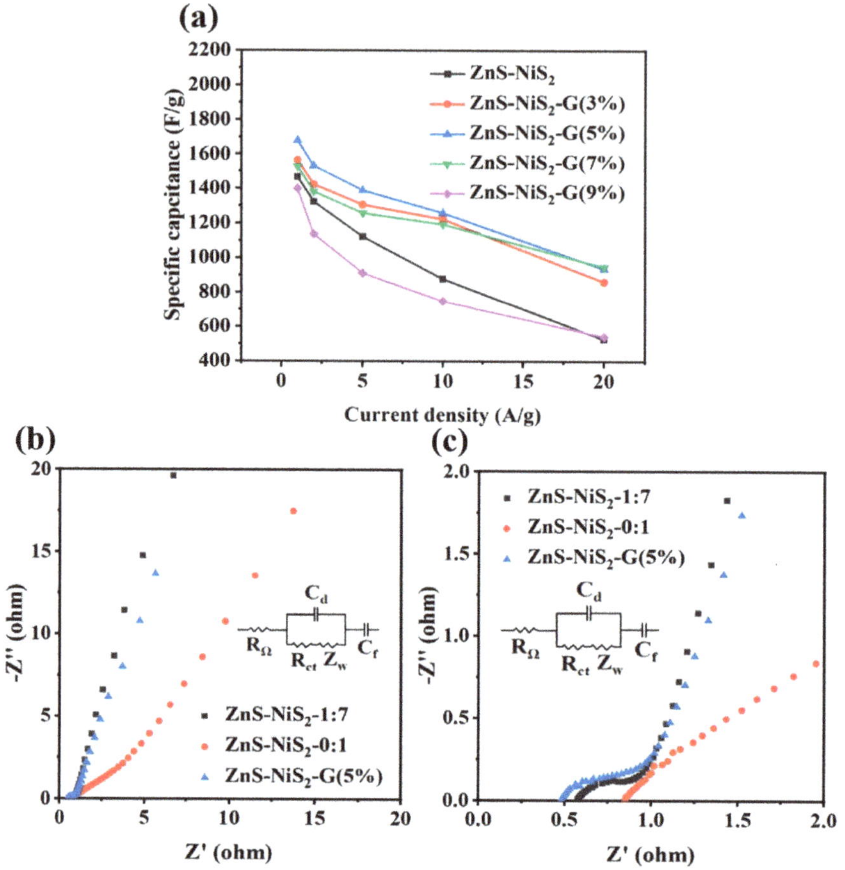

Figure 7. (**a**) Specific capacitance of ZnS-NiS_2-G samples with different graphene mass ratios at different current density, (**b**) EIS diagram of ZnS-NiS_2-1:7, 0:1 and ZnS-NiS_2-G (5%), (**c**) enlarged diagram of high-frequency region of EIS diagram.

Electrochemical impedance spectroscopy (EIS) analysis was performed to study the internal resistance and ion transfer properties of the electrodes. Figure 7b shows the results of the EIS test conducted on ZnS-NiS_2-G (5%), ZnS-NiS_2-1:7, and ZnS-NiS_2-0:1. In general, each of the EIS curves consists of an arc in the high frequency region and a roughly straight line in the low frequency region [38–40]. A possible equivalent circuit diagram is displayed in Figure 7c. The equivalent circuit comprises an internal resistance RΩ, a charge transfer resistance Rct, a double layer capacitor Cd, a Faradic pseudo-capacitor Cf, and a Warburg impedance Zw [41]. Furthermore, the value of the internal resistance RΩ can be determined by identifying the intersection point of the EIS curve and the horizontal axis. The RΩ values of the ZnS-NiS_2-G (5%), ZnS-NiS_2-1:7, and ZnS-NiS_2-0:1 samples are 0.48 Ω, 0.57 Ω, and 0.85 Ω, respectively. Notably, the ZnS-NiS_2-1:7 sample shows a lower RΩ value compared to the ZnS-NiS_2-0:1 sample. In the low-frequency region, the EIS lines of the ZnS-NiS_2-1:7 sample exhibit a greater gradient (3.7) compared to that of the ZnS-NiS_2-0:1 sample (0.6). This observation provides additional evidence indicating that the ZnS-NiS_2-1:7 composite material possesses a lower charge transfer resistance Rct. It indicates that the addition of ZnS nanosheets may promote the charge transfer property of the ZnS-NiS_2-1:7 sample, which can help to achieve a higher specific capacitance of the sample. On the basis of the ZnS-NiS_2-1:7 sample, the addition of graphene with high conductivity achieves a relatively lower RΩ value than that of the ZnS-NiS_2-G (5%) sample. However, the charge transfer resistance of the ZnS-NiS_2-G (5%) sample is slightly larger than that of ZnS-NiS_2-1:7. This might be due to the fact that graphene nanosheets slightly hinder the diffusion of ions in the electrolyte. In the future, we would like to analyze the EIS data by employing the Kramers–Kronig (KK) test and plan to address all relevant fitting parameters after fitting the EIS data. By employing the KK test and fitting suitable parameters, we want to enhance the accuracy and validity of our results, therefore improving the standard and significance of our research outcomes.

It is important to study the reaction kinetics of the electrode materials. According to the following formula, the relationship between the peak current i and scanning rate v in the CV curves can be used to analyze the reaction kinetics of the charging and discharging processes of electrode materials.

$$i = av^b \tag{7}$$

where i is the current density, v is the scanning rate, a and b are constants, while b is the slope of log i vs. log v with linear fitting. If b = 0.5, it indicates that the energy storage process is controlled by diffusion, and the electrode materials belong to the pseudo-capacitive materials. On the other hand, if b = 1, it means that the electrochemical reaction exhibits a capacitive behaviour. Therefore, if the value of b is between 0.5 and 1, it signifies that the electrochemical reaction is composed of both capacitive behaviour and diffusion-controlled processes. Figure 8a shows the linear relationships between log i and log v of cathodic and anodic processes in CV curves. The measured b values for cathodic and anodic processes were found to be 0.53 and 0.51, respectively. These findings suggest the coexistence of capacitive behaviour and diffusion-controlled processes in both cathodic and anodic processes. However, the diffusion-controlled process takes up the majority proportion in the electrode processes. The contribution of capacitive behaviour and diffusion-controlled processes to the total specific capacity can be further calculated by the following formulas [42,43]:

$$i = k_1v + k_2v^{1/2} \tag{8}$$

$$i/v^{1/2} = k_1v^{1/2} + k_2 \tag{9}$$

where k_1 and k_2 are the two coefficients and can be measured by fitting linear curves of $v^{1/2}$ versus $i/v^{1/2}$. Figure 8b shows the values k_1v and $k_2v^{1/2}$ for the ZnS-NiS_2-1:7 sample at 40 mv s^{-1}. The contributions of the capacitive behavior and the diffusion-controlled process under 40 mV s^{-1} were calculated to be 12% and 88%, respectively. In addition, we calculated the storage contribution at 5, 10, 20, and 30 mV s^{-1} as shown in Figure 8c. This is consistent with the earlier result that the diffusion-controlled process played a major role in the energy storage process. On the other hand, the capacitive contributions to the intercalation function increases as the scan rate increases [42].

In order to further investigate the practical application potential of the ZnS-NiS_2-1:7 sample, a supercapacitor device was fabricated and tested. The ZnS-NiS_2-1:7 was chosen as the positive electrode, while the AC was selected as the negative electrode. The mass loadings for the anode and cathodes were 6.6 mg and 1 mg, respectively. A Teflon wafer was used as a separator. The electrochemical performance of the ZnS-NiS_2-1:7//AC device was tested in a two electrode system with a 3 M KOH solution as the electrolyte. Figure 9a shows the CV curves of the ZnS-NiS_2-1:7 and AC electrodes at 20 mV s^{-1}. Notably, these positive and negative electrodes show pseudo-capacitance and EDLC behavior, respectively. There is some controversy about the term pseudo-capacitance in the field, which was discussed in our earlier paper [44]. The CV curve of ZnS-NiS_2-1:7 in Figure 9a originates from Figure 6a. To maintain data consistency, a voltage window of −0.25–0.65 V was established. Due to the relatively low scan rate, polarization was observed in the high-voltage region [45]. Figure 9b shows the CV curves of the ZnS-NiS_2-1:7//AC device at scan rates ranging from 5 mV s^{-1} to 40 mV s^{-1}. The CV curves exhibit minimal changes in shape as the scan rate increases, indicating excellent electron and ion transfer properties of the device [46,47]. In the case of the ZnS-NiS_2-G (5%)//AC device, no obvious polarization is observed, as shown in Figure S6. Figure 9c presents the GCD curves of the ZnS-NiS_2-1:7//AC device, with a potential range of 0–1.7 V. These curves closely resemble a symmetrical triangle with visible shoulders, which is due to the combined effect of the electrode materials. The ZnS-NiS_2-1:7//AC device exhibited specific capacitance values of 127.8, 116.2, 95.7, and 57.5 F g^{-1} at current densities of 1, 2, 5, and 10 A g^{-1}, respectively. Similarly, the electrochemical performance of the ZnS-NiS_2-G (5%)//AC device was measured under the same conditions, yielding specific capacitance values of 138.6, 121.4, 107.2, and 65.2 F g^{-1} at 1, 2, 5, and 10 A g^{-1}, respectively, as illustrated in Figure S6c.

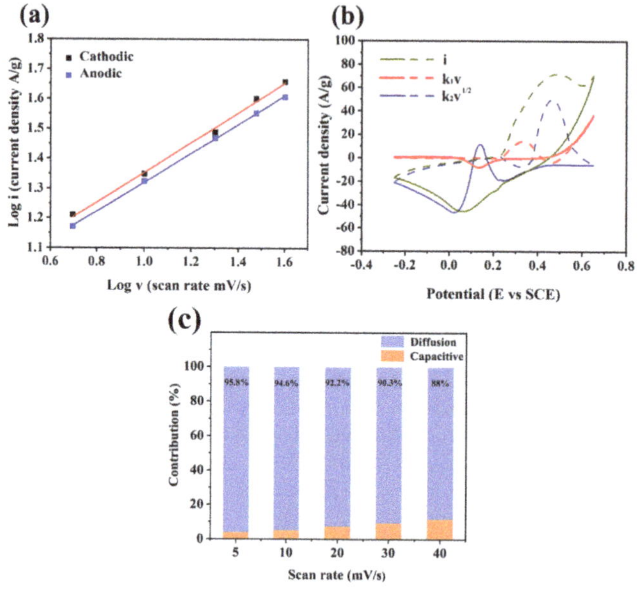

Figure 8. (**a**) Linear relationships between log i and log v in CV curves. (**b**). Storage contribution from the capacitive behavior and diffusion-controlled processes at 40 mV s^{-1}. (**c**) Percentage contribution of capacitive and diffusion-controlled components at varying scan rates for ZnS-NiS_2-1:7 sample.

Figure 9d shows the comprehensive rate performance of the above-mentioned devices. Additionally, Figure 9e displays the Ragone plot comparing the ZnS-NiS_2-1:7//AC and ZnS-NiS_2-G (5%)//AC devices with the existing literature. The ZnS-NiS_2-1:7//AC device exhibits a high energy density of 51.3 Wh kg^{-1} at a power density of 820.76 W kg^{-1}. Moreover, the ZnS-NiS_2-G (5%)//AC device demonstrates a higher energy density of 58.9 Wh kg^{-1} at 914.8 W kg^{-1}, and even at 9076.1 W kg^{-1} power density, it still maintains a 27.7 Wh kg^{-1} energy density, as shown in Figure S6d. The exceptional energy storage performance of both the ZnS-NiS_2-1:7//AC and ZnS-NiS_2-G (5%)//AC devices can be attributed to the high specific capacitance exhibited by the ZnS-NiS_2-1:7//AC device. The cycling stability of the ZnS-NiS_2-1:7//AC device was evaluated under a current density of 10 A g^{-1}. As illustrated in Figure 9f, the specific capacitance of the device remained at 94.9% even after 5000 cycles. Significantly, in the first 500 cycles, there was a notable increase observed in the specific capacitance. This phenomenon may be due to the better infiltration of the electrolyte into the electrode during the initial cycling process, which allows more active sites to participate in the redox reaction [48]. In the remaining cycles, a gradual and minimal decrease was observed in the specific capacitance, thereby indicating the excellent cycling stability of the ZnS-NiS_2-1:7//AC device.

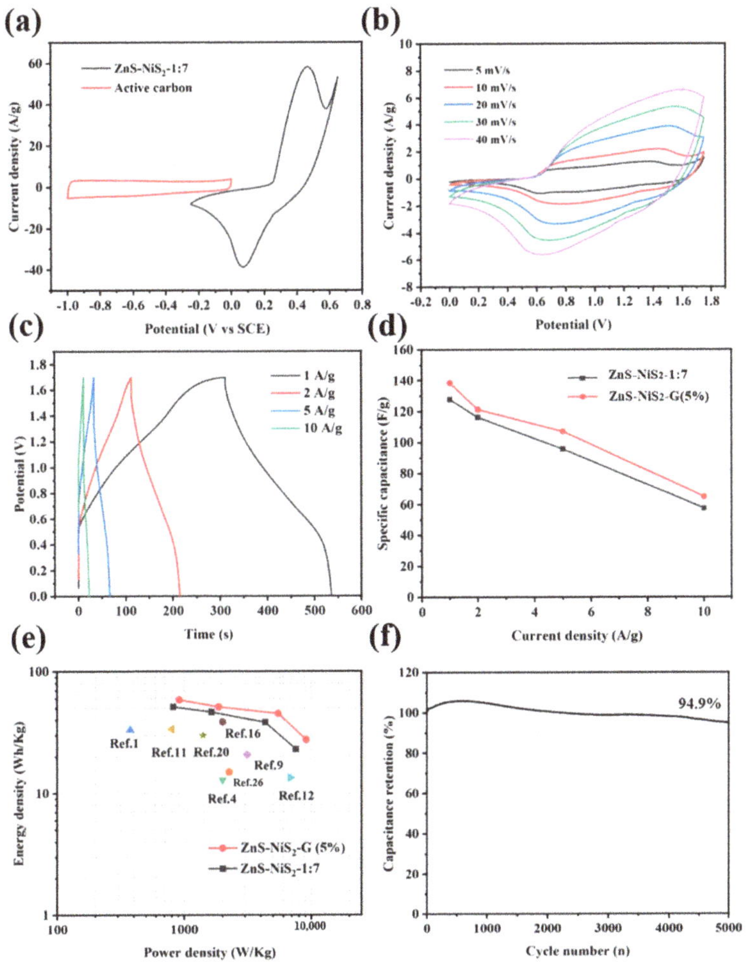

Figure 9. (**a**) CV curves of ZnS-NiS_2-1:7 and AC at 20 mV s^{-1}. (**b**) CV curves of ZnS-NiS_2-1:7//AC. (**c**) GCD curves of ZnS-NiS_2-1:7//AC. (**d**) Comprehensive rate performance of ZnS-NiS_2-1:7//AC and ZnS-NiS_2-G(5%). (**e**) Ragone plot of devices assembled by ZnS-NiS2-1:7 and ZnS-NiS_2-G(5%) [1,4,9,11,12,16,20,26]. (**f**) Cyclic stability of ZnS-NiS_2-1:7//AC.

4. Conclusions

A series of composite energy storage materials made of ZnS-NiS_2 were successfully prepared. A colloidal chemical method was carried out to synthesize ZnS nanosheets, and a solvothermal method was used to achieve the composite of ZnS nanosheets and NiS_2nanospheres. The optimized sample (ZnS-NiS_2-1:7) showed the highest specific capacitance of 1467.9 F g^{-1} at 1 A g^{-1}. A supercapacitor device was assembled by ZnS-NiS_2-1:7 and AC. The ZnS-NiS_2-1:7//AC device exhibited a high energy density of 51.3 Wh kg^{-1} at a power density of 820.8 W kg^{-1}. After 5000 cycles of charge and discharge, the specific capacitance of the ZnS-NiS_2-1:7//AC device remained at 94.9%, showing outstanding cyclic stability. The ZnS-NiS_2-1:7 sample is not only good on its own, but it may also be composited with other materials, such as graphene, to achieve even better electrochemical performances. The optimized ZnS-NiS_2-G (5%) sample showed a specific capacitance of 1681.0 F g^{-1} at 1 A g^{-1} due to the combined benefits of the superior conductivity of graphene and the high specific capacitance of ZnS-NiS_2-1:7. Furthermore, the ZnS-

NiS_2-G (5%)//AC device had an energy density of 59.0 Wh kg^{-1} at a powder density of 914.8 W kg^{-1}. The ZnS-NiS_2-based composite electrode materials have a good development prospect for supercapacitor applications. Further detailed analysis of the composite material nanostructure and its relationship with electrode performance will be performed in the future.

Supplementary Materials: The following supporting information can be downloaded at: https://www.mdpi.com/article/10.3390/batteries10010016/s1, Figure S1: (a,b) FESEM images of ZnS-NiS_2-G (5%); Figure S2. (a,b) TEM images of ZnS-NiS_2-G (5%), (c) SAED image of ZnS-NiS_2-G (5%). Figure S3. (a) XPS spectrum of ZnS-NiS2-1:7 sample, (b), (c), (d) high-resolution spectra of Zn, Ni, and S element regions. Figure S4. (a–c) CV curves of ZnS-NiS_2-0:1, 1:9, 1:5 samples, (d–f) GCD curves of ZnS-NiS_2-0:1, 1:9, 1:5 samples. Figure S5. (a–c). CV curves of ZnS-NiS_2-G (3%, 7%, 9%) electrodes, (d–f) GCD curves of ZnS-NiS_2-G (3%, 7%, 9%) electrodes.; Figure S6. (a) CV graphs of ZnS-NiS_2-G (5%) and AC electrode, (b) CV curve of ZnS-NiS_2-G(5%)//AC device, (c) GCD curve of ZnS-NiS_2-G (5%)//AC device, (d) comparison of ragone diagram of ZnS-NiS_2-G (5%) and ZnS-NiS_2-1:7 devices [49–56] (e) cyclic stability of ZnS-NiS_2-G(5%)//AC device.

Author Contributions: M.J.: writing-original draft, methodology, data curation, software. M.A.: writing—review and editing, formal analysis, resources. X.C.: methodology, writing—original draft, formal analysis, supervision. Y.E.: validation. L.T.: data curation. W.Y.: data curation, writing—review and editing. Y.L.: resources, data curation. W.J.: validation. All authors have read and agreed to the published version of the manuscript.

Funding: This work was supported by the National Natural Science Foundation of China (21875066); Shanghai Leading Academic Discipline Project (B502); and the Shanghai Key Laboratory Project (08DZ2230500).

Data Availability Statement: The data presented in this study are available on request from the corresponding author. The data are not publicly available due to privacy.

Conflicts of Interest: The authors declare that they have no known competing financial interests or personal relationships that could have appeared to influence the work reported in this paper.

References

1. Hua, Y.; Krishna, B.V.; Wang, T.; Ankinapalli, O.R.; Yu, J.S. Green synthesis and electrochemical properties of $A_3(PO_4)_2$ (A = Mn, Zn, and Co) hydrates for supercapacitors with long-term cycling stability. *J. Power Sources* **2022**, *552*, 232245. [CrossRef]
2. Li, S.; Yang, Y.; Hu, Z.; Li, S.; Ding, F.; Xiao, X.; Si, P.; Ulstrup, J. Hetero-structured NiS_2/CoS_2 nanospheres embedded on N/S co-doped carbon nanocages with ultra-thin nanosheets for hybrid supercapacitors. *Electrochim. Acta* **2022**, *424*, 140604. [CrossRef]
3. Wang, Q.; Yang, Y.; Chen, W.; Zhang, C.; Rong, K.; Gao, X.; Fan, W. Reliable coaxial wet spinning strategy to fabricate flexible MnO_2-based fiber supercapacitors. *J. Alloys Compd.* **2023**, *935*, 168110. [CrossRef]
4. Thalji, M.R.; Ali, G.A.; Liu, P.; Zhong, Y.L.; Chong, K.F. $W_{18}O_{49}$ nanowires-graphene nanocomposite for asymmetric supercapacitors employing $AlCl_3$ aqueous electrolyte. *Chem. Eng. J.* **2020**, *409*, 128216. [CrossRef]
5. Ng, W.; Yang, Y.; van der Veen, K.; Rothenberg, G.; Yan, N. Enhancing the performance of 3D porous N-doped carbon in oxygen reduction reaction and supercapacitor via boosting the meso-macropore interconnectivity using the "exsolved" dual-template. *Carbon* **2018**, *129*, 293–300. [CrossRef]
6. Liu, X.; Wang, J.; Hu, N.; Liao, J.; Zong, N.; Wei, J.; Li, M.; Wang, L.; Xu, R.; Yang, L. Facile synthesis of neuronal nickel–cobalt-manganese sulfide for asymmetric supercapacitors with excellent energy density. *J. Electroanal. Chem.* **2023**, *932*, 117262. [CrossRef]
7. Bahnasawy, N.; Sayed, D.M.; Allam, N.K. Engineered self-standing monolithic Cu-Zn-Ni oxide nanocubes as battery materials for durable all-solid-state hybrid supercapacitor devices. *J. Energy Storage* **2023**, *63*, 106997. [CrossRef]
8. El-Gendy, D.; Allam, N.; El Sawy, E. Novel high energy density electrodes based on functionalized/exfoliated molybdenum oxide nanoflakes for high-performance supercapacitors. *Mater. Today Chem.* **2023**, *29*, 101414. [CrossRef]
9. Zhang, C.; Sui, Q.; Lu, L.; Zou, Y.; Xu, F.; Sun, L.; Cai, D.; Xiang, C. Hollow core–shell $CuCo_2O_4$@MoNi-layered double hydroxides as an electrode material for supercapacitors. *J. Energy Storage* **2023**, *61*, 6106691. [CrossRef]
10. Attia, S.Y.; Bedir, A.G.; Barakat, Y.F.; Mohamed, S.G. A two-dimensional nickel-doped bismuth-layered double hydroxide structure as a bifunctional efficient electrode material for symmetric supercapacitors. *Sustain. Mater. Technol.* **2023**, *36*, e00595. [CrossRef]
11. Khan, M.W.; Shaheen, M.; Siddique, S.; Aftab, S.; Alzaid, M.; Iqbal, M.J. Evaluation of d-block metal sulfides as electrode materials for battery-supercapacitor energy storage devices. *J. Energy Storage* **2022**, *55*, 105418. [CrossRef]

12. Dong, J.; Li, S.; Ding, Y. Anchoring nickel-cobalt sulfide nanoparticles on carbon aerogel derived from waste watermelon rind for high-performance asymmetric supercapacitors. *J. Alloys Compd.* **2020**, *845*, 155701. [CrossRef]
13. Hussain, I.; Mohapatra, D.; Dhakal, G.; Lamiel, C.; Sayed, M.S.; Sahoo, S.; Mohamed, S.G.; Kim, J.S.; Lee, Y.R.; Shim, J.-J. Uniform growth of ZnS nanoflakes for high-performance supercapacitor applications. *J. Energy Storage* **2021**, *36*, 102408. [CrossRef]
14. Xie, X.; Wu, D.; Feng, X.; Ni, C.; Sun, X.; Kimura, H.; Du, W. A novel (α-β)NiS/Ni_3S_4-rGO electrode material for supercapacitors. *Colloid Interface Sci. Commun.* **2021**, *43*, 100453. [CrossRef]
15. Wang, H.; Tian, L.; Zhao, X.; Ali, M.; Yin, K.; Xing, Z. In situ growth MoS_2/NiS composites on Ni foam as electrode materials for supercapacitors. *Mater. Today Commun.* **2023**, *34*, 105041. [CrossRef]
16. Dar, M.A.; Dinagaran, S.; Govindarajan, D.; Ahamed, S.R.; Habib, F.; Siva, C.; Moholkar, A.V.; Ahmad, Z.; Yatoo, M.A. Snx-0MnxS nanomaterial based electrodes for future-generation supercapacitor and data storage devices. *J. Alloys Compd.* **2023**, *958*, 170523. [CrossRef]
17. Zhang, J.; Zhang, D.; Yang, B.; Shi, H.; Wang, K.; Han, L.; Wang, S.; Wang, Y. Targeted synthesis of NiS and NiS_2 nanoparticles for high−performance hybrid supercapacitor via a facile green solid−phase synthesis route. *J. Energy Storage* **2020**, *32*, 101852. [CrossRef]
18. Fischereder, A.; Martinez-Ricci, M.L.; Wolosiuk, A.; Haas, W.; Hofer, F.; Trimmel, G.; Soler-Illia, G.J.A.A. Mesoporous ZnS Thin Films Prepared by a Nanocasting Route. *Chem. Mater.* **2012**, *24*, 1837–1845. [CrossRef]
19. Olasanmi, O.O.; Anthony, M. Variation of ZnS deposition time on chemically prepared Cd1-xZnxS ternary compound from CdS/ZnS bilayers. *Results Opt.* **2023**, *11*, 100419. [CrossRef]
20. Yan, J.; Wang, S.; Chen, Y.; Yuan, M.; Huang, Y.; Lian, J.; Qiu, J.; Bao, J.; Xie, M.; Xu, H.; et al. Construction of a z-scheme $CdIn_2S_4$/ZnS heterojunction for the enhanced photocatalytic hydrogen evolution. *J. Alloys Compd.* **2023**, *948*, 169692. [CrossRef]
21. Zhang, W.; Qi, J.; Cao, T.; Lei, Z.; Ma, Y.; Liu, H.; Zhu, L.; Feng, X.; Wei, W.; Zhang, H. A facile method synthesizing marshmallow ZnS grown on Ti_3C_2 MXene for high-performance asymmetric supercapacitors. *J. Energy Storage* **2022**, *50*, 104652. [CrossRef]
22. Palanisamy, P.; Thangavel, K.; Murugesan, S.; Marappan, S.; Chavali, M.; Siril, P.F.; Perumal, D.V. Investigating the synergistic effect of hybridized WO_3-ZnS nanocomposite prepared by microwave-assisted wet chemical method for supercapacitor application. *J. Electroanal. Chem.* **2018**, *833*, 93–104. [CrossRef]
23. Yu, F.; Tiong, V.T.; Pang, L.; Zhou, R.; Wang, X.; Waclawik, E.R.; Ostrikov, K.; Wang, H. Flower-like $Cu_5Sn_2S_7$/ZnS nanocomposite for high performance supercapacitor. *Chin. Chem. Lett.* **2019**, *30*, 1115–1120. [CrossRef]
24. Teli, A.M.; Beknalkar, S.A.; Mane, S.M.; Chaudhary, L.S.; Patil, D.S.; Pawar, S.A.; Efstathiadis, H.; Shin, J.C. Facile hydrothermal deposition of Copper-Nickel sulfide nanostructures on nickel foam for enhanced electrochemical performance and kinetics of charge storage. *Appl. Surf. Sci.* **2021**, *571*, 151336. [CrossRef]
25. Ji, Z.; Chen, L.; Liu, K.; Ma, D.; Zhang, S.; Zhu, G.; Shen, X.; Song, P.; Premlatha, S. Nickel sulfide and cobalt sulfide nanoparticles deposited on ultrathin carbon two-dimensional nanosheets for hybrid supercapacitors. *Appl. Surf. Sci.* **2021**, *574*, 151727. [CrossRef]
26. Madhavi, J.; Prasad, V. ZnS and ZnS/CdS core-shell Nano particles: Synthesis, properties and Perspectives. *Surfaces Interfaces* **2020**, *21*, 100757. [CrossRef]
27. Yan, J.; Wang, S.; Chen, Y.; Yuan, M.; Huang, Y.; Lian, J.; Qiu, J.; Bao, J.; Xie, M.; Xu, H.; et al. Smart in situ construction of NiS/MoS_2 composite nanosheets with ultrahigh specific capacity for high-performance asymmetric supercapacitor. *J. Alloys Compd.* **2019**, *811*, 151915. [CrossRef]
28. Wang, B.; Williams, G.R.; Chang, Z.; Jiang, M.; Liu, J.; Lei, X.; Sun, X. Hierarchical NiAl layered double hydroxide/multiwalled carbon nanotube/nickel foam electrodes with excellent pseudocapacitive properties. *ACS Appl. Mater. Interfaces* **2014**, *6*, 16304–16311. [CrossRef]
29. Arif, M.; Shah, M.Z.U.; Ahmad, S.A.; Hussain, I.; Song, P.; Sajjad, M.; Huang, T.; Yi, J.; Shah, A. Exploring the potential of a ZnS/CNFs cloudy-like architecture for high-performance asymmetric supercapacitors. *J. Mater. Sci. Mater. Electron.* **2023**, *34*, 1340. [CrossRef]
30. Barazandeh, M.; Kazemi, S.H. High-performance freestanding supercapacitor electrode based on polypyrrole coated nickel cobalt sulfide nanostructures. *Sci. Rep.* **2022**, *12*, 4628. [CrossRef]
31. Wang, Q.; Tian, X.; Gao, Q.; Zhang, X.; Rong, K.; Cao, M.; Han, L.; Fan, W. Diversified constructions and electrochemical cycling stability of metal oxide fiber supercapacitors. *J. Mater. Res. Technol.* **2023**, *24*, 909–917. [CrossRef]
32. Salah, N.; Shehab, M.; El Nady, J.; Ebrahim, S.; El-Maghraby, E.; Sakr, A.-H. Polyaniline/ZnS quantum dots nanocomposite as supercapacitor electrode. *Electrochim. Acta* **2023**, *449*, 142174. [CrossRef]
33. Wu, X.; Yu, X.; Zhang, Z.; Liu, H.; Ling, S.; Liu, X.; Lian, C.; Xu, J. Anisotropic ZnS Nanoclusters/Ordered Macro-Microporous Carbon Superstructure for Fibrous Supercapacitor toward Commercial-Level Energy Density. *Adv. Funct. Mater.* **2023**, *33*, 2300329. [CrossRef]
34. Xu, Z.; Zhang, Z.; Yin, H.; Hou, S.; Lin, H.; Zhou, J.; Zhuo, S. Investigation on the role of different conductive polymers in supercapacitors based on a zinc sulfide/reduced graphene oxide/conductive polymer ternary composite electrode. *RSC Adv.* **2020**, *10*, 3122–3129. [CrossRef]
35. Si, L.; Xia, Q.; Liu, K.; Guo, W.; Shinde, N.; Wang, L.; Hu, Q.; Zhou, A. Hydrothermal synthesis of layered NiS_2/Ti_3C_2Tx composite electrode for supercapacitors. *Mater. Chem. Phys.* **2022**, *291*, 126733. [CrossRef]

36. Hu, J.; Sun, L.; Xie, F.; Qu, Y.; Tan, H.; Zhang, Y. An urchin-like Co-doped NiS_2/C nanorod array with enriched sulfur vacancies for asymmetric supercapacitors. *J. Mater. Chem. A* **2023**, *11*, 8380–8391. [CrossRef]
37. Beemarao, M.; Periyannan, P.; Ravichandran, K. High-performance supercapacitor electrode obtained by directly bonding 2D materials: Hierarchal NiS2 on reduced graphene oxide. *J. Mater. Sci. Mater. Electron.* **2023**, *34*, 1423. [CrossRef]
38. Rasappan, A.S.; Thangamuthu, V.; Natarajan, M.; Velauthapillai, D. Kirkendall effect induced NiFe: WS_2 core-shell nanocubes for Dye-sensitized solar cell and battery-type Supercapacitor applications. *J. Energy Storage* **2023**, *63*, 106964. [CrossRef]
39. Diao, Y.; Lu, Y.; Yang, H.; Wang, H.; Chen, H.; D'Arcy, J.M. Direct conversion of Fe_2O_3 to 3D nanofibrillar PEDOT microsupercapacitors. *Adv. Funct. Mater.* **2020**, *30*, 202003394. [CrossRef]
40. Okhay, O.; Tkach, A.; Gallo, M.J.H.; Otero-Irurueta, G.; Mikhalev, S.; Staiti, P.; Lufrano, F. Energy storage of supercapacitor electrodes on carbon cloth enhanced by graphene oxide aerogel reducing conditions. *J. Energy Storage* **2020**, *32*, 101839. [CrossRef]
41. Shen, X.; Wei, X.; Wang, T.; Li, S.; Li, H. Polypyrrole embedded in nickel-cobalt sulfide nanosheets grown on nickel particles passivated silicon nanowire arrays for high-performance supercapacitors. *Chem. Eng. J.* **2023**, *461*, 141745. [CrossRef]
42. Thalji, M.R.; Ali, G.A.; Shim, J.-J.; Chong, K.F. Cobalt-doped tungsten suboxides for supercapacitor applications. *Chem. Eng. J.* **2023**, *473*, 145341. [CrossRef]
43. Pandey, P.; Bhowmick, S.; Qureshi, M. Magnetically Triggered Interplay of Capacitive and Diffusion Contributions for Boosted Supercapacitor Performance. *ACS Appl. Mater. Interfaces* **2023**, *15*, 39435–39447. [CrossRef]
44. Li, X.F.; Liu, J.; Banis, M.N.; Lushington, A.; Li, R.Y.; Cai, M.; Sun, X.L. Atomic layer deposition of solid-state electrolyte coated cathode materials with superior high-voltage cycling behavior for lithium ion battery application. *Energy Environ. Sci.* **2014**, *7*, 768–778. [CrossRef]
45. Runfa, L.; Chen, X.; Hongliang, C.; Wei, Y.; Yuanfang, Z.; Siyu, C.; Wenrui, J.; Qi, Z.; Yi, E.; Meng, J.; et al. Facile synthesis of $Ni_3Se_4/Ni_{0.6}Zn_{0.4}O/ZnO$ nanoparticle as high-performance electrode materials for electrochemical energy storage device. *Nanotechnology* **2023**, *34*, 185401. [CrossRef]
46. Cui, Y.; Zhao, L.; Zhao, C.; Yu, H.; Zhao, B.; Gu, X.; Wang, J.; Meng, L.; Gao, X. Preparation of GO-based Cr-Zn bimetallic layered porous sulfide by ZIF template method for high performance supercapacitors. *J. Mol. Struct.* **2023**, *1275*, 134643. [CrossRef]
47. Wang, Y.; Zhang, Y.-Z.; Dubbink, D.; Elshof, J.E.T. Inkjet printing of δ-MnO_2 nanosheets for flexible solid-state micro-supercapacitor. *Nano Energy* **2018**, *49*, 481–488. [CrossRef]
48. Dong, Y.; Yue, X.; Liu, Y.; Zheng, Q.; Cao, Z.; Lin, D. Hierarchical core–shell-structured bimetallic nickel–cobalt phosphide nanoarrays coated with nickel sulfide for high-performance hybrid supercapacitors. *J. Colloid Interface Sci.* **2022**, *628*, 222–232. [CrossRef]
49. Fang, S.; Chen, Y.; Wang, S.; Xu, J.; Xia, Y.; Yang, F.; Wang, Y.; Lao, J.; Xiang, C.; Xu, F.; et al. Modified CNTs interfacial anchoring and particle-controlled synthesis of amorphous cobalt-nickel-boron alloy bifunctional materials for $NaBH_4$ hydrolysis and supercapacitor energy storage. *J. Alloys Compd.* **2023**, *936*, 167990. [CrossRef]
50. Sudhakaran, M.S.P.; Raju, R.; Youk, J.H. Polypyrrole-derived N-doped CNT nanocomposites decorated with CoNi alloy nanoparticles for high-performance supercapacitor electrodes. *Appl. Surf. Sci.* **2023**, *619*, 156796. [CrossRef]
51. Sun, J.; Zhang, J.; Shang, M.; Zhang, M.; Zhao, X.; Liu, S.; Liu, X.; Liu, S.; Yi, X. N, O co-doped carbon aerogel derived from sodium alginate/melamine composite for all-solid-state supercapacitor. *Appl. Surf. Sci.* **2023**, *608*, 155109. [CrossRef]
52. Chen, Y.; Muthukumar, K.; Leban, L., II.; Li, J. Microwave-assisted high-yield exfoliation of vanadium pentoxide nanoribbons for supercapacitor applications. *Electrochim. Acta* **2020**, *330*, 135200. [CrossRef]
53. Bulakhe, R.N.; Alfantazi, A.; Lee, Y.R.; Lee, M.; Shim, J.J. Chemically synthesized copper sulfide nanoflakes on reduced graphene oxide for asymmetric supercapacitors. *J. Ind. Eng. Chem.* **2021**, *101*, 423–429. [CrossRef]
54. Xu, H.; Lei, Z.; Xu, M.; Zhu, J.; Song, X.; Jin, X. Free-standing reduced graphene oxide/carboxymethylcellulose-polyaniline (RGO/CMC-PANI) hybrid film electrode for high-performance asymmetric supercapacitor device. *Int. J. Biol. Macromol.* **2023**, *236*, 123934. [CrossRef] [PubMed]
55. Sun, B.; Yao, M.; Chen, Y.; Tang, X.; Hu, W.; Pillai, S.C. Facile fabrication of flower-like γ-Fe_2O_3@PPy from iron rust for high-performing asymmetric supercapacitors. *J. Alloys Compd.* **2022**, *922*, 166055. [CrossRef]
56. Zhang, F.; Ma, J.; Qin, Y.; Wang, Y.; Xu, S.; Li, R. Rational construction of Ni–Co–S nanosheets encapsulated NiS_2 micro-blocks on carbon cloth as high-performance supercapacitor electrode. *Mater. Lett.* **2020**, *262*, 127021. [CrossRef]

batteries

Article

Influence of Acetonitrile on the Electrochemical Behavior of Ionic Liquid-Based Supercapacitors

Boryana Karamanova, Luybomir Soserov, Elefteria Lefterova, Toma Stankulov and Antonia Stoyanova *

Institute of Electrochemistry and Energy Systems, Bulgarian Academy of Sciences, G. Bonchev Str. 10, 1113 Sofia, Bulgaria; boriana.karamanova@iees.bas.bg (B.K.); l_stefanov@iees.bas.bg (L.S.); edl@iees.bas.bg (E.L.); tstankulov@iees.bas.bg (T.S.)
* Correspondence: antonia.stoyanova@iees.bas.bg

Abstract: The creation of supercapacitors with superior energy density and power capabilities is critical for advanced energy storage solutions. Ionic liquid electrolytes offer a promising alternative in this respect. However, improving their cycle stability and efficiency is a complex task requiring extensive research and significant effort. The high viscosity of ionic liquids (ILs) limits their lifetime, but this can be mitigated by increasing the temperature or adding solvents. In this research, the electrochemical performance of symmetric activated carbon supercapacitors with 1-Ethyl-3-methylimidazolium tetrafluoroborate ($EMIMBF_4$) and different ratios of acetonitrile (ACN) as electrolytes were investigated. Long-term galvanostatic charge/discharge tests, impedance studies, and cyclic voltammetry were performed at temperatures between 24 to 60 °C. The addition of ACN to the ionic liquid increased electrochemical stability and reduced internal resistance, with the best performance observed at a 1:2 volume ratio of $EMIMBF_4$ to ACN. This supercapacitor exhibited 87% cyclic stability after 5000 charge/discharge cycles in the voltage range of 0.05–2.8 V and a current rate of 1 Ag^{-1}. It also achieved an energy density of 23 $Whkg^{-1}$ and a power density of 748 Wkg^{-1}. The supercapacitors were stable at elevated temperatures up to 60 °C, showing no degradation after operation under various thermal conditions.

Keywords: carbon-based supercapacitors; ionic liquids; acetonitrile; temperature influence; cycling stability

1. Introduction

Energy storage is crucial for managing the variable nature of renewable energy sources and meeting the energy needs of rapidly advancing electronic devices and electric transport [1]. Supercapacitors (SCs) have attracted considerable interest due to their high specific capacitance, rapid energy storage, and long cycle life [2]. The performance of electric double-layer capacitors (EDLCs) is influenced by how ions are distributed near the electrode surface. This sensitivity means that both the structure of the electrode pores and the composition of the electrolyte play crucial roles in determining their characteristics [3]. Additionally, the electrochemical stability window (ESW) is crucial as it limits energy density, emphasizing the need for electrolytes with wider electrochemical windows to design high-energy-density supercapacitors [4]. The ESW is influenced by the properties of ions and solvents, as well as interactions between ions, solvents, and the electrode–electrolyte interface [5]. Additionally, the purity of the electrolytes plays a substantial role in SC operation [6].

Ionic liquids (ILs) are organic salts with melting points below 100 °C, dissolving into free ions in a solvent. They are composed of an asymmetric organic cation combined with either an inorganic or organic anion. ILs offer high solubility, low volatility, a conductivity of 10^{-3} to 10^{-2} $S{\cdot}cm^{-1}$, and non-flammability, with benefits varying by cation–anion combination. Due to their unique properties, ILs are promising electrolytes for energy storage devices [7]. Integrating ILs with nanostructured carbon could create advanced supercapacitors with high energy density and broad temperature range viability [8]. ILs have wider potential windows than water-based electrolytes, allowing higher energy densities despite higher viscosity. Strong ionic bonds also enhance stability and prevent evaporation.

ILs can be pure ILs, mixed, or quasi-solid-state electrolytes. Pure ILs are solvent-free, while mixtures improve energy storage potential. When selecting electrolytes, it is necessary to consider the support capabilities of the supercapacitor, such as ionic conductivity, thermal stability, operating voltage, specific energy, and power [9]. Using ILs alone is difficult due to high viscosity and cost. A practical approach involves integrating ILs into the electrolyte to improve ion mobility [10].

ILs are often combined with organic solvents to lower their viscosity and enhance the rate of ion transport [11], often quantified by the diffusion coefficient [12]. However, adding organic solvents reduces ionic association and alters the redox potential of ILs, narrowing the electrolyte's electrochemical stability window (ESW) and limiting the maximum working voltage (MWV) of EDLCs. The optimal concentration of ions and dissociation of ion pairs in IL-based electrolytes improve the mobility of cations and anions and ensure sufficient ion accessibility for charge transfer in carbon porous electrodes, resulting in the highest specific capacitance, with a tendency for it to increase with the MWV [13–15]. Consequently, pure ILs are preferred for achieving the highest energy density and specific capacitance in EDLCs. Nonetheless, when high-voltage devices are not required, a diluted IL electrolyte at an optimal concentration can be employed [16]. Adding organic solvents to ILs to diminish viscosity is an effective strategy for achieving desirable capacitive performance in EDLCs [17,18]. Acetonitrile (ACN), a common aprotic solvent, significantly enhances conductivity (up to 1.5 orders of magnitude) due to favorable interactions between ions and ACN, making conductivities less affected by the cation's size, shape, and mass [19]. Diluting ILs with AN generally boosts the ionic conductivity of a pure IL by improving ion transport rates (lower viscosity). Ionic conductivity is influenced by ionic density and the free volume of the electrolyte mixture [20,21].

ACN in $EMIMBF_4$ promotes ion solvation, reduces ion pairing, weakens Coulomb interactions, and facilitates ion mobility, thereby reducing IL viscosity [22,23]. The belief that lowering the electrolyte's viscosity enhances the specific capacitance is only valid when there are enough charge-carrying ions available. For example, graphene nanosheet electrodes demonstrate superior capacitive performance in a 2.0 M $EMIMBF_4$/acetonitrile electrolyte, featuring a wide voltage window (2.3 V), minimal infrared drop, and high specific capacitance (128.2 F g^{-1}) [24]. However, studies on the optimal IL concentration are lacking, and further studies are needed to enable the practical application of IL solvent electrolytes [4].

The practical application of ILs in energy storage systems requires an understanding of the electrolyte's physical properties at various temperatures, which significantly affect their viscosity and ionic conductivity and hence their long-term cycle life [25,26]. Therefore, understanding the temperature dependence for a specific IL–electrolyte system is crucial. For instance, it has been observed that a supercapacitor using the $EMIMBF_4$/acetonitrile electrolyte demonstrates long-term cyclic stability in charge/discharge cycles at both low and high operating temperatures for up to 1500 cycles. Beyond this point, galvanostatic charge/discharge curves noticeably deviate from ideal symmetric forms, leading to a significant decrease in specific capacitance [27].

In the present work, the electrochemical performance of symmetric activated carbon-based supercapacitors with $EMIMBF_4$ ionic liquid-based electrolytes and different ACN ratios were investigated through long-term galvanostatic charge/discharge tests, impedance studies, and cyclic voltammetry at different heating modes from 24 to 60 °C. In order to clarify the reasons for the ongoing processes in the investigated systems, electrochemical studies were also carried out after the prolonged regimes.

2. Materials and Methods

2.1. Electrolyte Preparation

The electrolyte contained a neat ionic liquid (1-Ethyl-3-methylimidazolium tetrafluoroborate, Fluorochem 96.3% (HPLC)) and mixtures of ionic liquid and acetonitrile (Spectrophotometric Grade, 99.7+% Alfa Aesar) in various vol.% ratios: EMIMBF4:ACN = 9:1 (9IL/ACN), EMIMBF4:ACN = 2:1 (2IL/ACN), $EMIMBF_4$:ACN = 1:1 (IL/ACN), and EMIMBF4:ACN = 1:2 (IL/2ACN). For comparison, neat ACN was also examined.

The electrolytes were prepared in a glove box mBraun–Labmaster Pro.

2.2. Preparation of the Carbon Electrodes and Supercapacitor Cell Assembly

Commercial activated carbon (YP-50F, "Kuraray Europe" GmbH) was used as the active material to prepare the electrodes for supercapacitor cells. The composition consisted of 80 wt.% activated carbon, 10 wt.% poly(vinylidene fluoride-co-hexafluoropropylene), and 10 wt.% graphite fibers.

Before use, the binder poly(vinylidene fluoride-co-hexafluoropropylene, Sigma-Aldrich, St. Louis, MI, USA) in the form of granules was dissolved in N,N-dimethylacetamide (3% solution, 99% Alfa Aesar), to obtain a homogeneous mixture. All materials were stirred for 15 min.

The electrode layers were prepared using a casting technique. This involved pouring the electrode mixture onto a glass plate, followed by several drying stages: 5 h at 40 °C to evaporate the solvent, then drying for 12 h at 70 °C. The forming free-standing flexible film electrode was detached with distilled water and thermally treated at 120 °C for 1 h to remove the traces of water, and then finally heated for 20 min at 160 °C to improve the mechanical strength [28].

Before assembly, the electrodes were cut with a circular area of 0.64 cm^2, dried under a vacuum at 80 °C for 12 h in order to outgas the electrode, and then mounted in a two-electrode coin Swagelok-type cell with a Whatman separator. The cell was filled with the electrolyte in a dry box under an argon atmosphere (H_2O и O_2 < 0.01 ppm).

2.3. Electrochemical Characterization

The performance of the supercapacitors was investigated through a series of electrochemical measurements using cyclic voltammetry (CV) at different scan rates ranging from 10 to 50 mVs^{-1}, within a voltage range of 0.05–3.2 V, depending on the electrolyte composition. These measurements were conducted using a Multi PalmSens system (model 4, Houten, The Netherlands). Electrochemical impedance spectroscopy (EIS) investigations were performed using the same apparatus, with frequencies ranging from 10 MHz to 1 mHz.

Cyclic galvanostatic charge–discharge (GCD) and long-term tests were performed using an Arbin LBT21084 measurement system. The GCD tests were conducted over the voltage range of 0.05–2.8 V and 0.05–3.0 V while continuously charging the cells with a current of 200 to 4000 mAg^{-1} for 100 cycles per step. Long-term tests were carried out for 5000 charge and discharge cycles at 200 mAg^{-1}.

The specific capacitance determined from the cyclic voltammetry curves is calculated using Equation (1) and the specific capacitance is calculated from the charge/discharge curves according to Equation (2):

$$Cs = [4(I/(dV/dt)/m] \tag{1}$$

where dV/dt is the voltage scan rate, I is the current, and m is the mass of the active carbon material.

$$C = [4(I \times \Delta t)/(m \times \Delta V)] \tag{2}$$

where Δt (s), m (g), ΔV (V), and I (A) indicate the discharge time, mass of active material, voltage window, and discharge current, respectively. The value 4 is necessary to denote the capacitance of a single electrode [28].

The energy density (E) and power density (P) are calculated from Equations (3) and (4), using the specific discharge capacitance:

$$E = C\,\Delta V^2/7.2 \quad (3)$$

and

$$P = 3600\,E/t \quad (4)$$

3. Results and Discussion

The different characteristics of ILs necessitate careful selection based on the specific operating conditions and requirements of the device being developed. Additionally, the structure of electrodes affects how ionic liquids (ILs) are arranged at the interface. Mixing ILs with organic electrolytes can enhance conductivity and durability under harsh conditions. However, this method narrows the voltage window of IL electrolytes, which in turn limits the energy density of supercapacitors [8].

Conductivity and viscosity are key parameters that significantly influence electrolyte characteristics. Establishing a relationship between these parameters and molar concentration is crucial, especially for ionic liquids due to their high viscosity. Although partial data exist for some systems, comprehensive studies aimed at determining the optimal concentration of ionic liquids are still lacking [19]. In this study, a pure electrolyte of $EMIMBF_4$ and its mixtures with ACN in various volume ratios were investigated. $EMIMBF_4$ was selected because it is one of the most commonly used ionic liquids in experimental studies on supercapacitors [29] and offers high electrochemical stability [30]. On the other hand, acetonitrile was chosen as the organic solvent due to its outstanding oxidative stability and high dielectric constant, which ensures its compatibility with electrolyte salts. [31]. ACN has a much lower viscosity compared to $EMIMBF_4$ (0.3 centipoises versus 41 centipoises, respectively), which is likely to reduce charge transfer resistance and consequently improve the rate performance of the SC [32].

Various methods have been employed to accurately investigate the electrochemical characteristics of SCs with different ACN ratios. To monitor changes in the investigated systems, CV and EIS curves were recorded at the beginning of the experiments and after conducting a long-term galvanostatic charge/discharge test. The resulting CV curves, recorded at 10 mVs^{-1} and different voltage windows, are presented in Figure 1. For comparison, the voltammograms of pure ACN are also presented.

The initial voltammograms exhibit the typical characteristics of symmetrical supercapacitor systems. Supercapacitors with IL electrolytes containing ACN show nearly rectangular shapes, unlike those using pure $EMIMBF_4$. This difference can be explained by the lower resistance in the supercapacitor due to the reduced viscosity of the electrolyte. This result is expected and is most likely due to the reduced ionic association that occurs when organic solvents are introduced into pure IL. This change alters the redox potential, resulting in a narrowing of the electrochemical stability window of the electrolyte. Although the potential window does not reach the electrolyte's decomposition voltage, local high voltages in porous carbon can cause decomposition. Additionally, the strong polarity of ionic liquids makes it difficult to eliminate trace water, which may also lead to electrolyte decomposition and affect the potential window [33].

Galvanostatic charge/discharge tests were conducted at different current loads and long-term studies were performed. Based on the results shown in Figure 1, the voltage limits for the investigated supercapacitors were selected, ensuring that the electrolyte remained stable and its efficiency and longevity were not compromised by undesirable chemical reactions. The voltage limits were set at 0.05–3.0 V for neat IL and the electrolyte mixture containing a small amount of 9IL/ACN, and at 0.05–2.8 V for the other studied solutions.

A significant drawback of ionic liquids is their low stability during prolonged cycling, which has focused our efforts on conducting long-term tests. The results of these tests at a current load of 200 mAg^{-1} for a total of 5000 cycles are shown in Figure 2.

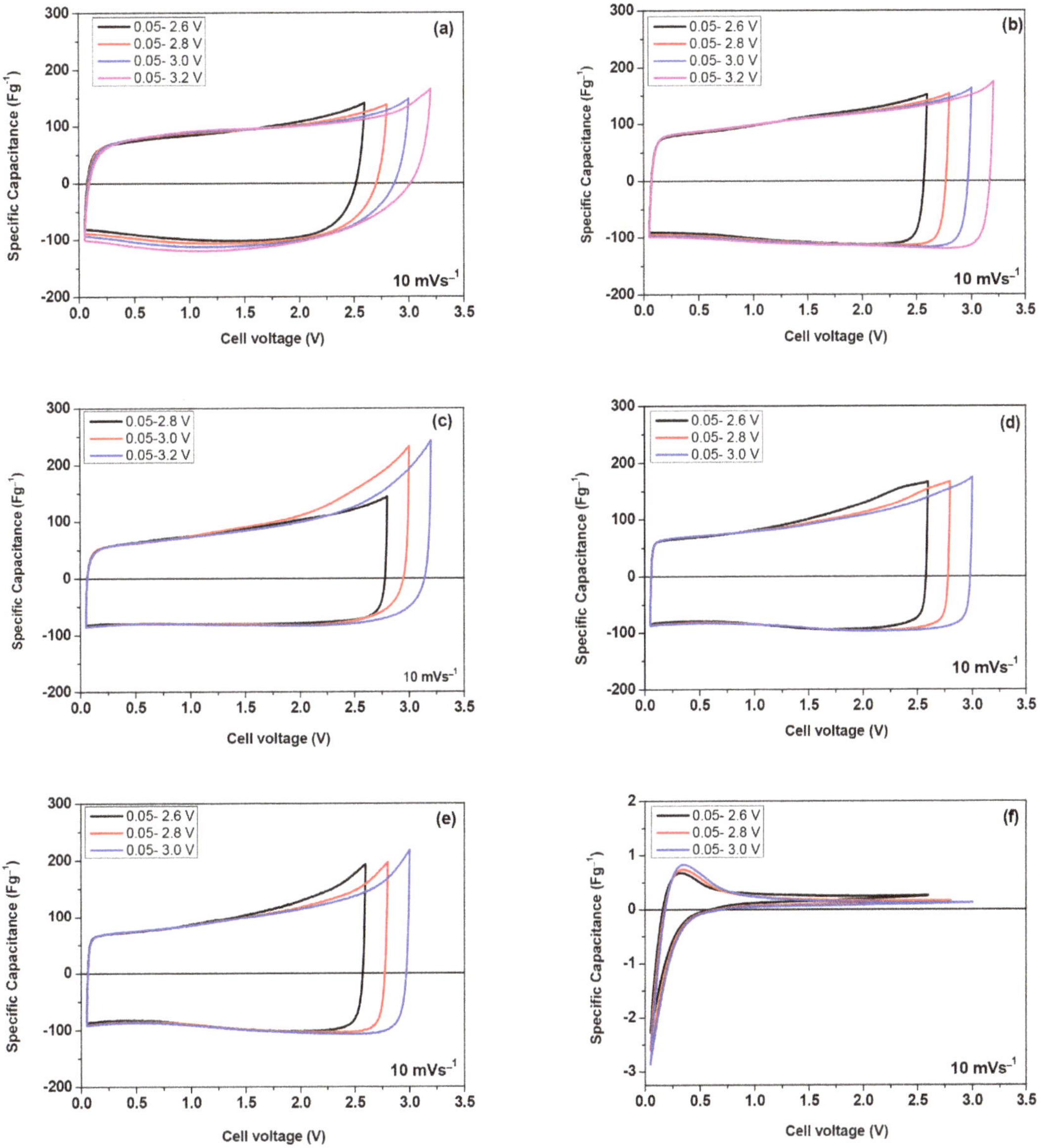

Figure 1. Cyclic voltammogram curves measured at a scan rate of 10 mVs^{-1} at different voltage ranges from 0.05 to 3.2 V for symmetric supercapacitors with varying ratios of ACN to $EMIMBF_4$: IL: (**a**), 9IL/ACN (**b**), IL/ACN (**c**), 2IL/ACN (**d**), IL/2ACN (**e**), and ACN (**f**).

The results show that increasing the concentration of ions in pure ionic liquids (ILs) does not significantly enhance the capacitance of nanoporous carbon-based EDLCs. This is due to the increased difficulty in separating ions with opposite charges in pure ILs. The mechanism of EDLCs changes from the adsorption of opposite ions in dilute ACN to an ion exchange in pure ILs. Furthermore, the interface structure is influenced by both the electrode surface and the IL electrolytes [34].

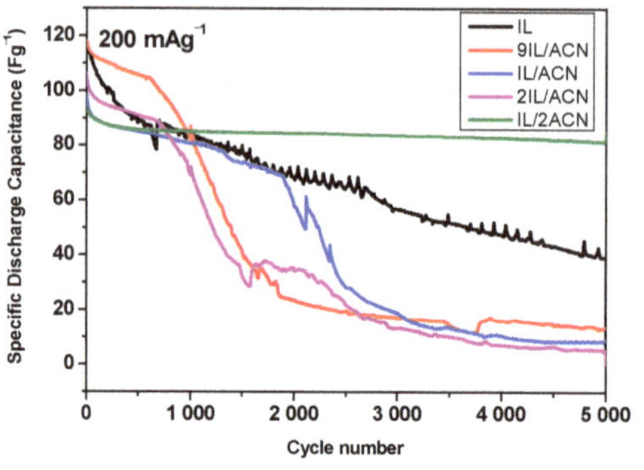

Figure 2. Long-term cycling stability at 200 mAg^{-1} of the symmetrical supercapacitors with different electrolyte ratios of ACN to $EMIMBF_4$ up to 5000 GCD cycles.

A detailed look of the cyclic stability curves reveals distinct electrochemical characteristics. Up to 1000 cycles, the discharge capacitance of the SC with pure ionic liquid is the highest, but it has low stability, reaching a value of 65% loss at 5000 cycles. After 1000 cycles, a sharp decline in the stability of the SC with electrolytes is observed, where the IL/ACN ratio favors IL. By 1900 charge/discharge cycles, similar behavior is noted with the SC using electrolytes containing IL/ACN. An impressive result, however, is observed with the cyclic stability of the SC with IL/2ACN. The discharge capacitance remains nearly constant, with capacitance loss not exceeding 13% after 5000 cycles. In order to clarify the results obtained, paying particular attention to the increased stability upon cycling of the SC with electrolytes containing IL/2ACN, CV curves were also performed after prolonged cycling (Figure 3a,b), as well as impedance measurements. In addition, voltage profiles before and after cycling tests at 200 mAg^{-1} are presented in Figure 3c,d, which aids in the overall analysis of the results.

For a better understanding of the long-term test results, Table 1 presents the calculated values for the discharge capacitance retention after 5000 cycles and for the ohmic drop before and after the GCD test.

Table 1. Capacitance retention after 5000 charge/discharge cycles and ohmic drops before and after cycling.

Electrolyte Composition, IL:AN	Capacitance Retention after 5000 GDC, %	iR Drops—Initial, V	iR Drops—after 5000 GDC, V
IL	33	0.087	0.452
9Il/ACN	12	0.025	1.101
2IL/ACN	5	0.021	1.943
IL/ACN	9	0.019	1.975
IL/2ACN	87	0.016	0.082

The results in Table 1 show that during the first cycle, the ohmic drop of the cell with pure IL is the highest. As the ACN content in the electrolyte increases, the ohmic drop in the cell decreases. Considering that the 'IR drop' includes the resistance of the electrolyte solution, the contact resistance at the interface between the active material and the receiver, and the internal resistance of the active material, with the electrolyte solution's resistance usually being predominant, the results obtained are probably related to an increase in the electrolyte's conductivity, as impedance studies have shown [35]. However, after 5000 cycles, the cell with a 2:1 ratio exhibits the lowest ohmic drop and maintains almost constant capacitance (only a 13% decrease). For the other cells, the ohmic drop increases significantly, leading to a sharp decline in their discharge capacitance. A possible reason for this is the degradation of the electrolyte and the formation of a passivation layer on the electrodes, as discussed by K. Karuppasamy et al. [33].

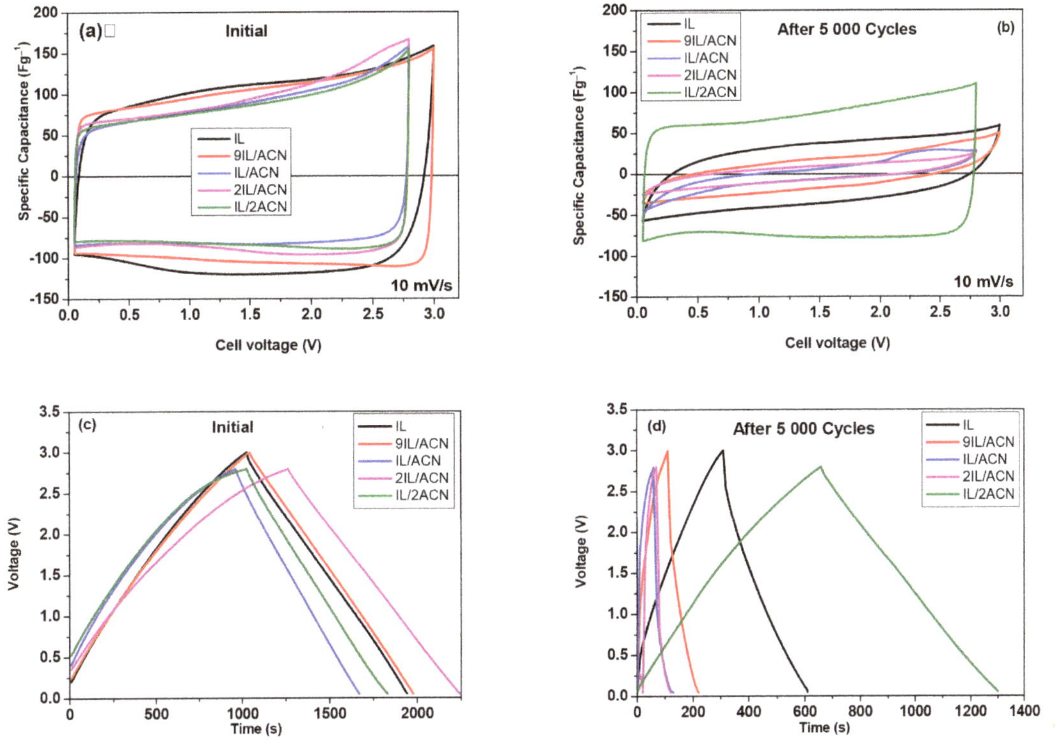

Figure 3. Cyclic voltammograms at scan rate of 10 mV s^{-1} (**a**,**b**) and galvanostatic charge/discharge curves at 200 mAg^{-1} (**c**,**d**) for symmetric SCs before and after 5000 GCD cycles.

As can be seen, the rectangular shape of the IL/2ACN supercapacitor remains unchanged, with area shrinkage observed. This result confirms the good stability of this supercapacitor. Conversely, the shape of the CV curves in other investigated systems changes, which is in agreement with the results of the long-term galvanostatic test (Figure 3).

The electrochemical properties of the studied carbons were assessed using Ragon plots for supercapacitor cells with different electrolyte compositions before and after the long-term test (Figure 4).

The energy density of the SCs before the start of the long-term test was highest for SCs with pure ionic liquid and those with an electrolyte containing acetonitrile in an IL/ACN ratio of 9:1 (35 and 36 Whkg^{-1} at power outputs of 140 and 137 Wkg^{-1}, respectively). However, after 5000 charge/discharge cycles at 200 mAg^{-1}, the trend changes. The SC with an electrolyte containing an IL/ACN ratio of 1:2 shows the highest energy density (21 Whkg^{-1} at a power output of 120 Wkg^{-1}), while the energy density of the SC with pure ionic liquid decreases drastically (12 Whkg^{-1} at a power output of 141 Wkg^{-1}), even with a wider voltage window.

Rate response is a crucial performance parameter for electric double-layer capacitors, evaluated by charging and discharging them at different current densities. The discharge capacitance and their dependence on the current rate were compared for the supercapacitor showing the most stable behavior in an electrolyte with IL/2ACN, and for the SC with an electrolyte containing neat IL (Figure 5).

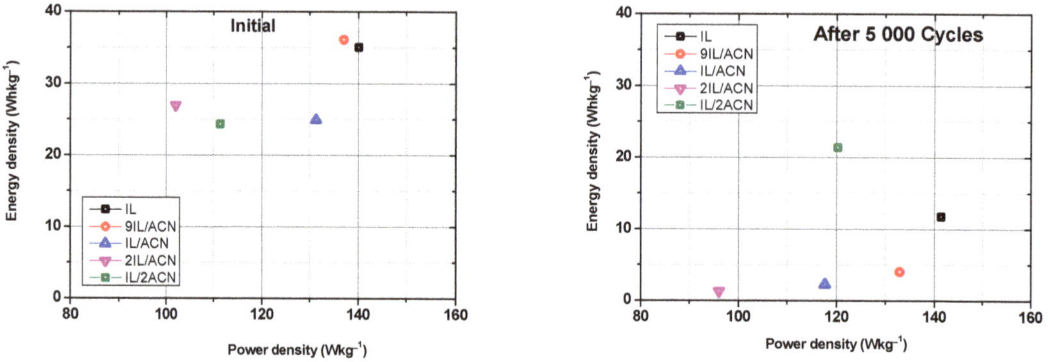

Figure 4. Energy density versus power density (Ragone plot) for symmetric supercapacitors with different IL electrolyte compositions.

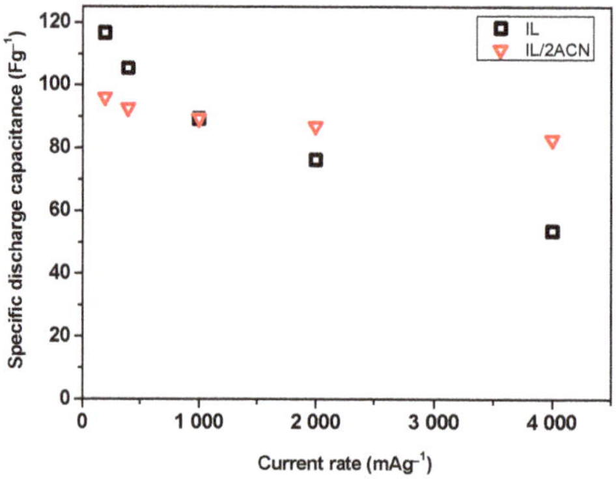

Figure 5. Specific discharge capacitance as a function of current load for symmetric supercapacitors with electrolytes containing IL and IL/2ACN.

As shown in Figure 5, the SC using IL/2ACN has the most stable behavior, maintaining a constant discharge capacity at different current rates. Conversely, the SC with pure $EMIMBF_4$ shows a noticeable drop in capacitance with the increasing current load. These findings indicate that the IL/2ACN electrolyte improves the rate response in terms of rate, providing more reliable operation under different discharge conditions. This improvement can be attributed to the significant influence of electrolyte viscosity and conductivity on the rate response.

Figure 5 demonstrates that the supercapacitor (SC) utilizing IL/2ACN exhibits the most stable behavior, maintaining a consistent discharge capacitance across various current rates. In contrast, the SC employing pure $EMIMBF_4$ experiences a notable decline in capacitance as the current load increases. These observations suggest that the IL/2ACN electrolyte enhances the rate response, ensuring more dependable operation under diverse discharge conditions. This enhancement is attributed to the considerable impact of electrolyte viscosity and conductivity on the rate response. This trend is reflected in the Ragone plot (Figure 6), illustrating a slight reduction in energy density (from 25.5 to 21.5 Whkg^{-1}) with increasing power density (from 138 to 2750 Wkg^{-1}) for the IL/2ACN SC.

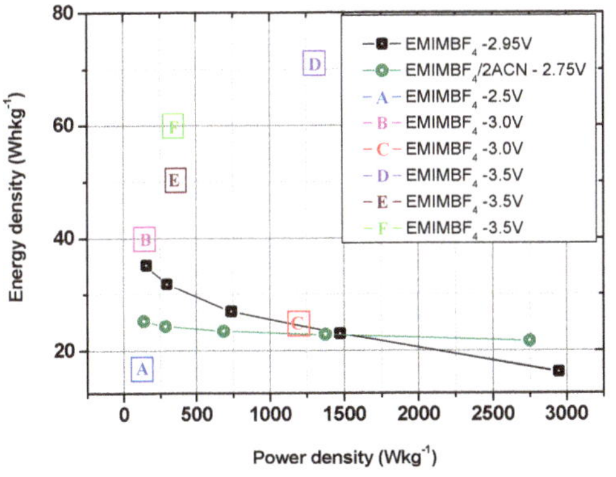

Figure 6. Energy density versus power density (Ragone plot) for symmetric supercapacitors with $EMIMBF_4$:ACN = 1:2 (IL/2ACN). The literature data for carbon-based SCs with neat $EMIMBF_4$ are presented for comparison: A [36], B [37], C [38], D [39], E [40], F [41].

The study of supercapacitors using ionic liquids at elevated temperatures is of major importance due to their distinctive chemical composition and characteristics. ILs are more viscous than conventional electrolytes due to the large size of the ions, resulting in reduced capacitance and power density at room temperature. However, studies have shown that optimal IL performance is achieved at temperatures around 60 °C [42]. Furthermore, temperature has a significant impact on both the lifetime estimation and the efficiency enhancement of the supercapacitor [24].

Our studies at elevated temperatures (40 and 60 °C) were mainly focused on the investigation of supercapacitors with neat ionic liquid and with the IL/2ACN electrolyte, which showed the highest cycling stability at room temperature. The results are presented in Figures 7 and 8 and show that the SC with the IL/2ACN electrolyte has a stable behavior at 40 °C.

Electrochemical impedance spectroscopy (EIS) was performed to analyze the impedance characteristics with different ratios of ACN in the electrolyte, both at the beginning of the tests and after 5000 cycles. Figure 9 presents the Nyquist plot and the capacitance versus frequency dependence, while Table 2 shows the calculated values.

Table 2. The values of electrolyte resistance (Rel), electrode resistance (Rp), and cell resistance (Ri) determined from Nyquist plots.

Electrolyte Composition, IL/AN	Temperature	Rel Ω	Rp Ω	Ri Ω
IL	room temperature	2.2	2.3	11.5
	after 5000 cycles	6.8	-	52
	40 °C	1.7	3.7	8.9
	60 °C	1.3	3.3	6.78
Il/2ACN	room temperature	1.13	0.35	3.2
	after 5000 cycles	1.7	0.9	6.2
	40 °C	1.02	0.53	2.9
	60 °C	0.94	0.48	-

The Nyquist plots are inherent to the porous electrodes in the carbon bilayer capacitors. The semicircle at high frequencies is due to the pore resistance/resistance of the electrodes (Rp is the diameter of the semicircle), the line with a slope $\leq 45°$at medium to low frequencies is due to the diffuse region, and the nearly vertical line at low frequencies is the capacitive part is due to the capacitance of DL. From the intersection of its continuation with Z', the cell resistance (Ri) is determined. The left intercept of Z'' at the Z' axis (the origin of the semicircle) represents the resistance (Rel) of the electrolyte in the cell. As can be seen from Table 2 the addition of ACN causes a decrease in Ri. The same role is played by the increase in operating temperatures as can be seen from the Nyquist plots of SCs with IL and IL/2ACN (Figure 10). In both cases, the decrease in Ri is a result of a decrease in Rel and a shortening of the diffuse portion (i.e., a decrease in diffusion resistance).

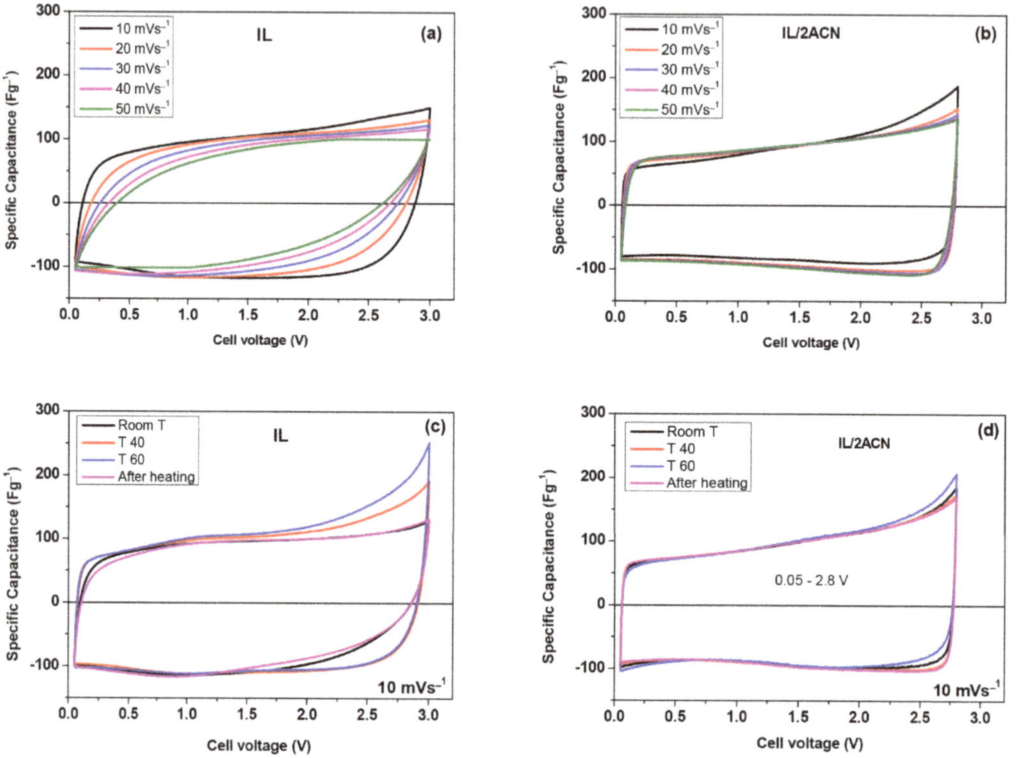

Figure 7. Cyclic voltammograms for SCs with IL and IL/2ACN electrolyte at room temperature and different scan rates (**a**,**b**) and at scan rates of 10 mV s^{-1} and 40 °C and 60 °C (**c**,**d**).

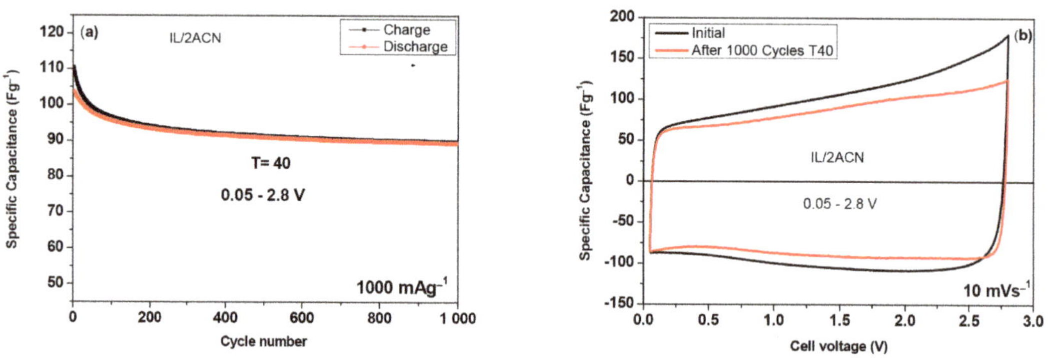

Figure 8. Long-term cycling stability at 1000 mAg^{-1} of the symmetrical supercapacitor with the IL/2ACN electrolyte (**a**) and cyclic voltammograms before and after 1000 GCD cycles (**b**).

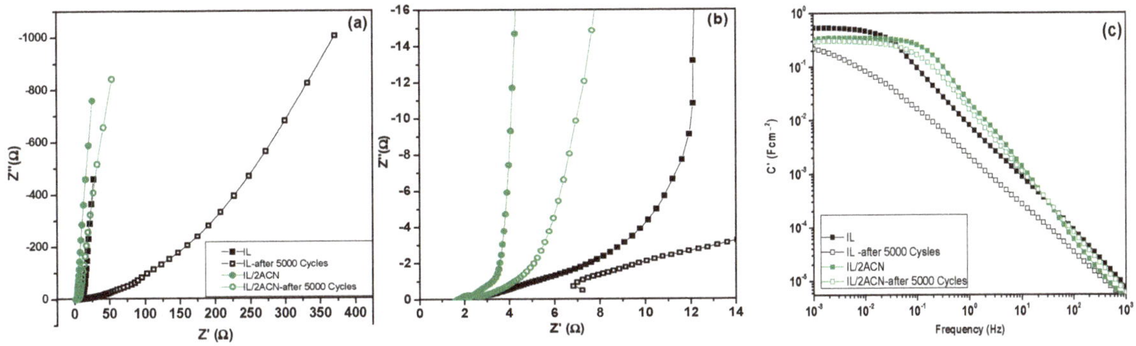

Figure 9. EIS of SCs with neat $EMIMBF_4$ and IL/2ACN initially and after 5000 GCD: Nyquist plots (**a**) full scale and (**b**) zoomed; C′ vs. frequency of the cells (**c**).

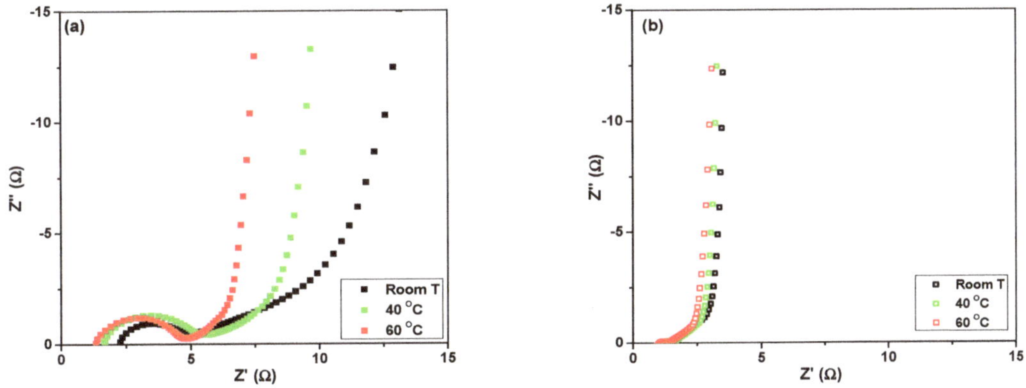

Figure 10. Nyquist plots of SCs with IL (**a**) and IL/2ACN (**b**) at different temperatures.

The addition of ACN has a contradictory effect on the stability of the cell under cycling. In relatively small amounts, prolonged cycling leads to a decrease in Rel and a change in the nature of the AC response, namely from capacitive to diffusive (Figures 9b and 10). At a ratio of 1:2, the increase in Ri is small, the capacitive character is preserved, and the capacitance is close to the initial sample. Our assumption of good stability is that with enough ACN, it plays a dual role—on the one hand, the viscosity decreases sharply, thereby facilitating ion transport, and on the other hand, it prevents IL degradation.

Additional information can be obtained from the diffusion coefficient (D), which can be calculated using Equations (5) and (6) that are valid in the diffusion region [43–45]:

$$-Z'' = \frac{1}{C_l \omega}, \tag{5}$$

and

$$IZI = \frac{L}{C_l (D\omega)^{-1/2}} \tag{6}$$

where L and C_l are the thickness and capacitance, respectively.

Therefore, C_l and D can be calculated from the slopes of the $-Z''$ vs. ω^{-1} and IZI vs. $\omega^{-1/2}$ linear plots, as illustrated in Figure 11, while Table 3 presents the calculated values.

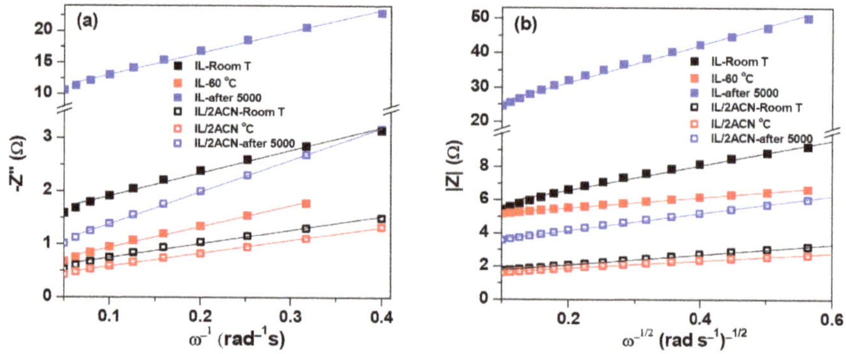

Figure 11. Plots of $-Z''$ vs. ω^{-1} (**a**) and IZI vs. $\omega^{-1/2}$ (**b**) for diffusion coefficient determination.

Table 3. Calculated values of diffusion coefficient (D) of pure IL and IL/2AN at different temperatures.

Electrolyte Composition, IL/AN	Temperature	D m^2s^{-1}
IL	room temperature	7.01×10^{-8}
	after 5000 cycles	1.05×10^{-7}
	60 °C	3.85×10^{-7}
Il/2ACN	room temperature	1.06×10^{-7}
	after 5000 cycles	1.64×10^{-7}
	60 °C	2.38×10^{-7}

The resulting values for D correlate with those reported by [46]. No decrease in the value of the diffusion coefficient of pure IL was observed after 5000 charge/discharge cycles despite the deteriorated cell characteristics. This is a surprising result that deserves further investigation.

4. Conclusions

Studies on symmetric supercapacitors using an $EMIMBF_4$-based electrolyte with various ratios of acetonitrile (ACN) demonstrate that adding ACN significantly enhances electrochemical stability and reduces internal resistance. These improvements are due to ACN's dual role at higher concentrations: it significantly lowers viscosity, facilitates ion transport, and prevents the degradation of the ionic liquid (IL). Optimal performance is observed at a volumetric ratio of 1:2 for $EMIMBF_4$ to ACN. The supercapacitor shows 87% cyclic stability over 5000 charge/discharge cycles within a voltage range of 0.05–2.8 V and at a current density of 1 Ag^{-1}, achieving an energy density of 23 $Whkg^{-1}$ and a power density of 748 Wkg^{-1}. Moreover, these supercapacitors exhibit stability at elevated temperatures up to 60 °C, showing no degradation under varying thermal conditions.

Adding solvents such as ACN to $EMIMBF_4$ is a practical approach to reduce costs in IL-supercapacitors, which are promising due to their relatively high energy density and power characteristics. These findings provide a basis for future research aimed at developing more efficient solutions using compatible and environmentally friendly materials to produce safe, efficient, and reliable IL-based devices capable of operating under a variety of conditions.

Author Contributions: Conceptualization, A.S. and T.S., methodology, A.S., T.S. and. B.K.; validation, B.K. and L.S.; investigation, B.K. and L.S.; data curation, B.K., E.L. and L.S.; writing—original draft preparation, B.K. and T.S.; writing—review and editing, A.S. and E.L.; visualization, B.K.; project administration, B.K.; funding acquisition, B.K. All authors have read and agreed to the published version of the manuscript.

Funding: This research was funded by the Bulgarian Ministry of Education and Science under the National Program "Young Scientists and Postdoctoral Students-2".

Data Availability Statement: The original contributions presented in the study are included in the article.

Acknowledgments: The authors gratefully acknowledge the support of the Bulgarian Ministry of Education and Science under the National Program "Young Scientists and Postdoctoral Students-2".

Conflicts of Interest: The authors declare no conflicts of interest. The funders had no role in the design of the study; in the collection, analyses, or interpretation of the data; in the writing of the manuscript; or in the decision to publish the results.

References

1. González, A.; Goikolea, E.; Barrena, J.A.; Mysyk, R. Review on supercapacitors: Technologies and materials. *Renew. Sustain. Energy Rev.* **2016**, *58*, 1189–1206. [CrossRef]
2. Czagany, M.; Hompoth, S.; Keshri, A.K.; Pandit, N.; Galambos, I.; Gacsi, Z.; Baumli, P. Supercapacitors: An Efficient Way for Energy Storage Application. *Materials* **2024**, *17*, 702. [CrossRef]
3. Liu, K.; Wu, J. Boosting the Performance of Ionic-Liquid-Based Supercapacitors with Polar Additives. *J. Phys. Chem. C* **2016**, *120*, 24041–24047. [CrossRef]
4. Wang, Y.; Xue, K.; Yan, C.; Li, Y.; Zhang, X.; Su, K.; Ma, P.; Wan, S.; Lang, J. Tuning of Ionic Liquid–Solvent Electrolytes for High-Voltage Electrochemical Double Layer Capacitors: A Review. *Batteries* **2024**, *10*, 54. [CrossRef]
5. Venâncio, R.; Vicentini, R.; Pinzón, C.M.J.; Corrêa, D.A.; Miranda, A.N.; Queiroz, A.C.; Degasperi, F.T.; Siqueira, L.J.; Da Silva, L.M.; Zanin, H. Combining electrochemical, molecular simulation and operando techniques to investigate the stability of electrodes and organic electrolytes used in EDLCs. *Energy Storage Mater.* **2023**, *62*, 102943. [CrossRef]
6. Mendhe, A.; Panda, H.S. A review on electrolytes for supercapacitor device. *Discov. Mater.* **2023**, *3*, 29. [CrossRef]
7. Xiong, W.; Yin, Z.; Zhang, X.; Tu, Z.; Hu, X.; Wu, Y. Ionic Liquids Endowed with Novel Hybrid Anions for Supercapacitors. *ACS Omega* **2022**, *7*, 26368–26374. [CrossRef]
8. Pan, S.; Yao, M.; Zhang, J.; Li, B.; Xing, C.; Song, X.; Su, P.; Zhang, H. Recognition of Ionic Liquids as High-Voltage Electrolytes for Supercapacitors. *Front. Chem.* **2020**, *8*, 261. [CrossRef]
9. Rahmi, V.A.; Zunita, M. Ionic liquid-based electrolyte in supercapacitors. *AIP Conf. Proc.* **2024**, *3073*, 060001. [CrossRef]
10. Eftekhari, A. Supercapacitors utilising ionic liquids. *Energy Storage Mater.* **2017**, *9*, 47–69. [CrossRef]
11. Chaban, V.V.; Voroshylova, I.V.; Kalugin, O.N.; Prezhdo, O.V. Acetonitrile Boosts Conductivity of Imidazolium Ionic Liquids. *J. Phys. Chem. B* **2012**, *116*, 7719–7727. [CrossRef]
12. Yang, H.; Yang, J.; Bo, Z.; Chen, X.; Shuai, X.; Kong, J.; Yan, J.; Cen, K. Kinetic-Dominated Charging Mechanism within Representative Aqueous Electrolyte-based Electric Double-Layer Capacitors. *J. Phys. Chem. Lett.* **2017**, *8*, 3703–3710. [CrossRef]
13. Wu, G.; Tan, P.; Wang, D.; Li, Z.; Peng, L.; Hu, Y.; Wang, C.; Zhu, W.; Chen, S.; Chen, W. High-performance Supercapacitors Based on Electrochemical-induced Vertical-aligned Carbon Nanotubes and Polyaniline Nanocomposite Electrodes. *Sci. Rep.* **2017**, *7*, 43676. [CrossRef]
14. Yu, C.; Masarapu, C.; Rong, J.; Wei, B.; Jiang, H. Stretchable Supercapacitors Based on Buckled Single-Walled Carbon-Nanotube Macrofilms. *Adv. Mater.* **2009**, *21*, 4793–4797. [CrossRef]
15. Lei, Z.; Liu, Z.; Wang, H.; Sun, X.; Lu, L.; Zhao, X.S. A high-energy-density supercapacitor with graphene–CMK-5 as the electrode and ionic liquid as the electrolyte. *J. Mater. Chem. A* **2013**, *1*, 2313–2321. [CrossRef]
16. Wong, S.I.; Lin, H.; Sunarso, J.; Wong, B.T.; Jia, B. Optimization of ionic-liquid based electrolyte concentration for high-energy density graphene supercapacitors. *Appl. Mater. Today* **2020**, *18*, 100522. [CrossRef]
17. Frackowiak, E. Supercapacitors based on carbon materials and ionic liquids. *J. Braz. Chem. Soc.* **2006**, *17*, 1074–1082. [CrossRef]
18. Zhang, S.; Brahim, S.; Maat, S. High-voltage operation of binder-free CNT supercapacitors using ionic liquid electrolytes. *J. Mater. Res.* **2018**, *33*, 1179–1188. [CrossRef]
19. Chaban, V.V.; Voroshylova, I.V.; Kalugin, O.N.; Prezhdo, O.V. Acetonitrile Drastically Boosts Conductivity of Ionic Liquids. *arXiv* **2012**. [CrossRef]
20. Wang, H.; Liu, S.; Huang, K.; Yin, X.; Liu, Y.; Peng, S. BMIMBF4 Ionic Liquid Mixtures Electrolyte for Li-ion Batteries. *Int. J. Electrochem. Sci.* **2012**, *7*, 1688–1698. [CrossRef]
21. Kwon, H.-N.; Jang, S.-J.; Kang, Y.C.; Roh, K.C. The effect of ILs as co-salts in electrolytes for high voltage supercapacitors. *Sci. Rep.* **2019**, *9*, 1180. [CrossRef]
22. Ruiz, V.; Huynh, T.; Sivakkumar, S.R.; Pandolfo, A.G. Ionic liquid–solvent mixtures as supercapacitor electrolytes for extreme temperature operation. *RSC Adv.* **2012**, *2*, 5591–5598. [CrossRef]
23. Deshpande, A.; Kariyawasam, L.; Dutta, P.; Banerjee, S. Enhancement of Lithium Ion Mobility in Ionic Liquid Electrolytes in Presence of Additives. *J. Phys. Chem. C* **2013**, *117*, 25343–25351. [CrossRef]
24. Liu, W.; Yan, X.; Lang, J.; Xue, Q. Effects of concentration and temperature of EMIMBF4/acetonitrile electrolyte on the supercapacitive behavior of graphene nanosheets. *J. Mater. Chem.* **2012**, *22*, 8853–8861. [CrossRef]

25. Bahaa, A.; Alhammadi, A.; Lethesh, K.C.; Susantyoko, R.A.; Bamgbopa, M.O. Ionic liquid electrolyte selection for high voltage supercapacitors in high-temperature applications. *Front. Chem.* **2024**, *12*, 1349864. [CrossRef]
26. Kötz, R.; Hahn, M.; Gallay, R. Temperature behavior and impedance fundamentals of supercapacitors. *J. Power Sources* **2006**, *154*, 550–555. [CrossRef]
27. Likitchatchawankun, A.; Whang, G.; Lau, J.; Munteshari, O.; Dunn, B.; Pilon, L. Effect of temperature on irreversible and reversible heat generation rates in ionic liquid-based electric double layer capacitors. *Electrochim. Acta* **2020**, *338*, 135802. [CrossRef]
28. Thomas, M.; Veleva, S.; Karamanova, B.; Brigandì, A.; Rey-Raap, N.; Arenillas, A.; Stoyanova, A.; Lufrano, F. Highly stable and reliable asymmetric solid-state supercapacitors with low self-discharge rates. *Sustain. Mater. Technol.* **2023**, *38*, e00770. [CrossRef]
29. Nishida, T.; Tashiro, Y.; Yamamoto, M. Physical and electrochemical properties of 1-alkyl-3-methylimidazolium tetrafluoroborate for electrolyte. *J. Fluor. Chem.* **2003**, *120*, 135–141. [CrossRef]
30. Kazemiabnavi, S.; Zhang, Z.; Thornton, K.; Banerjee, S. Electrochemical Stability Window of Imidazolium-Based Ionic Liquids as Electrolytes for Lithium Batteries. *J. Phys. Chem. B* **2016**, *120*, 5691–5702. [CrossRef]
31. Yamada, Y.; Furukawa, K.; Sodeyama, K.; Kikuchi, K.; Yaegashi, M.; Tateyama, Y.; Yamada, A. Unusual Stability of Acetonitrile-Based Superconcentrated Electrolytes for Fast-Charging Lithium-Ion Batteries. *J. Am. Chem. Soc.* **2014**, *136*, 5039–5046. [CrossRef]
32. Kim, E.; Han, J.; Ryu, S.; Choi, Y.; Yoo, J. Ionic Liquid Electrolytes for Electrochemical Energy Storage Devices. *Materials* **2021**, *14*, 4000. [CrossRef]
33. Karuppasamy, K.; Theerthagiri, J.; Vikraman, D.; Yim, C.-J.; Hussain, S.; Sharma, R.; Maiyalagan, T.; Qin, J.; Kim, H.-S. Ionic Liquid-Based Electrolytes for Energy Storage Devices: A Brief Review on Their Limits and Applications. *Polymers* **2020**, *12*, 918. [CrossRef]
34. Salanne, M. Ionic Liquids for Supercapacitor Applications. *Top. Curr. Chem.* **2017**, *375*, 63. [CrossRef]
35. Xu, C.; Li, B.; Du, H.; Kang, F.; Zeng, Y. Supercapacitive studies on amorphous MnO_2 in mild solutions. *J. Power Sources* **2008**, *184*, 691–694. [CrossRef]
36. Sudhan, N.; Subramani, K.; Karnan, M.; Ilayaraja, N.; Sathish, M. Biomass-Derived Activated Porous Carbon from Rice Straw for a High-Energy Symmetric Supercapacitor in Aqueous and Non-aqueous Electrolytes. *Energy Fuels* **2017**, *31*, 977–985. [CrossRef]
37. Karnan, M.; Subramani, K.; Sudhan, N.; Ilayaraja, N.; Sathish, M. Aloe vera Derived Activated High-Surface-Area Carbon for Flexible and High-Energy Supercapacitors. *ACS Appl. Mater. Interfaces* **2016**, *8*, 35191–35202. [CrossRef]
38. Subramani, K.; Sudhan, N.; Karnan, M.; Sathish, M. Orange Peel Derived Activated Carbon for Fabrication of High-Energy and High-Rate Supercapacitors. *ChemistrySelect* **2017**, *2*, 11384–11392. [CrossRef]
39. Boujibar, O.; Ghamouss, F.; Ghosh, A.; Achak, O.; Chafik, T. Activated carbon with exceptionally high surface area and tailored nanoporosity obtained from natural anthracite and its use in supercapacitors. *J. Power Sources* **2019**, *436*, 226882. [CrossRef]
40. Wang, Y.; Hao, L.; Zeng, Y.; Cao, X.; Huang, H.; Liu, J.; Chen, X.; Wei, S.; Gan, L.; Yang, P.; et al. Three-dimensional hierarchical porous carbon derived from resorcinol formaldehyde-zinc tatrate/poly(styrene-maleic anhydride) for high performance supercapacitor electrode. *J. Alloys Compd.* **2021**, *886*, 161176. [CrossRef]
41. Wang, Y.; Zeng, Y.; Zhu, J.; Yang, C.; Huang, H.; Chen, X.; Wang, R.; Yan, P.; Wei, S.; Liu, M.; et al. From dual-aerogels with semi-interpenetrating polymer network structure to hierarchical porous carbons for advanced supercapacitor electrodes. *Colloids Surf. A Physicochem. Eng. Asp.* **2022**, *649*, 12935. [CrossRef]
42. Zaccagnini, P.; Serrapede, M.; Armandi, M.; Bianco, S.; Carminati, S.; Zampato, M.; Melis, G.; Pirri, C.F.; Lamberti, A. A high-temperature high-pressure supercapacitor based on ionic liquids for harsh environment applications. *Electrochim. Acta* **2023**, *447*, 142124. [CrossRef]
43. Penner, R.; Martin, C. Electrochemical investigations of electronically conductive polymers. 2. Evaluation of charge-transport rates in polypyrrole using an alternating current impedance method. *Phys. Chem.* **1989**, *93*, 984–989. [CrossRef]
44. Ahuja, P.; Sahu, V.; Ujjain, S.K.; Sharma, R.K.; Singh, G. Performance evaluation of Asymmetric Supercapacitor based on Cobalt manganite modified graphene nanoribbons. *Electrochim. Acta* **2014**, *14*, 429–436. [CrossRef]
45. Ahuja, P.; Ujjain, S.K.; Kanojia, R. Electrochemical behaviour of manganese & ruthenium mixed oxide@ reduced graphene oxide nanoribbon composite in symmetric and asymmetric supercapacitor. *Appl. Surf. Sci.* **2018**, *427 Pt A*, 102–111. [CrossRef]
46. KVan Aken, K.L.; McDonough, J.K.; Li, S.; Feng, G.; Chathoth, S.M.; Mamontov, E.; Fulvio, P.F.; Cummings, P.T.; Dai, S.; Gogotsi, Y. Effect of cation on diffusion coefficient of ionic liquids at onion-like carbon electrodes. *J. Phys. Condens. Matter* **2014**, *26*, 284104. [CrossRef]

batteries

MDPI

Review

Empowering the Future: Cutting-Edge Developments in Supercapacitor Technology for Enhanced Energy Storage

Mohamed Salaheldeen [1,2,3,*], Thomas Nady A. Eskander [4], Maher Fathalla [5], Valentina Zhukova [1,2], Juan Mari Blanco [2], Julian Gonzalez [1,2], Arcady Zhukov [1,2,6] and Ahmed M. Abu-Dief [4,7,*]

1 Department of Polymers and Advanced Materials, Faculty of Chemistry, University of the Basque Country, UPV/EHU, 20018 San Sebastián, Spain; valentina.zhukova@ehu.eus (V.Z.); julianmaria.gonzalez@ehu.eus (J.G.); arkadi.joukov@ehu.eus (A.Z.)
2 Department of Applied Physics I, EIG, University of the Basque Country, UPV/EHU, 20018 San Sebastián, Spain; juanmaria.blanco@ehu.eus
3 Physics Department, Faculty of Science, Sohag University, Sohag 82524, Egypt
4 Department of Chemistry, Faculty of Science, Sohag University, Sohag 82524, Egypt; thomas_nady_post@science.sohag.edu.eg
5 Department of Chemistry, Faculty of Science, Islamic University of Madinah, Madinah P.O. Box 42351, Saudi Arabia; maherfathalla80@iu.edu.sa
6 IKERBASQUE, Basque Foundation for Science, 48011 Bilbao, Spain
7 Department of Chemistry, College of Science, Taibah University, Madinah P.O. Box 344, Saudi Arabia
* Correspondence: mohamed.salaheldeenmohamed@ehu.eus (M.S.); amamohammed@taibahu.edu.sa (A.M.A.-D.)

Abstract

The accelerating global demand for sustainable and efficient energy storage has driven substantial interest in supercapacitor technology due to its superior power density, fast charge–discharge capability, and long cycle life. However, the low energy density of supercapacitors remains a key bottleneck, limiting their broader application. This review provides a comprehensive and focused overview of the latest breakthroughs in supercapacitor research, emphasizing strategies to overcome this limitation through advanced material engineering and device design. We explore cutting-edge developments in electrode materials, including carbon-based nanostructures, metal oxides, redox-active polymers, and emerging frameworks such as metal–organic frameworks (MOFs) and covalent organic frameworks (COFs). These materials offer high surface area, tunable porosity, and enhanced conductivity, which collectively improve the electrochemical performance. Additionally, recent advances in electrolyte systems—ranging from aqueous to ionic liquids and organic electrolytes—are critically assessed for their role in expanding the operating voltage window and enhancing device stability. The review also highlights innovations in device architectures, such as hybrid, asymmetric, and flexible supercapacitor configurations, that contribute to the simultaneous improvement of energy and power densities. We identify persistent challenges in scaling up nanomaterial synthesis, maintaining long-term operational stability, and integrating materials into practical energy systems. By synthesizing these state-of-the-art advancements, this review outlines a roadmap for next-generation supercapacitors and presents novel perspectives on the synergistic integration of materials, electrolytes, and device engineering. These insights aim to guide future research toward realizing high-energy, high-efficiency, and scalable supercapacitor systems suitable for applications in electric vehicles, renewable energy storage, and next-generation portable electronics.

Keywords: batteries; electrode–electrolyte interface materials; nanostructured materials; carbon-based electrodes; renewable energy

1. Introduction

Energy storage systems (ESSs) are critical for addressing efficiency, power quality, and reliability, and they are vital for contemporary power systems, particularly within the context of direct current (DC) and alternating current (AC) systems. Their role in maintaining grid stability and facilitating the integration of renewable energy sources (RESs) is indispensable [1]. Beyond traditional grid applications, ESSs are also critical in specialized domains such as aircraft, shipboard systems, and electric vehicles, where they manage peak load demands and improve overall system reliability and efficiency [1]. In the realm of microgrids, which are gaining prominence as decentralized and resilient energy solutions, ESSs serve to mitigate the intermittency challenges posed by renewables. Their ability to store excess energy during high generation periods and release it during peak demand or low renewable output enhances the reliability and resilience of these systems, thereby ensuring a consistent power supply and optimizing the utilization of locally generated green energy [2–5]. Microgrids, particularly those utilizing DC power, require careful consideration of the power exchange dynamics influenced by load conditions and power generation characteristics. These dynamics are categorized into low-frequency and high-frequency components, with the former reflecting the gradual variations inherent to renewable sources and regular energy consumption patterns [6]. High-frequency components, on the other hand, are influenced by rapid fluctuations that can impact system stability and necessitate advanced energy storage solutions to maintain equilibrium. Electrochemical capacitors, or supercapacitors, are emerging as a significant energy storage solution due to their high-power density and ultrahigh cyclic stability [7]. These characteristics position them favorably in the quest for high-performance energy devices capable of supporting the growing demands of technology. Despite their lower energy density compared to batteries, supercapacitors are the subject of extensive research aimed at pushing the boundaries of charge storage capabilities. Advancements in the understanding of electrical double-layer formation and pseudocapacitive and intercalation-type (akin to batteries) behaviors have significantly improved the electrochemical performance of supercapacitors. The introduction of innovative nanostructured active materials, such as carbon-, metal-, redox-polymer-, metal–organic framework-, and covalent organic framework-based electrodes, coupled with advanced electrolytes that offer superior stability in conventional aqueous and novel systems, has further bolstered their capabilities [7]. Additionally, the meticulous examination of processes occurring at the electrode–electrolyte interface has led to refined techniques and a roadmap for the next generation of high-performance supercapacitors. The increasing demand for energy and the need for sustainable energy systems have made ESSs an area of renewed interest. The development of sustainable energy systems is critical, due to factors such as declining fuel prices, geopolitical conflicts, and pandemics. Decreasing fuel prices have slowed investment in renewable energy, while geopolitical conflicts and pandemics highlight the need for resilient energy systems that can operate independently. ESSs are essential for providing reliable backup power and enabling microgrids to operate independently of the larger grid. Therefore, developing effective ESS technologies is crucial for creating sustainable energy systems that meet modern society's demands while mitigating the impact of external factors [8]. A multitude of energy storage solutions are available, categorized into electrochemical, mechanical, electrical, and hybrid systems. Within the electrochemical domain, we find batteries, hydrogen storage fuel cells, and flow batteries as prominent examples. On the mechanical side, pumped hydroelectric energy storage (PHES), gravity energy storage (GES), compressed air energy storage (CAES), and flywheel systems are notable. Additionally, supercapacitor energy storage (SES) and superconducting magnetic energy storage (SMES) represent distinct electrical storage technologies. Hybrid configurations often integrate a combination of these methods, such as merging batteries with PHES or pairing supercapacitors with thermal storage. The selection of the most appropriate technology hinges on various factors unique to each application, such as system requirements, economic viability, and the performance metrics of the chosen solution. These factors are essential to ensure an optimal match between the storage technology and its intended use, which can significantly influence the overall efficiency and effectiveness of the system in question [9]. Batteries are widely used energy storage devices that meet the requirements of different industrial and consumer applications. They offer high energy density, making them suitable for applications requiring long discharge times. However, batteries have limitations such as lower power density, slower charge–discharge speeds, and limited cycle life compared to supercapacitors [10]. Lithium-ion batteries are a common type of battery used in high-power storage applications. Supercapacitors (SCs) are energy storage devices that offer superior power density, faster charge–discharge speeds, and longer cycle life compared to batteries [11]. They store energy through the accumulation of electric charge at the interface between an electrode and an electrolyte [12]. SCs are suitable for applications requiring high-speed energy delivery, such as hybrid vehicles and wearable electronic devices [13]. However, SCs have lower energy storage capability compared to batteries, which has driven research efforts to increase their energy density. Hybrid energy storage systems (HESSs) combine the advantages of batteries and supercapacitors to achieve high energy and power density [14]. A battery–supercapacitor HESS (BS-HESS) is widely used in renewable energy integration, smart grids, and microgrids. HESSs require sophisticated energy management systems (EMS) to coordinate energy flows between the battery and the supercapacitor, ensuring efficient and safe operation [15]. This review is particularly timely in light of several converging developments in the field of energy storage. Recent breakthroughs in hybrid supercapacitor systems, combining battery-like and capacitor-like behaviors, have opened new frontiers for achieving both high energy and power densities. Simultaneously, there is a growing emphasis on sustainability-driven research, including the use of environmentally benign materials and green electrolytes, aligning with global efforts to reduce carbon emissions and enhance energy efficiency. Furthermore, the increasing demand for next-generation applications—such as flexible and wearable electronics, electric vehicles, and grid-scale renewable energy integration—necessitates a comprehensive update that captures the latest innovations in materials, device architectures, and performance optimization strategies. This review aims to bridge these emerging needs by offering a holistic synthesis of current advancements and future directions in supercapacitor technology.

2. Energy Storage Technologies

2.1. Lithium-Ion Batteries

Lithium-ion batteries are vital to a range of contemporary applications, from portable electronic devices to the aerospace sector and electric vehicles, due to their proficient ability to interconvert chemical and electrical energy during discharge and charge cycles [16]. Their design typically features a series or parallel arrangement of electrochemical cells, each encompassing a negative electrode (anode) and a positive electrode (cathode), separated by an electrolyte and a separator within the cell itself [17,18]. The choice of electrolyte and separator has a substantial impact on the battery's performance and security profile. Solid-state electrolytes represent a burgeoning technology that promises to enhance charge and discharge rates while bolstering safety [19]. However, the challenge with polymer-based electrolytes is their limited electrochemical stability, which restricts the selection of materials that can effectively conduct ions. In contrast, liquid electrolytes present a broader spectrum of options, as solvents boast varying viscosity and dielectric constants that influence performance in distinct ways [20]. In the realm of cathode materials, commonly utilized substances include lithiated metal oxides, with notable examples being $LiCoO_2$, Li-Mn-O, and $LiFePO_4$, along with lithium-layered metal oxides [21]. Each of these materials showcases distinct electrochemical properties, which in turn have significant implications for energy density, power density, and safety aspects, as outlined in Table 1. For instance, while lithium cobalt oxide boasts high energy density, it is associated with thermal stability and safety concerns. Conversely, lithium iron phosphate offers improved safety and thermal stability, albeit at the cost of a lower energy density. Therefore, the selection of cathode materials requires careful consideration to achieve an optimal balance between various performance metrics and safety requirements. During the charging phase, the cathode material undergoes a transformation into lithium ions, which migrate through the lithium salt electrolyte toward the anode. Upon reaching the anode, these ions combine with incoming electrons from an external circuit. The electrolyte, primarily consisting of organic carbonates of lithium such as LiPF6, plays a critical role in facilitating this reaction [22]. It is essential to note that the properties of the electrolyte, such as the ionic conductivity, viscosity, flammability, and thermal stability, are significantly influenced by the choice of solvent and additives. Consequently, tailoring the electrolyte formulation is crucial for optimizing the battery performance, safety, and longevity. This is achieved by enhancing the efficiency of the ion transfer and thermal stability, which are key factors in prolonging the service life and reducing the risk of failure or degradation.

Table 1. Comparison between three cathode materials [23].

Parameters	Li-Ion Manganese	Li-Ion Cobalt	Li-Ion Phosphate
Specific energy density (Wh/kg)	100–135	150–190	90–120
Cell voltage (nominal V)	3.8	3.6	3.3
Cycle life (80% discharge)	500–1000	500–1000	1000–2000
Internal resistance (mΩ)	25–75	150–300	25–50
Fast charge time (Hours)	<1	2–4	<1

Lithium-ion batteries are celebrated for their exceptional energy density, swift responsiveness, prolonged cyclic endurance, and remarkable efficiency. Notably, the discharge voltage characteristics of Li-Mn and Li-phosphate variants display a significantly flat profile, with approximately 80% of the total energy capacity being accessible within this range. This feature greatly simplifies the design process for various applications. Their prevalence is particularly evident in the realms of portable electronic devices and as promising components within electric and hybrid vehicles. Despite these benefits, the integration of lithium-ion batteries on a large-scale faces hurdles, primarily due to the substantial costs stemming from the requirement for specialized packaging and sophisticated internal overcharge safeguarding mechanisms. Moreover, the introduction of these batteries into power systems and the intricate management of battery energy storage systems connected to the grid present additional complexities that must be addressed [24–27].

2.2. Pumped Hydro Energy Storage (PHES)

Pumped hydroelectric storage (PHS) is an established and widely-implemented method in the renewable energy sector for addressing the storage of electrical power. This technology serves a critical function in minimizing the variability inherent in renewable energy sources by accumulating electrical energy during off-peak periods and discharging it during times of heightened demand, as depicted in (Figure 1). The PHS operates through a mechanism that employs two water basins at varying altitudes. The excess electricity generated during low-demand intervals is harnessed to transfer water from the lower basin to the higher one. Conversely, during peak demand, the water is released from the upper reservoir, descending through turbines to produce power as it returns to the lower basin. This process effectively bridges the gap between supply and demand, enhancing the reliability and efficiency of renewable energy systems.

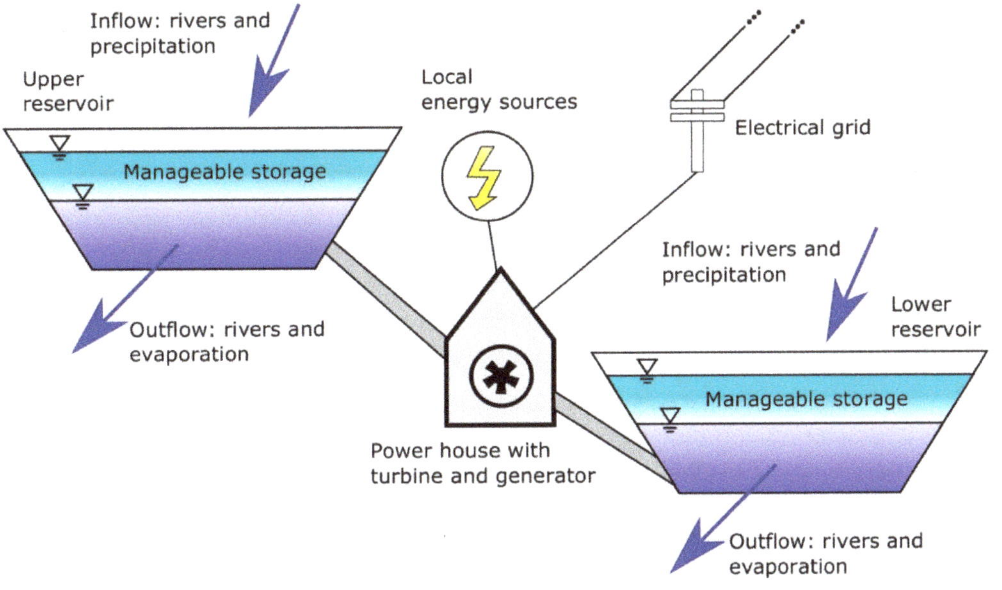

Figure 1. Structure of pumped hydro energy storage [28], licensed under CC BY 4.0.

The capacity for energy storage in PHES is contingent upon the water volume retained and the altitudinal disparity between the two interconnected reservoirs. This system plays a critical role in bolstering the stability of electrical grids and enhancing their overall reliability [29,30].

2.3. Compressed Air Energy Storage (CAES)

CAES, or compressed air energy storage, represents a progressive solution for storing energy efficiently. This technology operates by compressing air and storing it in subterranean caverns or designated tanks during off-peak electricity consumption hours. The compressed air is later released during peak demand, undergoing a controlled expansion that propels turbines to produce power. The system is characterized by a two-phase cycle, comprising compression and expansion. During the compression phase, excess electrical energy is astutely utilized to compact and stockpile atmospheric air. In the subsequent expansion phase, the confined high-pressure air is gradually released, facilitating its expansion to drive turbines for electricity generation. It is noteworthy that, despite boasting an energy efficiency spectrum of 40–70%, CAES remains a pivotal component in managing fluctuations in energy demand. Its contributions to the electrical grid's stability are substantial, illustrating its significance in the broader context of energy management [31–33].

2.4. Supercapacitor Energy Storage

Supercapacitors, or ultracapacitors, stand out as a unique category of energy storage devices, bridging the characteristics of typical capacitors and batteries. They leverage electrochemical mechanisms to store energy, which is fundamentally different from the electrostatic approach employed by conventional capacitors. This advanced technology provides supercapacitors with substantially greater energy storage potential. These high-performance devices are known for their exceptional power and specific capacitance properties [34]. The energy storage in supercapacitors is facilitated through a combination of electrical double-layer capacitance (EDLC) and pseudocapacitance. EDLC is characterized by the buildup of ions at the juncture of the electrode and electrolyte, which generates an electrical double layer without involving charge transfer across the boundary [35]. In contrast, pseudocapacitance is a Faradaic process that relies on redox reactions to store charge across the electrode–electrolyte interface. The effectiveness of supercapacitors is contingent upon various factors such as the nature of the electrode material, the composition of the electrolyte, and the intricacies of the device's architecture [36]. For optimal performance, the electrode material should possess a high surface area to enhance capacitance, excellent electrical conductivity to support efficient charge flow, and robust stability in electrochemical environments [37]. Additionally, the selection of an electrolyte with high ionic conductivity and a broad electrochemical window is crucial for rapid ion movement and elevated operational voltages. In terms of device design, minimizing the internal resistance is vital for reducing energy losses, while optimizing the pathways for ion diffusion is key to achieving superior overall performance.

2.4.1. Types of Supercapacitors

Supercapacitors, also known as ultracapacitors, are distinguished into three primary categories: electrochemical double-layer capacitors (EDLC), hybrid capacitors, and pseudocapacitors, as illustrated in Figure 2. These distinctions arise from variations in their material composition, manufacturing techniques, and specific design features of their electrodes, which in turn influence their suitability for different application domains. Electrochemical double-layer capacitors (EDLCs) are a subset that can be further divided into three types based on the materials utilized for constructing their electrodes. These include capacitors that employ activated carbon, carbon nanotubes, and graphene as the foundation for their energy storage capabilities. Activated carbon is widely recognized for its high surface area and excellent electrical conductivity, making it a popular choice for EDLC electrodes. Carbon nanotubes, on the other hand, offer unique structural properties that enhance the performance of these devices, while graphene-based supercapacitors leverage the exceptional characteristics of graphene, such as high surface area and excellent mechanical and electrical properties, to achieve higher energy densities and power performance [38,39]. Pseudocapacitors, a separate class, are categorized into polymeric and metal oxide/hydroxide types. Polymer pseudocapacitors rely on the redox reactions of conductive polymers, which contribute to their energy storage mechanism. In contrast, metal oxide/hydroxide pseudocapacitors derive their capacitance from the reversible Faradaic interactions between the metal oxide or hydroxide electrodes and the electrolyte, leading to the formation or dissolution of surface species and the movement of electrons or ions. These variations in material selection and the underlying electrochemical processes within pseudocapacitors determine their specific capacities and rate capabilities. Each type of supercapacitor offers distinct advantages and limitations, guiding their application in various fields such as automotive, renewable energy storage, and portable electronics. The choice of supercapacitor type is contingent upon specific performance requirements, such as energy density, power density, cyclic stability, and cost-effectiveness. Understanding these classifications allows for a more informed decision when selecting a suitable technology to meet the demands of a given application [40,41].

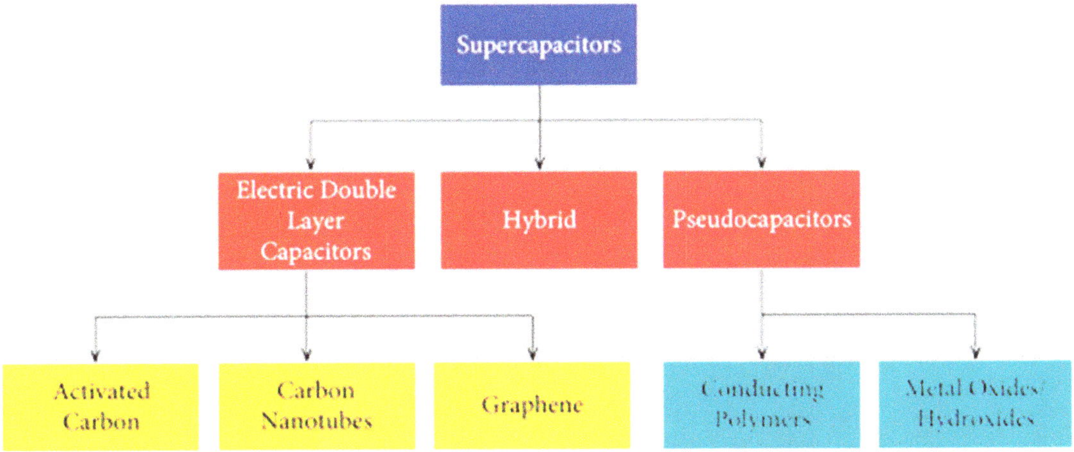

Figure 2. Overview of the three main types of supercapacitors and their material classifications.

Electrical Double-Layer Capacitors (EDLCs)

Electric double-layer capacitors (EDLCs) operate on the principle of charge separation, much like traditional capacitors, but they boast significantly higher capacitance values. This enhanced capacitance is achieved through the utilization of high-surface-area porous materials, such as activated carbon [42], which contrasts with the flat plates found in conventional capacitors. The term "electric double layer" is derived from the mechanism by which EDLCs store energy—electric charge accumulates at the interface between the electrode and the electrolyte, creating a double layer of charge. These capacitors are particularly advantageous in stationary and mobile systems that demand high power. They excel in scenarios requiring rapid energy capture, for instance, during regenerative braking in vehicles, attributed to their low time constant of less than a minute. The storage of energy in EDLCs is an electrostatic process, facilitated by the reversible adsorption of electrolyte ions onto the electrode material. The polarization at the electrolyte/electrode junction results in capacitance that is characteristic of a double layer.

The reasons for EDLCs' superior energy storage capabilities, when juxtaposed with standard capacitors, are two-fold:

1. The extensive surface area of the electrode material, riddled with numerous pores, enables a substantial enhancement in the storage capacity for electric charge. This intricate structure allows for a dramatically increased volume for the formation of electric double layers, which are crucial for energy storage [43].
2. The ultrathin electrical double layers that are established at each electrode–electrolyte interface contribute to the elevated capacitance of these devices.

The construction of EDLC supercapacitors is analogous to batteries, comprising two electrodes immersed in an electrolyte and separated by an ion-permeable separator to prevent direct electrical contact (Figure 3a). Upon charging, the anions and cations in the electrolyte migrate towards the respective positive and negative electrodes, forming two distinct double layers. This arrangement can be conceptualized as a series of capacitors within the cell, with each electrode–electrolyte interface acting as an individual capacitor. The separation of ions across the double layers results in a potential difference throughout the cell, which is a fundamental aspect of its energy storage functionality.

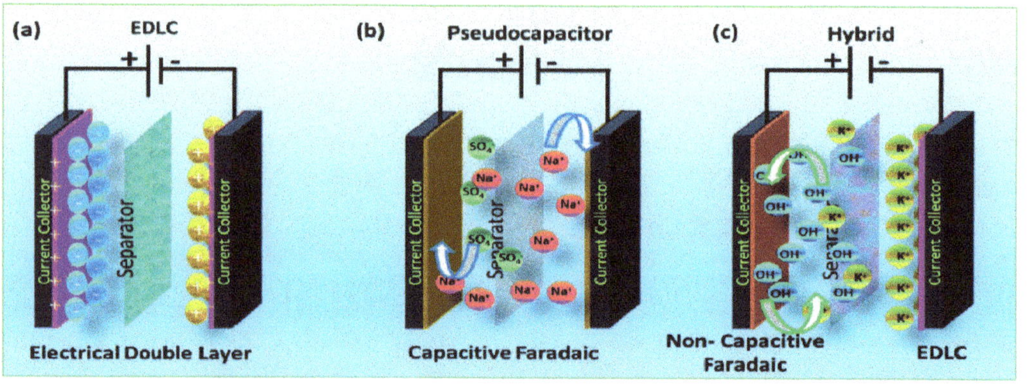

Figure 3. Supercapacitors' charge storage mechanisms include (**a**) EDLCs, (**b**) pseudocapacitors, (**c**) and hybrid supercapacitors. Reproduced with approval from [44], copyright (2021), Royal Society of Chemistry.

Pseudocapacitors

Pseudocapacitors operate on the principle of Faradaic reactions for energy storage, where the electrode–electrolyte interface facilitates the electrostatic storage of charge. The electrochemical process in these devices involves reversible reduction–oxidation (redox) reactions that occur at the electrode surface when a voltage is introduced. These reactions are characterized by a rapid flow of Faradic current, which is the result of charge transfer across the double layer (Figure 3b). In contrast to electrochemical double-layer capacitors (EDLCs), pseudocapacitors exhibit an enhanced rate of electrochemical processes, contributing to higher specific capacitance, and consequently, greater energy densities [45,46].

Hybrid Supercapacitors

Hybrid supercapacitors have gained significant attention in recent times due to their remarkable ability to improve energy density while preserving power density. These advanced devices surpass the capacitance levels of traditional electric double-layer capacitors (EDLCs) and pseudocapacitors (PCs) (Figure 3c), offering a unique solution to the quest for superior energy storage. The asymmetry inherent in their design, which involves the integration of an EDLC and a PC, enhances the overall capacitance values by leveraging the complementary characteristics of each component. This innovative approach opens up a promising avenue for the development of sustainable and high-performance energy storage systems, which are crucial for the growing demand in the field of hybrid electric vehicles (HEVs). The integration of different storage mechanisms in hybrid supercapacitors is a strategic move towards achieving more efficient energy storage. Specifically, the combination of EDLCs, which rely on electrostatic charge storage, and PCs, which utilize fast and reversible Faradaic reactions at the electrode–electrolyte interface, provides a synergistic effect that overcomes the individual limitations of each technology. This synergy is crucial for advancing the performance of these devices and is particularly advantageous for energy-demanding applications such as HEVs. To enhance the performance of hybrid supercapacitors for energy-efficient purposes, the introduction of innovative materials is essential. These materials should facilitate an increase in the surface-to-volume ratio, thereby contributing to enhanced energy storage capabilities. The fundamental concept behind these hybrid storage systems is the combination of the two distinct storage mechanisms: the double-layer capacitance from EDLCs and the pseudocapacitance from PCs. The architecture of hybrid supercapacitors can be tailored to be either symmetric or asymmetric, depending on the desired application and performance criteria. The choice between symmetric and asymmetric designs depends on the specific requirements of the system, such as the voltage window, energy density, and power density [47]. The comparison of various supercapacitor types reveals their relative advantages and disadvantages, as presented in (Figure 4). This comprehensive analysis helps in the selection of the most suitable hybrid configuration for different applications. Understanding these characteristics is pivotal for researchers aiming to design and develop advanced hybrid supercapacitors that cater to the evolving needs of the energy sector, especially in the context of transportation electrification and the push towards a greener future.

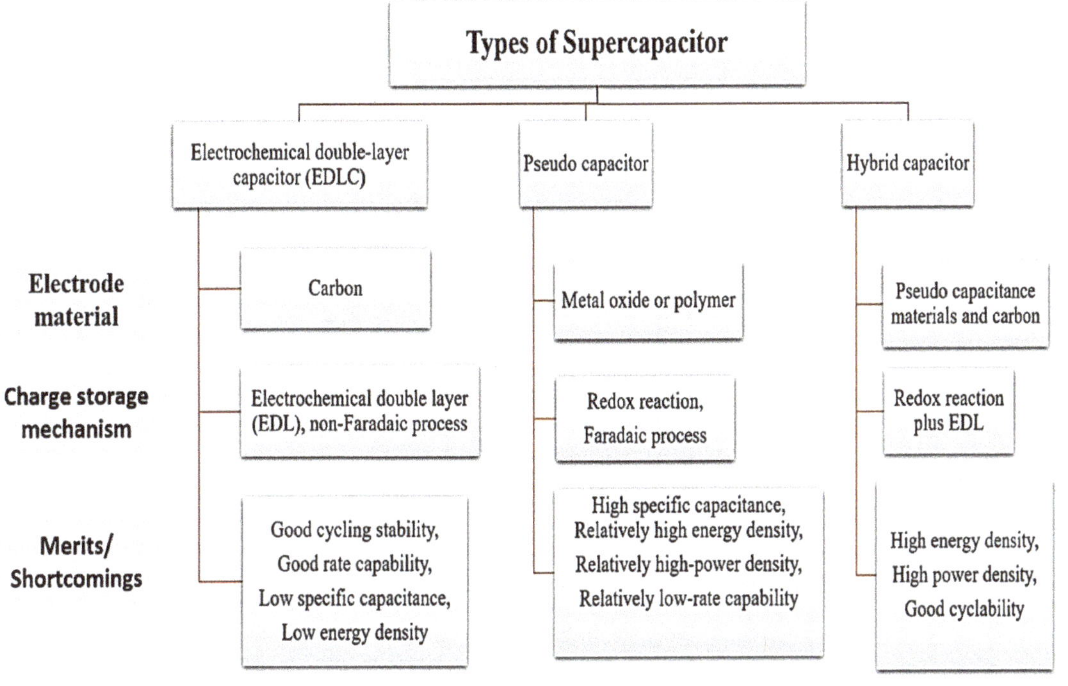

Figure 4. Comparative analysis of the key performance metrics of electrochemical double-layer capacitors (EDLCs), pseudocapacitors, and hybrid supercapacitors.

2.5. Key Materials in Electrochemical Energy Storage Devices

The performance of supercapacitors is highly dependent on the materials used for their construction, particularly the electrode and electrolyte materials. The electrode material should possess a high surface area, excellent electrical conductivity, good electrochemical stability, and low cost. The electrolyte should have high ionic conductivity, a wide electrochemical window, and good compatibility with the electrode material.

2.5.1. Electrode Materials

Carbon-based materials, transition metal oxides, conducting polymers, metal chalcogenides, and composite materials are some of the electrode materials utilized in supercapacitors [48].

Carbon-Based Materials

Various nanomaterials derived from carbon have been developed and incorporated into the design of energy conversion and storage systems, with a particular emphasis on their role as electrode components in supercapacitors (SCs). Notable examples of such materials include activated carbon (AC) [49], graphene [50], carbon nanotubes (CNT) [51], and carbon nanofibers (CNF) [52]. These materials are favored due to their exceptional properties, which include a substantial specific surface area (SSA), excellent electrical conductivity, tailorable porous architecture, economic viability for large-scale manufacturing, and the ease with which they can be modified through chemical and physical means to suit specific application requirements (Figure 5). Various techniques, such as carbonization, activation processes, functionalization, and doping, are frequently employed to optimize the physical and chemical characteristics of these materials. These methods are crucial in enhancing the electrochemical performance of supercapacitors when utilized as electrodes and are comprehensively discussed in the subsequent sections

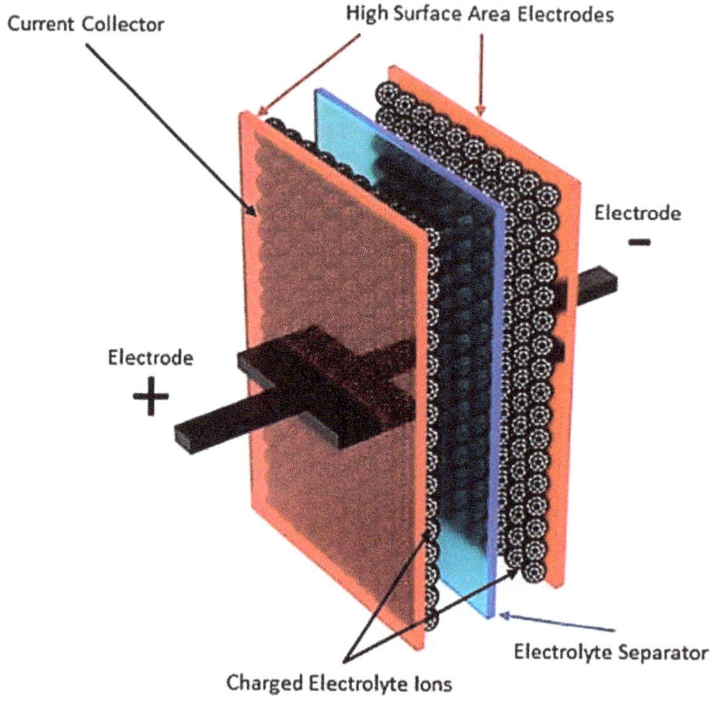

Figure 5. The schematic of the supercapacitor structure highlighting the role of carbon-based electrodes.

Transition Metal Oxides

The utilization of metal oxides in the realm of electrochemical capacitors has gained considerable attention due to their intrinsic merits, such as lower resistivity and improved specific capacitance. A multitude of studies have focused on the potential of various transition metal oxides, which generally outperform conductive polymers in terms of capacitance properties and present higher energy densities than their carbon-based counterparts [53]. Notable entries in this category are nickel oxide (NiO), manganese dioxide (MnO_2), ruthenium dioxide (RuO_2), and iridium oxide (IrO_2). However, while ruthenium dioxide is recognized for its exceptional performance, its high cost and environmental concerns impose significant barriers to widespread application [54]. On the other hand, materials such as MnO_2 and ZnO are naturally abundant and have showcased substantial capacitance values, thereby establishing themselves as appealing alternatives for supercapacitor technologies. These oxides offer the possibility of achieving high energy storage capacity and excellent cyclic stability, which are essential for developing advanced energy storage systems that can meet the growing demands of various industries.

Conducting Polymers

Pseudocapacitors have found notable promise in the realm of energy storage due to the potential of conductive polymers. These polymers stand out from electric double-layer capacitors in terms of their enhanced charge density, simplified manufacturing, and economic viability [55]. Among the key conductive polymers employed as electrodes in such devices are polyaniline (PANi), polypyrrole (Ppy), and polythiophene (PTh). These materials have gained considerable attention as they exhibit a range of desirable characteristics, such as superior conductivity, cost-effectiveness, and adaptability for various applications. Their distinctive properties are succinctly summarized in (Figure 6). Extensive research endeavors have been undertaken to explore the full scope of these polymers, resulting in numerous substantial advancements that have bolstered their performance [56].

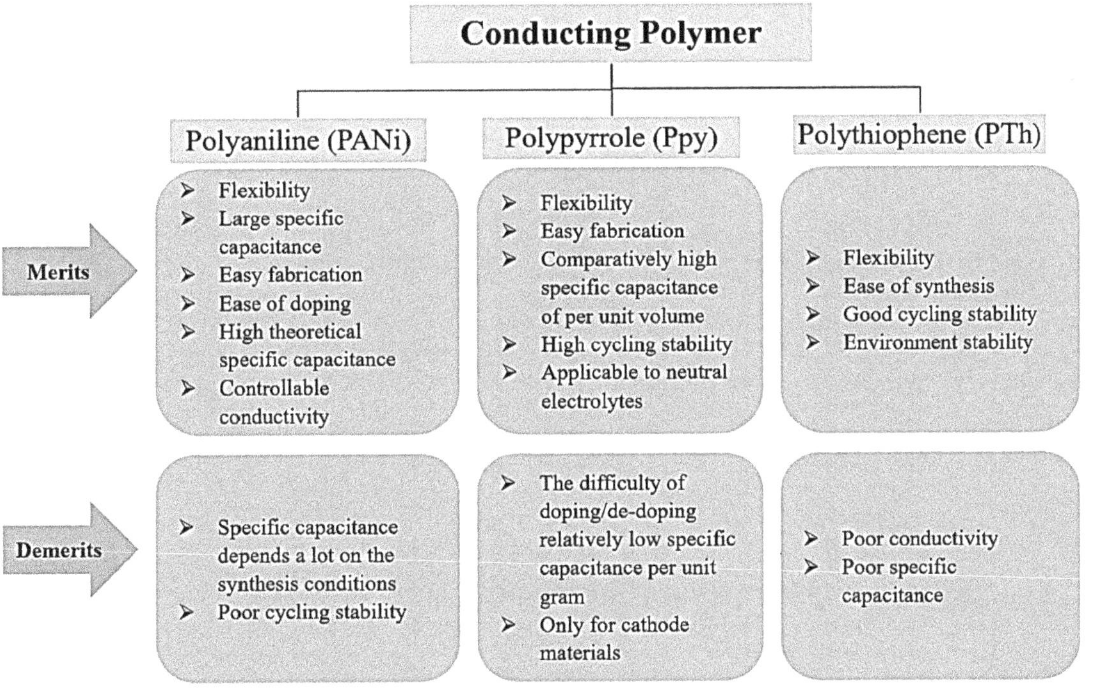

Figure 6. Comparison of polyaniline (PANi), polypyrrole (PPy), and polythiophene (PTh) as electrode materials for pseudocapacitors.

Polyaniline (PANi)

Polyaniline (PANi) is a promising conducting polymer that stands out for several favorable attributes, including its flexibility, substantial specific capacitance, cost-effectiveness, ease of synthesis, and adjustable conductivity, with a high doping level of 0.5. Despite these advantages, PANi faces challenges such as its susceptibility to rapid degradation during repeated charge/discharge cycles and the necessity for a protic solvent or ionic liquid to ensure optimal performance [57]. In an effort to mitigate these drawbacks, researchers have integrated PANi with carbon materials and metal oxides, yielding composites that show significant promise. Sivakumar and colleagues employed interfacial polymerization to produce PANi nanofibers, which exhibited a specific capacitance of 554 F/g at a current density of 1.0 A/g. Despite the initial success, it was noted that the specific capacitance value deteriorated rapidly, and the cyclic stability was less than ideal. To address this, they developed a composite of PANi with multi-walled carbon nanotubes (MWCNT), achieving an enhanced specific capacitance of 606 F/g [58]. This demonstrates the potential of combining PANi with other materials to improve its performance characteristics. Li and others conducted both theoretical and experimental studies on the specific capacitance of PANi in sulfuric acid, revealing a discrepancy where the experimental values were significantly lower than the theoretical predictions. The highest theoretical specific capacitance value determined was 2000 F/g, highlighting the substantial potential that remains untapped in PANi [59]. This disparity suggests that further research and optimization are essential to realize the full capacitive capabilities of this material.

Polypyrrole

Polypyrrole (PPy), a notable member of the conductive polymer family, is valued for its flexibility, straightforward fabrication, and enhanced cyclic stability, which contribute to its high volumetric capacitance. Its elevated density is reflected in its substantial capacitance per unit volume. However, PPy also presents certain challenges, namely the complex doping process and a comparatively low specific capacitance on a per gram basis. A notable synthesis approach for PPy-based supercapacitors involves the creation of free-standing films, as suggested by Yang and collaborators. The inclusion of a surfactant in this process has been found to yield more favorable outcomes, as it promotes the formation of smaller pores within the film's structure. This method showcased a capacitive retention of 75%, along with a peak specific capacitance of 261 F/g [60]. Moreover, the work conducted by Rajesh et al. [61] examined the electropolymerization of polypyrrole doped with phytic acid. This study reported impressive results with a maximum specific capacitance reaching 343 F/g and a capacitance retention rate of 91% at 10 A/g after subjecting the material to 4000 charge–discharge cycles. The use of phytic acid as a dopant in the PPy films not only facilitated the doping process but also significantly improved the film's electrochemical properties. These findings underscore the potential of PPy in the realm of supercapacitors, despite its limitations, and encourage further exploration into the optimization of its synthesis techniques and doping strategies.

Polythiophene (Pth)

Polythiophenes have been explored for their application in energy storage systems, particularly as electrode materials. These polymers exhibit several favorable properties, such as a high flexibility, ease of synthesis, and environmental compatibility. However, they are also associated with limitations, namely poor electrical conductivity and relatively low specific capacitance [62]. To address these issues, researchers have developed various strategies to enhance their performance. For instance, Laforgue et al. [63] have chemically synthesized poly(thiophene) (Pth) and poly(3,4-ethylenedioxythiophene) (PFPT), achieving specific capacitances of 7 mAh/g and 40 mAh/g, respectively. These values highlight the potential of these materials in the context of energy storage. Further research has focused on improving the capacitance of polythiophenes. Patil's group employed the successive ionic layer adsorption and reaction (SILAR) method to produce PTh thin films, which demonstrated a notable enhancement in capacitance up to 252 F/g when using $FeCl_3$ as an oxidizing agent [64]. Another study reported the preparation of PTh films through chemical bath deposition, which yielded an impressive maximum specific capacitance of 300 F/g [64]. These findings underscore the effectiveness of different synthesis techniques in optimizing the properties of polythiophenes for energy storage applications

Metal Chalcogenides

Transition metal sulfides (TMSs) have garnered significant interest as appealing candidates for supercapacitor electrodes, largely attributed to their high theoretical capacitance values, exceptional electrical conductivity, and advantageous redox characteristics. Specific TMSs, such as manganese (Mn), vanadium (V), cobalt (Co), iron (Fe), copper (Cu), nickel (Ni), molybdenum (Mo), zinc (Zn), tungsten (W), and tin (Sn), have exhibited substantial enhancements in their electrochemical performance upon undergoing compositional and structural optimization. Strategic approaches in composite formation and morphological manipulation play a pivotal role in amplifying the efficacy of these TMS electrodes [65].

Binary Composites as Electrode Materials

Composite substances, denoted by their dual- or tri-material structure, are formed by integrating two or three distinct components. Specifically, within this realm, composite materials encompass blends of metal oxides, carbon-based elements, and conductive polymers. These materials can be divided into two primary classifications: (a) binary composites, exemplified by combinations like NiO with Co_3O_4 or MnO_2 with CNT (carbon nanotubes), and (b) ternary composites, which involve more complex arrangements such as GO (graphene oxide) intertwined with MWCNT (multi-walled carbon nanotubes) and PANI (polyaniline) or Co_3O_4 in conjunction with NG (nanographene) and CNTs.

Binary compound electrodes are fabricated by integrating two different substances, which can be any combination of carbon-based materials, metal/metal oxides/metal hydroxides, and conductive polymers. These composite systems typically demonstrate enhanced electrochemical efficacy in comparison to their standalone constituents. This superiority arises from the concurrent engagement of both Faradaic and non-Faradaic charge storage processes within the composite structure. By combining these elements, the resulting material benefits from a synergistic interplay of properties, which can be optimized for various electrochemical applications. Let us delve into some illustrations of binary composites and assess their electrochemical performance characteristics.

Carbon–Carbon Composites

The performance of carbon materials in supercapacitors is largely contingent upon the accessible specific surface area (SSA) for electrolyte ions, which can enhance the energy and power density. To achieve a high SSA, researchers have developed non-covalent functionalized graphene. Despite its remarkable properties, graphene tends to disperse poorly, due to aggregation caused by Van der Waals forces. In a study by Zhang et al. [66], single-walled carbon nanotubes (SWCNTs) were introduced as spacers to prevent restacking and thus maintain the accessible surface area for the electrolyte. This strategy yielded a specific capacitance of 261 F/g and an energy density of 123 Wh/kg in a graphene-based supercapacitor with an ionic liquid electrolyte operating at 3.7 V. Another approach to optimize graphene's capacitive characteristics involves combining it with activated carbon (AC) to form a composite. Zheng et al. [67] synthesized such a graphene/AC nano-sheet electrode material for supercapacitors. The unique architecture of this composite provided two distinct types of porosity: micropores from AC and mesopores resulting from the interconnection of the graphene sheets. This combination led to specific capacitances of 103 F/g and 210 F/g in organic and aqueous electrolytes, respectively. Moreover, the material demonstrated excellent cyclic stability with a capacitance retention rate of 94.7% after undergoing 5000 charge–discharge cycles.

Carbon–MOs Composites

We are aware of the well-established drawbacks of carbon materials and transition metal oxides (TMOs), including their low specific capacitance, poor electrical conductivity, and restricted electrochemical stability. To overcome these challenges, researchers have turned to the fabrication of composite electrodes for supercapacitors (SCs) that incorporate both TMOs and carbon materials. A notable example is the synthesis of MnO_2/CNT composites through a hydrothermal method, which has been shown to exhibit a specific capacitance of 223 F/g, thereby surpassing the performance of CNT and MnO_2 electrodes used in isolation [68]. The integration of TMOs with carbon materials typically results in composites with higher surface areas compared to their individual components. Several research initiatives have focused on improving the capacitance of these composite materials. For instance, cobalt oxide has been combined with graphene [69] and carbon nanofibers (CNFs) to create Co_3O_4/CNFs [69]. These combinations have displayed enhanced surface areas and correspondingly increased electrical conductivity, which are essential for improved supercapacitor performance. Additionally, the exploration of ternary metal oxides (TMOs) has also shown promise. A research team led by Xiong et al. [70] utilized hydrothermal synthesis to produce nickel–cobalt–manganese triple hydroxide nanoneedles, which exhibited outstanding electrochemical properties. The synthesized ternary metal oxide demonstrated a noteworthy specific capacitance of 1400 F/g, an impressive energy density of 30 Wh/kg at a high-power density of 39 kW/kg, and an exceptional cyclic stability, maintaining 100% of its capacitance over an extensive 3000 cycles. This study highlights the efficacy of such a synthesis approach in creating advanced materials with excellent performance metrics suitable for energy storage applications.

Carbon–CPs Composites

The study conducted by Liangliang et al. [71] has highlighted the potential of nanocomposites comprising activated carbon cathodes and conductive polymer anodes in electric double-layer capacitors (EDLCs) to enhance energy and power density, surpassing alternative materials. Conductive polymers, like polyaniline (PANI), have garnered significant attention in the creation of supercapacitors (SCs) due to their high capacitance and excellent cyclic performance. However, PANI alone confronts issues with rate capability and cycle stability. The integration with carbon-based materials, such as single-walled carbon nanotubes (SWCNT), has been shown to significantly boost these properties. For instance, a PANI–SWCNT blend exhibited a 65% increase in capacitance and a threefold energy density improvement compared to the individual components. An intriguing development in the field is the flexible supercapacitor introduced by Yang et al. [72], which employs a reduced graphene oxide (rGO) aerogel–PANI composite fabricated via electrodeposition. This material boasts a specific capacitance of 432 F/g and an energy density of 25 Wh/kg, while maintaining 85% of its capacitance over 10,000 charge–discharge cycles. The high flexibility stability under various bending conditions underscores its potential for flexible energy storage solutions. Furthermore, researchers have explored ternary composites to optimize SC performance. An example is the work of Yan et al. [73], who synthesized a material through in situ polymerization combining graphene nanosheets (GNS), carbon nanotubes (CNT), and PANI. This GNS/CNT/PANI composite presented a specific capacitance of 1035 F/g, which, although marginally less than the GNS/PANI binary structure (1046 F/g), significantly outperformed the CNT/PANI binary system (780 F/g). Most importantly, the ternary composite displayed remarkable cyclic stability, retaining 94% of its capacitance after 1000 cycles. This suggests that despite the slightly lower specific capacitance compared to the GNS/PANI binary, the GNS/CNT/PANI ternary composition offers superior long-term performance reliability, making it an attractive candidate for supercapacitors that demand robustness and consistent function over numerous charge–discharge cycles.

MOs–CPs

The integration of metal oxides (MOs) with conducting polymers (CPs) is a notable advancement in the realm of electrode materials for supercapacitors (SCs). This synergistic pairing has been observed to offer substantially better electrochemical properties than their individual components. The MOs provide high specific capacitance and stability, while the CPs contribute to improved electrical conductivity and charge storage capabilities [74]. A prime example is the study by Liu and colleagues, who developed a MoO_3 electrode with a polypyrrole (PPy) coating through in situ polymerization [75]. The resultant material demonstrated a remarkable energy density of 20 Wh/kg at a power density of 75 W/kg, along with a specific capacitance of 110 F/g at a current density of 100 mA/g. When configured as an asymmetric supercapacitor employing a 0.5 M K_2SO_4 aqueous electrolyte, the energy density increased to 12 Wh/kg, with power density reaching 3 kW/kg. These findings underscore the potential of such composites for high-performance energy storage solutions. Taking this concept further, the formation of ternary nanocomposites that combine carbon-based substances, MOs, and CPs has yielded electrode materials with remarkable electrochemical performance enhancements. These hybrid materials exhibit a significant improvement in both energy and power density in comparison to their pure counterparts. The careful selection of components and synthesis techniques optimize the synergistic effects between the different materials, leading to enhanced charge storage and transport properties. This approach opens up a new avenue for the design of electrodes with tailored characteristics that can meet the increasing demands of various industrial and electronic applications. The integration of these advanced materials into SCs could pave the way for more efficient and reliable energy storage systems, thereby contributing to the advancement of sustainable energy technologies.

MOs–MOs

Binary metal oxide combinations, particularly the MO–MO configuration, have drawn substantial interest in the realm of supercapacitor research due to their advantageous attributes, including improved electrical conductivity, broad potential ranges, and notable energy and power densities. A prominent illustration is the core-shell structured NiO/Co_3O_4 electrode material introduced by Adhikari et al. [76]. This design, featuring NiO/Co_3O_4@NF as the cathode and activated carbon serving as the anode, showcased outstanding electrochemical properties, achieving a capacitance of 2760 F/g at a 2 A/g current density and an energy density of 81.45 Wh/kg at a power density of 4268 W/kg. The most remarkable aspect of their study was the electrode's cyclic durability, with the material retaining 95.5% of its initial capacitance throughout an impressive 12,000 charge–discharge cycles. This outcome underscores the promising potential of binary metal composites for the development of supercapacitors with exceptional performance indicators.

Ternary Composite Materials

Recent research has been increasingly focused on ternary composite materials for supercapacitors due to their ability to outperform binary composites. These advanced materials are constructed by combining three individual components to generate synergistic properties that improve the specific capacitance, cyclic stability, rate capability, and energy density in supercapacitor electrodes. Some noteworthy examples of such ternary composites include Mn_3O_4@$NiCo_2O_4$@NiO, NiO/PANI/CNT, CoO/NiO−Cu@CuO, and MnO_2-AgCNT-CC, each exhibiting distinct advantages that drive the progression of supercapacitor technology. A prime illustration of a ternary composite is the graphene/PANI/Co_3O_4

electrode developed by Lin and colleagues [77]. This material integrates the high stability and capacitance of Co_3O_4 within a 3D porous scaffold formed by graphene and polyaniline. The specific capacitance of this hydrothermally synthesized composite is 1247 F/g at a current density of 1 A/g, with an energy density reaching 190 Wh/kg. Furthermore, it demonstrates excellent cyclic stability, with no loss in capacitance observed after 3500 charge–discharge cycles. These combined features suggest that graphene/PANI/Co_3O_4 has significant potential for the future of supercapacitor electrodes. Another intriguing advancement involves metal–metal–carbon-based composites, such as the one reported by Yang et al. [78]. They developed a ternary system consisting of Co_3O_4, NiO, and graphene oxide (GO), which formed a sea urchin-inspired architecture on a GO substrate that served as the negative electrode in an asymmetric supercapacitor setup with activated carbon as the positive electrode. This design achieved a specific capacitance of 883 F/g at 1 A/g and an energy density of 50.2 Wh/kg. Impressively, it displayed remarkable capacitance retention, with only an 18% decrease after 3000 cycles. The complex microstructures of the sea urchin-like Co_3O_4-NiO and the Co_3O_4-NiO/GO composite can be examined using field-emission scanning electron microscope (FE-SEM) images, revealing their intricate patterns at different scales (Figure 7).

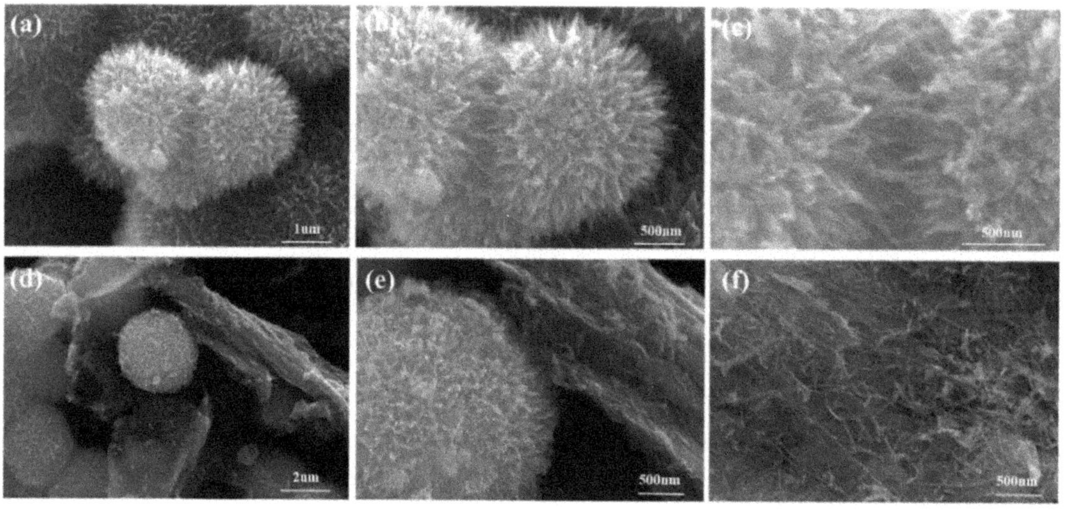

Figure 7. (**a–c**) Sea urchin-shaped Co_3O_4-NiO FE-SEM photographs at various magnifications and (**d–f**) Co_3O_4-NiO/GO FE-SEM photographs at various magnifications. Reproduced with approval from ref. [78] copyright 2020, *Journal of Material Science and Technology*, Elsevier.

Moreover, the cyclic voltammetry (CV) curves of the Co_3O_4-NiO/GO//AC electrode are presented at varying potentials in (Figure 8). Additionally, this figure showcases the galvanostatic charge–discharge (GCD) profiles of the same electrode material at distinct current densities. A cyclic stability test was also conducted on Co_3O_4-NiO/GO//AC at a rate of 10 A/g to evaluate its performance consistency over multiple charge–discharge cycles.

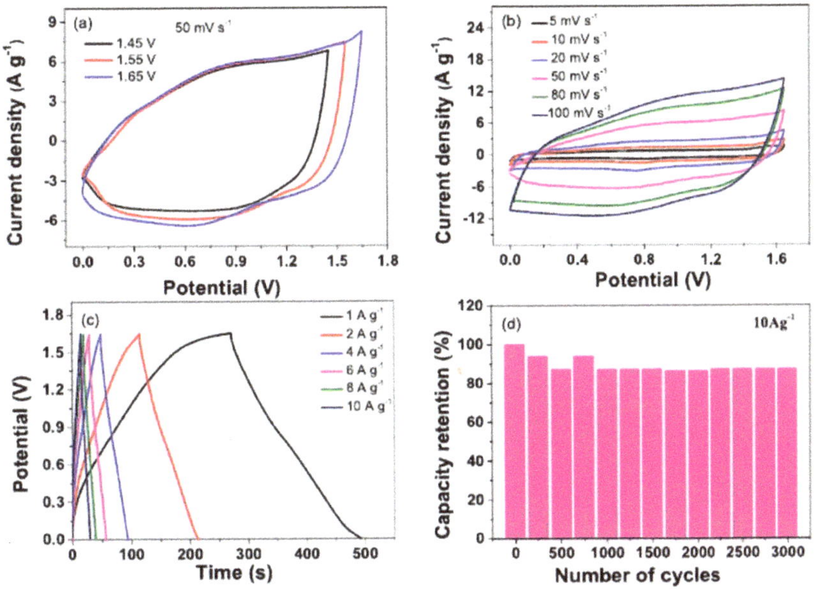

Figure 8. (**a**) Co_3O_4-NiO/GO/AC CV curves at various potentials; (**b**) CV curves and (**c**) Co_3O_4-NiO/GO/AC GCD curves at various current densities; and (**d**) cyclic stability test performed at 10 Ag^{-1}. Reproduced with approval from ref. [78].

A study spearheaded by Muhammad Usman and colleagues focused on enhancing the stability of polyaniline (PANI) by creating a ternary material named PANI and Fe–Ni co-doped Co_3O_4 (PANI@FNCO) using an in situ synthesis approach. With its high stability, high processability, and adjustable conducting and optical properties, polyaniline is the most prospective and studied conducting polymer. The concentration of the dopant determines the conductivity of polyaniline, which only exhibits metal-like conductivity at pH values below 3.23. They are classified as leucoemeraldine, emeraldine, and pernigraniline, by their oxidation state, i.e., leucoemeraldine exists in a sufficiently reduced state, and pernigraniline exists in a fully oxidized state. Only when polyaniline is partially oxidized does it become conductive; when it is fully oxidized, it functions as an insulator [79]. The resulting PANI@FNCO nanocomposite presented exceptional electrochemical characteristics, such as a substantial specific capacitance of 1171 F/g at 1 A/g and an energy density of 144 Wh/kg. Furthermore, the cyclic stability was maintained at 84% over an extensive 2000 cycles [80]. Hao and colleagues [81] engineered a ternary carbon-based composite for enhancing supercapacitor performance, which incorporated graphene oxides (GOs), multi-walled carbon nanotubes (MWCNTs), and polyaniline (PANI). This composite material, denoted as GO/MWCNT/PANI, served as the positive electrode in an asymmetrically designed supercapacitor with KOH-activated GO/MWCNT functioning as the negative electrode. The synergistic combination of these components led to an enhanced electrical conductivity and promoted efficient electron transfer. Consequently, the resulting supercapacitor exhibited an impressive specific capacitance of 696 F/g at a scan rate of 20 mV/s, as well as an energy density of 69 Wh/kg and power density of 6.4 kW/kg. Most importantly, the cyclic stability of this material was demonstrated by maintaining 89% of its capacitance over an extensive 3000 charge–discharge cycles, highlighting its robust performance and potential for practical applications. Xu et al. [82] introduced a ternary composite electrode made of graphene (GN), activated carbon (AC), and polypyrrole (PPy). Employing vacuum filtration and anodic constant current deposition techniques, the researchers crafted a

GN/AC/PPy material that displayed flexibility and robustness. The unique architecture, with AC intercalated between GN layers to prevent restacking, allowed for a maximum specific capacitance of 178 F/g at 0.5 mA/cm2. Additionally, the electrode material displayed commendable resilience, retaining 83.6% of its capacitance after 500 stretching and bending cycles, indicating its suitability for flexible energy storage devices. In a separate endeavor, a ternary composite of N-doped graphene nanosheets (NG), carbon nanotubes (CNTs), and Co_3O_4, denoted as Co_3O_4/NG/CNTs, was synthesized via hydrothermal methods. The synergy of these components led to a significant enhancement in the electrochemical properties, achieving a specific capacitance of 456 F/g. This metal–carbon–carbon combination showcased remarkable potential in supercapacitor applications.

Another study by Li et al. [83] presented a ternary composite of conductive polymers, namely PANI and PPy, supported by a carbon-based matrix derived from wood. This environmentally friendly and sustainable substrate contributed to the composite's excellent performance, with a specific capacitance of 360 F/g at 0.2 A/g, highlighting its prospects for eco-friendly supercapacitors.

Some Novel Emerging Materials

The inherent limitation of transition metal oxides (TMOs) and transition metal hydroxides (TMHs) often includes unsatisfactory electrical conductivity levels. MXenes, a novel class of materials, exhibit exceptional mechanical robustness and hydrophilic characteristics. Despite their potential, the production of MXenes faces hurdles primarily due to the insufficient interaction among the sheets and the scarcity of advanced assembly methods [84,85]. The difficulty in achieving proper alignment further complicates their fabrication. To tackle this, researchers have devised several composite approaches that focus on improving MXenes' electrical properties. Notable among these strategies are the wet-spinning assembly process, the utilization of self-propagating reduction methods, the construction of MXene/CNT sandwich structures, the creation of in situ polymerized CNT-PANI nanocomposites, and the development of self-healing 3D micro-supercapacitors [86]. These innovative techniques strive to overcome the challenges associated with MXene assembly and enhance their overall performance.

Organic Framework (COF, MOF)–Carbon Hybrids

Metal–organic frameworks (MOFs) are created when metal ions bond with organic ligands to produce porous coordination polymers, which can exhibit one-dimensional to three-dimensional structures. The nature of the metal ions and linkers employed significantly affects the pore dimensions, surface area, and electrical properties of these materials, with factors such as the metal ions' electronic characteristics, ionic radii, and coordination geometries contributing to the determination of MOF morphology and electrochemical behavior [87]. To address the challenge of graphene layers restacking due to van der Waals interactions, researchers have established two main approaches: First, they have developed methods to chemically modify graphene nanosheets (GNSs) to impart repulsive properties that hinder restacking. Second, they have incorporated physical spacers to either preserve or enhance the separation between GNS layers [88]. Among these spacers, covalent organic frameworks (COFs) have emerged as particularly effective, as they are constructed from regularly arranged small organic molecules that form coordination polymers. COFs offer several advantages over MOFs in electrochemical applications, including lower mass, reduced environmental impact, and improved stability. Furthermore, they can be tailored to possess adjustable in-plane porosity and high surface areas. For instance, 2D COFs composed of aminopyridine and substituted-phloroglucinol linkers, synthesized via solvothermal techniques, can achieve substantial surface areas of around 690 m^2/g and pore widths exceeding 1 nm [89]. In the realm of supercapacitors, MOFs have been successfully utilized as two-dimensional spacers to prevent graphene sheets from restacking. A notable example is the hybrid material, UPZCNs-K4, synthesized by Li and colleagues, which exhibits a high gravimetric capacitance of 402 F/g at a charge density of 1 A/g. When implemented in supercapacitors, these hybrid carbon sheets demonstrated energy densities of approximately 16 Wh/kg with potassium hydroxide (KOH) and 22 Wh/kg with sodium sulfate (Na_2SO_4) electrolytes [90].

2.5.2. Electrolytes

Electrolytes are vital for the functionality of supercapacitors as they facilitate the movement of ions, thereby influencing the device's overall performance and operational voltage range. Typically, supercapacitor electrolytes are categorized into three primary classes: aqueous solutions, organic solvent-based systems, and ionic liquids. Each category contributes distinct characteristics to the supercapacitor's performance profile, such as conductivity and voltage capabilities. These classifications are essential in optimizing the design and selection of electrolytes to meet specific supercapacitor requirements and enhance their efficiency in various applications.

Aqueous Electrolytes

The efficiency of aqueous electrolytes in energy storage surpasses that of solid or gel-like semi-solid alternatives, largely due to their liquid nature and associated low viscosity that facilitates rapid ionic movement. These electrolytes can be further categorized into three distinct groups based on their pH properties: (i) acidic, characterized by a pH value below 7; (ii) alkaline, with a pH above 7; and (iii) neutral, maintaining a pH of 7. This classification is crucial, as it impacts their interaction with electrodes and the overall system performance in various storage applications.

Acidic Electrolyte

H_2SO_4 is a commonly studied aqueous electrolyte, due to its remarkable ionic conductivity and small ionic dimensions. At 25 °C, it showcases the highest ionic conductivity among its counterparts. This high conductivity is advantageous, as it leads to greater specific capacitance and a lower equivalent series resistance (ESR), which are essential for achieving superior energy and power densities. Nevertheless, the elevated conductivity can induce corrosion concerns. To address this, Naik et al. combined 0.2 M H_2SO_4 with 1 M KNO_3 to create a solution that not only mitigates corrosion but also lowers the KNO_3 resistance and enhances the electrolyte's overall capacitance [91]. Research comparing the electrochemical properties of mesoporous MoS_2 with different electrolytes found that 1 M H_2SO_4 yielded the highest capacitance at 225 F/g, whereas 2 M KOH provided the lowest at 123 F/g, underscoring the substantial influence of the electrolyte on performance, which remained relatively consistent across various electrode studies [92,93]. In a different study, Bakhshandeh and associates used the partial reduction and functionalization of acylated graphene oxide to create a functionalized and partly reduced graphene oxide (FPRGO). The electrochemical analysis of FPRGO in 1 M H_2SO_4, as seen in the charge–discharge curves and cyclic voltammograms (Figures 9 and 10), revealed that the material exhibits a combination of electrochemical double-layer capacitance (EDLC) and pseudocapacitance properties, highlighting its adaptable nature within such an electrolyte environment [94].

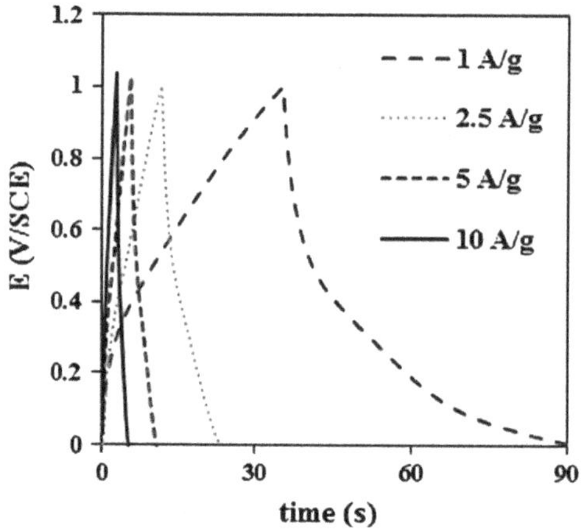

Figure 9. Charge–discharge curve for FPRGO at various densities of current in 1 M H_2SO_4 (Reproduced with permission from Ref. [94]).

The dimension of pores in a material significantly affects its ability to store charge, particularly in supercapacitors. Research has demonstrated that for materials interacting with acidic electrolytes, a pore size of approximately 3 nanometers optimizes the charge-storing capacity. The performance enhancement can be attributed to the superior accommodation of potassium ions in pores of these dimensions compared to hydrogen ions [95]. Phosphate- and nitrogen-doped carbons have showcased higher capacitance in concentrated potassium hydroxide (6 M KOH, reaching 160 F/g) than in 1 M sulfuric acid (138 F/g), which is likely due to their larger pore structure that allows more effective trapping of the larger potassium cations. Investigations into various electrode configurations have been undertaken using a range of electrolytes, with experiments conducted in both two-electrode [95] and three-electrode setups [96]. Despite these variations, it is generally observed that 1 M H_2SO_4 provides the most favorable conditions for achieving high capacitance. Figure 11 illustrates the specific capacitance and internal resistance fluctuations in a symmetric EDLC (electric double-layer capacitor) device utilizing 3 M H_2SO_4 as the electrolyte across different temperatures. A notable increase in equivalent series resistance (ESR) is discernible at −30 °C, which can be ascribed to the electrolyte's reduced stability in proximity to its freezing point.

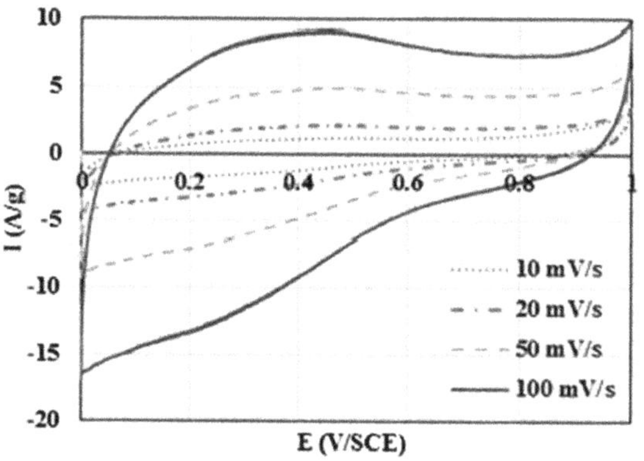

Figure 10. CV curves for FPRGO in 1 M H_2SO_4 solution at different scan rates (Reproduced with permission from Ref. [94]).

The study examining the reactivity of aqueous and organic electrolytes in response to temperature fluctuations determined that aqueous electrolytes demonstrate less reactivity compared to their organic counterparts. Specifically, it was found that while aqueous electrolytes experienced a 15% reduction in capacitance, organic electrolytes such as Et_4NBF_4/PC underwent a more substantial capacitance loss of 32% when exposed to varying thermal conditions [97]. The utilization of H_2SO_4 as an industrial electrolyte is constrained by its narrow potential window, which extends only to 0.6 V. This limitation leads to decreased energy density and an elevated risk of corrosion. On the other hand, sulfuric acid-based electrolytes exhibit higher capacitance in both electric double-layer capacitors (EDLCs) and redox systems, with H^+ ions significantly participating in the charge storage process and enhancing the overall performance.

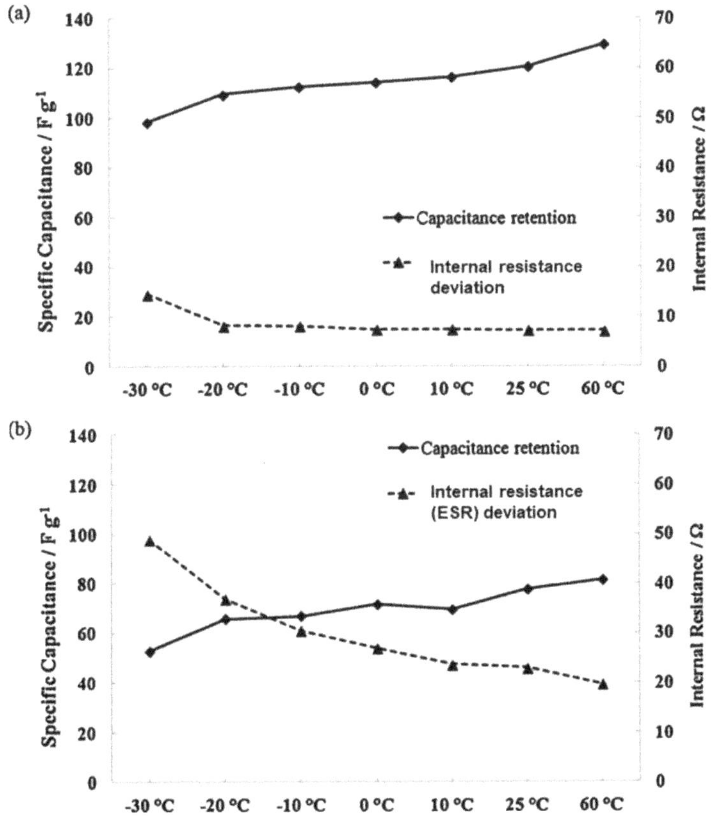

Figure 11. (**a**) Specific capacitance and (**b**) internal resistance deviation with variation in temperature for 3 M H_2SO_4 solution (Reproduced with permission from [98]).

Alkaline Electrolytes

Alkaline electrolyte solutions, such as 6 M potassium hydroxide (KOH), are frequently regarded as suitable candidates for enhancing the performance of supercapacitors, much like their acidic analogs. The appeal of 6 M KOH is attributed to its small ionic radius and exceptionally high ionic conductivity, which is measured at 0.6 S/cm under standard conditions of 25 °C. When the charge storage properties of freeze-dried reduced graphene oxide (rGO) are compared in 6 M KOH, 1 M sodium sulfate (Na_2SO_4), and a protic ionic liquid electrolyte (ILE), it becomes evident that KOH provides the most substantial capacitance, reaching 71.8 F/g at a scanning rate of 50 mV/s. Despite this, the protic ILE exhibits a superior energy density, which is a direct consequence of its broader potential window. This observation underscores the importance of considering both ionic conductivity and potential range when evaluating electrolyte options for supercapacitor applications [99]. Research conducted by Balaji et al. highlighted boron-doped graphene in 20% KOH with an impressive capacitance of 286 F/g and an energy density of 5.3 Wh/kg, whereas in 1-ethyl 3-methylimidazolium tetrafluoroborate ($EMIMBF_4$), the energy density soared to 43.1 Wh/kg despite a slightly lower capacitance of 138 F/g. This is attributed to the smaller ionic radius of KOH, its reduced viscosity, and enhanced ionic conductivity [100]. Interestingly, while alkaline electrolytes like KOH may offer lower energy density compared to organic variants, they exhibit higher capacitance in acidic environments. This discrepancy is largely due to the more confined potential window of alkaline electrolytes, which in turn limits their charge storage capacity. For instance, the capacitance of a porous carbon electrode in 6 M KOH was measured at 355.6 F/g using a three-electrode configuration at a current density of 1 A/g. When the setup was switched to a two-electrode configuration, the capacitance decreased to 78.15 F/g, underscoring the higher capacitance that KOH can achieve in electrochemical double-layer capacitor (EDLC)-type carbon-based electrodes [101]. Further exploration into the realm of 1-D $NiMoO_4/CoMoO_4$ nanorod arrays coated with 2-D Ni-Co-S nanosheets revealed that utilizing 1 M KOH as an electrolyte led to a maximum capacitance of 778.1 F/g at a current density of 0.5 A/g. This study also presented an impressive energy density of 33.1 Wh/kg at 0.25 A/g and power density of 3195 W/kg at 4 A/g for flexible asymmetric supercapacitors [102]. These findings suggest that KOH is not only suitable for EDLC electrodes but also for pseudocapacitive ones. Moreover, an alternative to KOH, namely KOH/PVA, has been employed as an electrolyte in Co_3O_4/rGo-based electrodes, resulting in notable capacitance and exceptional energy and power densities of 38.8 Wh/kg and 400 W/kg, respectively [103]. This underscores the potential of alkaline electrolytes in enhancing the performance of supercapacitors. While supercapacitors equipped with alkaline electrolytes like KOH can boast higher power densities, their integration with batteries in hybrid storage systems still requires further investigation. Another alkaline option, LiOH, has been proposed as a viable electrolyte substitute for KOH, warranting additional research in this domain [104].

Neutral Electrolytes

Sulfate-based neutral electrolytes, such as sodium sulfate (Na_2SO_4), are commonly favored in electrochemical systems for their broad operating range and minimal propensity for corrosion. Despite these advantages, they generally exhibit inferior electrochemical performance when compared to acidic and basic electrolytes. Research conducted by Qin et al. focused on the use of nitrogen-doped carbon nanotubes (CNTs) within a neutral electrolyte, achieving a substantial specific surface area of 2608 m^2/g and a significant energy density of 27.3 Wh/kg at a power density of 182 W/kg in a 1 M Na_2SO_4 solution. The study contrasted the performance of carbon paper collectors in sulfuric acid (H_2SO_4) with nickel foam collectors in potassium hydroxide (KOH), revealing that the latter's capacitance was lower. This is because KOH lacks H^+ ions, which are vital for the occurrence of redox reactions. Consequently, the nickel foam's charge storage is predominantly influenced by double-layer capacitance (DLC), which is characterized by rapid charge–discharge rates that surpass the redox mechanism. When analyzing the CNT/MnO_2 electrode combination, the researchers found that it demonstrated the highest capacitance in sulfuric acid, reaching 2523 F/g at a 5 mV/s scan rate. This high capacitance can be attributed to the presence of redox peaks, which signify substantial energy storage capabilities. In a neutral Na_2SO_4 electrolyte, however, the capacitance was observed to be 1676 F/g at the same scan rate, with the absence of redox peaks indicating that the charge storage was mainly attributed to electric double-layer capacitance (EDLC). While MnO_2 has higher solubility in H_2SO_4, which could potentially lead to decreased stability, the study showed that Na_2SO_4 offers superior cyclic stability, with the CNT/MnO_2 electrode maintaining 80% capacitance retention following 3000 cycles. This is due to the reduced solubility of MnO_2 in neutral environments, which enhances its stability and longevity in such conditions [105]. Further examination of MnO_2/CNT electrodes in various electrolytes revealed a capacitance of 43.2 F/g in a 0.5 M Na_2SO_4 solution, underscoring the importance of electrolyte selection [106]. An asymmetric device utilizing V_2O_5@3D graphene and Fe_3O_4@3D graphene as positive and negative electrodes, respectively, achieved an energy density of 54.9 Wh/kg and a power density of 898 W/kg in 1 M Na_2SO_4, with minimal ohmic resistance [107]. Different morphologies of MnO_2/carbon electrodes were studied in a 1 M Na_2SO_4 solution, and it was found that the peach-like structure of Mn_2O_3/C offered the highest surface area, which correlated with the highest capacitance [108]. Conversely, the electrode with the lowest surface area displayed the lowest capacitance. KCl is another neutral electrolyte that has been considered for electrochemical applications, with a potential window spanning from −0.2 to 0.3 V (vs. Ag/AgCl) [109]. Although it has a neutral pH, similar to Na_2SO_4, its capacitance is lower due to the less effective conductivity of its anions [110]. However, the cobalt oxide (Co_3O_4) electrode exhibited lower capacitance in KCl than in Na_2SO_4, despite both having a neutral pH [111]. This suggests that the ionic conductivity and size have more influence on the electrochemical capacitance than the electrolyte's pH. Moreover, while some materials may show high capacitance in acidic conditions, they may not sustain this performance over multiple cycles due to solubility or pH sensitivity issues [112]. Hodaei and colleagues reported a high capacitance of 311 F/g at 1 A/g for nitrogen-doped titanium dioxide (TiO_2) electrodes in 3 M KCl, which also maintained an impressive 98.9% capacitance retention after 4000 cycles (Figure 12). Cyclic voltammetry (CV) studies indicated that the specific capacitance of MnO_2 is dependent on the concentration of cations, with the capacitance increasing with higher cation concentrations across all tested electrolytes. These findings collectively emphasize the significance of ionic conductivity and material stability [113].

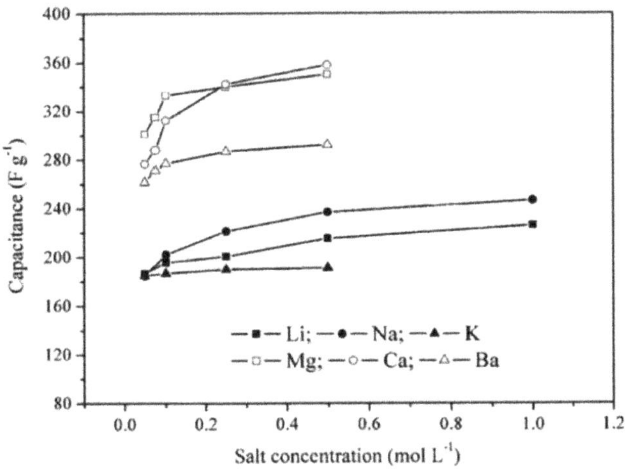

Figure 12. Capacitance and salt concentration dependence (Reproduced with permission from Ref. [113]).

Organic Electrolytes

Currently, there is a growing interest in exploring organic electrolytes for various applications, due to their higher energy densities, which are attributed to their extensive potential windows. Organic electrolytes, while beneficial in certain aspects such as high stability potential windows, often suffer from lower ionic conductivity and a higher propensity for ion solvation, which in turn affects their dielectric properties. Furthermore, the larger ionic dimensions of organic electrolytes hinder their diffusion through the narrow pores of typical electrode materials, thereby limiting their overall performance. To address the limitations of organic electrolytes, particularly their lower pseudocapacitive response compared to alternative systems, researchers have focused on optimizing the pore size within the electrode material. This modification is essential to enhance the interaction between the electrolyte and the electrode, thereby improving the electrochemical efficiency. Among the commonly studied organic salts is tetraethylammonium tetrafluoroborate ($TeABF_4$), which has been examined in combination with solvents such as propylene carbonate (PC) and acetonitrile (ACN). Despite the lower energy density of $TeABF_4$ in PC due to its higher viscosity and boiling point, it is generally preferred over $TeABF_4$ in ACN, due to the latter's elevated toxicity concerns. It has been observed that at a concentration of 1.1 M, $TeABF_4$ faces solubility issues [114]. To better comprehend the behavior of organic electrolytes, researchers have turned to simulation studies, which provide insights into the underlying mechanisms governing their performance. For instance, Yang and colleagues have investigated the Gogotsi–Simon effect, where they observed an increase in the specific capacitance as the pore width decreased [115]. A study was undertaken to assess the mass ratio and full utilization of the electrolyte potential window. The findings suggest that 1 M TEABF4/ACN can achieve a maximum voltage of 2.9 V, while 1 M TEABF4/PC reaches 2.7 V, indicating a potential enhancement in device operation efficiency [116]. The energy density of the resulting 2D N-doped carbon nanosheets in 1 M TEABF4 was determined to be 34 Wh/kg [117], which is on par with the 29.8 Wh/kg reported by Zou et al. for polymer nanobelt screw electrodes in TEABF4/ACN [118]. Despite the higher capacitance of acidic and alkaline electrolytes, organic electrolytes are preferred for their cost-effectiveness and environmental benefits [119]. To optimize the performance of these electrolytes, researchers have explored various models and experimental methods to determine the suitable pore size for organic electrolytes. By altering the solvents, the interlayer distance of graphene oxide can be adjusted, which in turn affects the capacitance. For instance, when using methanol, the interlayer distance increased to 8.84 Å, whereas in acetone, it expanded to 8.99 Å, and in propanol, it reached 9.5 Å. This tailoring is crucial for ion transport [120]. A notable investigation by Vijaykumar et al. demonstrated a 25% enhancement in gravimetric capacitance when employing the TEABF4/ACN electrolyte [121]. The potential window significantly influences the energy density of supercapacitors, and controlling it through structural manipulation is essential. Researchers examined the areal capacitance of 28 carbon samples with pores sizes ranging from 0.7 to 1.5 nm in 1 M TEABF4/ACN. Interestingly, no substantial variation in capacitance was observed across different pore sizes, possibly due to a diminishing relative permittivity of organic electrolytes, as the pore sizes decreased from mesopores to sub-nanometer dimensions [122]. Furthermore, EMIMBF4 salt has also been employed in organic electrolytes. Through molecular modeling, scientists have compared EMIMBF4 and EMIMBF4/ACN, discovering that the latter exhibits a 55–60% greater capacitance [123]. Another computational study focused on the impact of porosity and decoordination on capacitance and found that monodispersed pores generally yield higher capacitance values compared to polydispersed ones [124]. These findings collectively underscore the importance of understanding and optimizing the relationship between pore structure and electrolyte properties to enhance the performance of supercapacitors. The aforementioned findings cast doubt on prevailing theories concerning the role of pores in contributing to total capacitance, highlighting the need for further exploration of phenomena at the microscale level of the electrode/electrolyte interface [125]. An investigation conducted by Ue and collaborators demonstrated that certain salts, namely 1-ethyl-1-methylpyrrolidinium ($MEPYBF_4$), triethylmethylammonium ($TeMABF_4$), and tetramethylene–pyrrolidinium ($TMPYBF_4$), exhibit superior electrolytic conductivity at practical concentrations around 0.65 M, attributed to their minute ionic dimensions while preserving strong dissociative characteristics. These salts are highly soluble in propylene carbonate (PC), which enables conductivities exceeding 2 M, unlike $TEABF_4$ that is confined to 1 M solubility [126]. The compact nature of the Li^+ ionic radius renders organic electrolytes based on lithium salts a prevalent selection for pseudocapacitors and hybrid energy storage systems such as Li-ion capacitors (LICs). This preference is reflected in the exploration of their application within electric double-layer capacitors (EDLCs). The emergence of lithium-ion batteries (LIBs) has significantly shaped the utilization of these salts in various electrochemical storage systems. In this context, Kim and their team engineered a flexible micro-supercapacitor utilizing an organic gel electrolyte constituted of PMMA-PC-$LiClO_4$-HQ. The incorporation of hydroquinone (HQ) as a redox species within this electrolyte significantly enhanced the capacitance and energy density of the supercapacitor device [127]. The highest specific capacitance for porous activated carbon derived from pea skin was observed in 1 M H_2SO_4, with a notable energy density of 17.86 Wh/kg in 1 M $LiClO_4$ [128]. Comparative analysis of PANI-NFs electrodes in 1 M H_2SO_4, 1 M Na_2SO_4, and 1 M $LiClO_4$ in PC showed that $LiClO_4$ exhibited superior capacitive behavior, potentially due to the smaller ClO_4^- anion compared to the SO_4^{2-} anion [129]. Moreover, the thermal properties of $LiClO_4$ in acetamide displayed remarkable capacitance and a eutectic temperature beneath its melting point (234 °C), which decreased with the increasing acetamide concentration [130]. An electrolyte of 1 M $LiClO_4$ combined with a vanadium pentoxide (V_2O_5) nanofiber (VNF) electrode achieved an impressive energy density of 78 Wh/kg, and in the context of the Na6V10O28 electrode, 1 M $LiClO_4$ yielded a commendable energy density of 73 Wh/kg [131,132]. These outcomes suggest that $LiClO_4$ is a highly suitable electrolyte for vanadium-based electrodes (Figure 13).

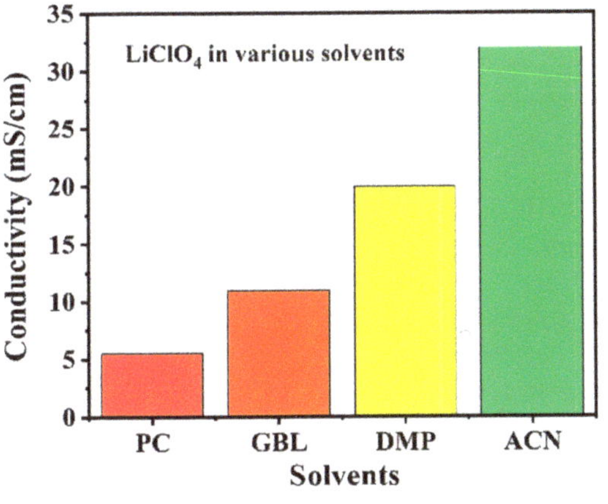

Figure 13. Conductivity of 1 M $LiClO_4$ at 25 °C in different solvents. PC (propylene carbonate), GBL (gamma-butyrolactone), DMP (N,N-dimethylpropionamide), and ACN (acetonitrile). Under Creative Commons Attribution-NonCommercial 3.0 Unported License.

$LiPF_6$ is commonly used in organic electrolytes alongside $LiClO_4$. When implemented in an electrolyte at a concentration of 1 M for activated carbon/graphene, it yielded a specific capacitance of 19.45 F/g, accompanied by an electrolyte resistance of 4.3 Ω [133]. This was shown in prior research. For vertically aligned carbon nanotube (CNT) electrodes [134]., the capacitance reportedly reached 101 F/g. A hierarchical porous carbon electrode, when paired with a 1 M $LiPF_6$/EC/DEC electrolyte, demonstrated an energy density of 288 W/kg, with a capacitance of 625 F/g at a current density of 1 A/g [135]. Notably, graphene-deposited mesocarbon nanobead electrodes have shown a high power density of 10,000 W/kg under the influence of $LiPF_6$ [136]. The conductivity profile of 1 M LiPF6 across diverse solvents at room temperature can be visualized in Figure 14.

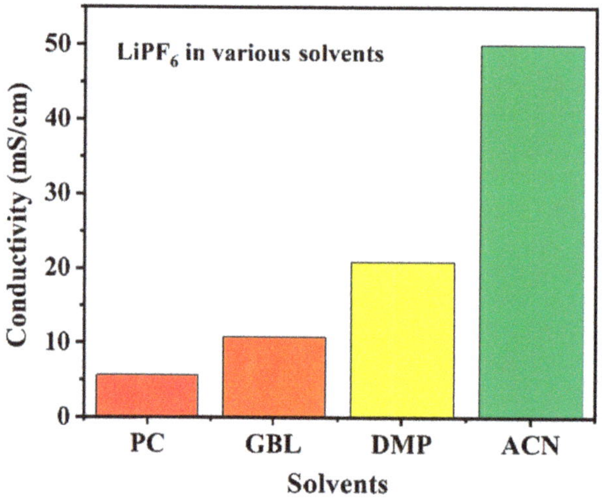

Figure 14. Conductivity of 1 M $LiPF_6$ in various solvents at 25 °C. Under Creative Commons Attribution-NonCommercial 3.0 Unported License.

Ionic Liquid-Based Electrolytes

Ionic liquids, often known as room-temperature molten salts, are characterized by their ability to dissociate into free ions at temperatures below 100 °C without the formation of discrete molecules. These unique substances have been proposed as viable alternatives to conventional electrolytes in energy storage applications due to their exceptional properties. Typically, an ionic liquid is composed of an asymmetrical organic cation paired with an inorganic or organic anion, which imparts a variety of advantageous features. Some of these include high thermal and chemical stability, significant electrochemical stability, and a broad range of conductivity, generally spanning from 10^{-3} to 10^{-2} S/cm. These attributes are contingent upon the specific combination of cations and anions within the liquid. Multiple modeling investigations have been undertaken to elucidate the relationship between the concentrations of anions and cations within ionic liquids and their resulting physical and chemical properties [137,138]. Despite their inherently high viscosity, ionic liquids have found favor in the production of supercapacitors due to their capacity to yield higher energy densities than aqueous counterparts. This is largely attributed to the extended potential window that ionic liquids offer compared to water-based electrolytes. Furthermore, the robust nature of the ions and the strong interactions they form with solvents confer a heightened stability to ionic liquids, effectively resisting volatility and evaporation, which is a prevalent issue with water-based solutions. These factors contribute significantly to the growing interest in ionic liquids as a superior choice for enhancing the performance and durability of energy storage technologies.

Ionic liquids are generally categorized into three major types according to their ionic constitution: aprotic, protic, and zwitterionic. As depicted in (Figure 15), these classifications are based on the nature of their ions. Aprotic ionic liquids are particularly suitable for use in supercapacitors and lithium-ion batteries due to their specific charge-carrier properties. Conversely, protic ionic liquids are typically employed in the realm of fuel cells where their unique ionic interactions play a crucial role. Lastly, zwitterionic ionic liquids find their niche in ion–liquid membranes, which exploit their distinct structural features for optimal performance. The most frequently utilized cations and anions in the formulation of these substances are presented in Figures 16 and 17, highlighting the versatility and tailorability of these materials for various applications [139,140].

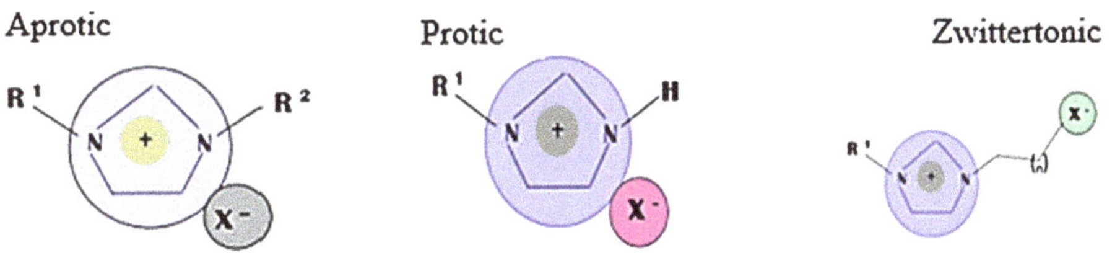

Figure 15. Ionic liquid types according to their components.

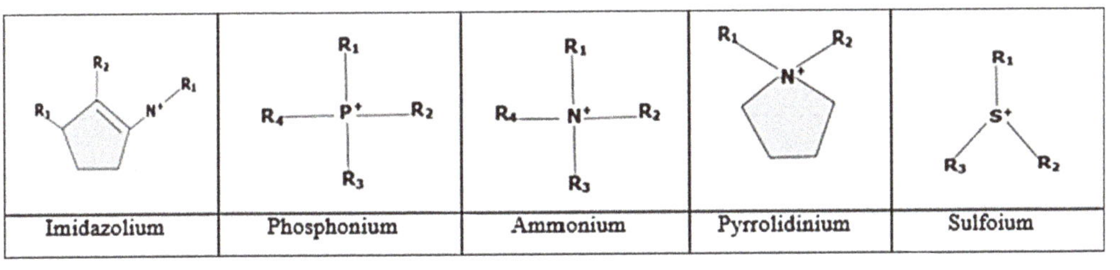

Figure 16. Ionic liquid cations that are frequently utilized in supercapacitors.

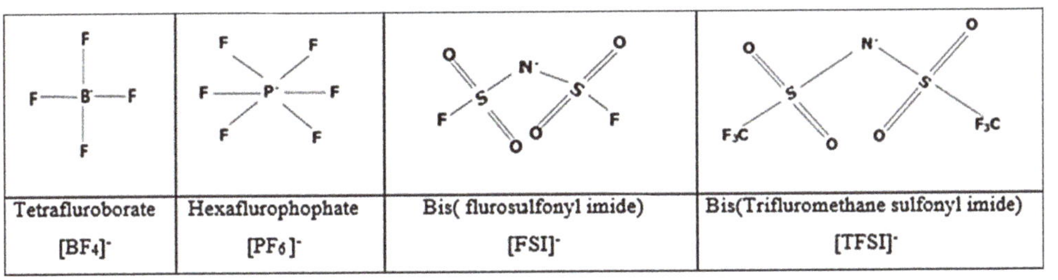

Figure 17. Frequently employed ionic liquid anions in supercapacitors.

When comparing the effects of carbon nanofiber electrodes in aqueous KOH, Na_2SO_4, and (1-ethylimidazolium bis(trifluoromethanesulfonyl)imide) IL-based electrolytes, it was observed that the KOH electrolyte provided a superior specific capacitance of 368.8 F/g at 0.5 A/g, attributed to the elevated K^+ ion mobility. The performance of these electrolytes in device configurations was then evaluated, with the IL-based device showcasing a noteworthy 24 Wh/kg energy density at a power density of 750.3 W/kg, surpassing the others. Table 2 compiles a list of prominent ILs that have been explored as electrolytes in contemporary supercapacitor research [141].

Table 2. The performance of IL electrolyte-based supercapacitors [141].

IL Electrolyte	Electrode	Capacitance	References
[EMIm][SeCN]	Activated carbon	49 F/g for Al 56 F/g for stainless steel	[142]
[EDMF]BF$_4$	GC electrode	126 F/g	[143]
[EMIM]BF$_4$	Nonporous gold	12.1 to 6.6 F/cm^3	[144]
EMIM-TFSI	"Peppered"-activated carbon		[145]
[EMIM][TFSI]/FS	MnO$_2$@CF//FeOOH/PPy@CF		[146]
EMIM-TFSI	Mesoporous reduced graphene oxide	104.3 F/g	[147]
[EMIM][TFSI]/LiCl/Al$_2$O$_3$	(MWNT)/V$_2$O$_5$ nanowires (NWs)	10.6 mF/cm^2 at 0.5 mA/cm^2	[148]
(EMIMTFSI)/ acetonitrile	Manganese oxide (MnO$_x$)-decorated carbonized porous silicon nanowire	635 F/g	[149]
[BMPyr$^+$] [DCA$^-$]	α-Co(OH)$_2$	28.6 F/g @0.15 mA/cm^2	[139]
EMIMBF$_4$	Boron-doped graphene	138 F/g	[100]
EMIMBF$_4$	Highly conductive mesoporous activated carbon fiber	204 F/g @0.5 A/g	[150]
(1-Ethylimidazolium bis(trifluoromethane sulfonyl) Imide	Carbon nanofiber	77.1 F/g @0.5 A/g	[99]

The aforementioned progression underscores the vital function of ionic liquids (ILs), a subset of organic electrolytes, in the context of advanced energy storage technologies. Despite certain drawbacks such as limitations associated with large cationic and anionic species, these properties can be cleverly harnessed to augment the energy density of capacitive energy storage solutions, particularly within the realm of supercapacitors that integrate bi-redox ILs [151]. As the demand for high-performance energy storage systems continues to surge, the appeal of supercapacitors utilizing IL-based electrolytes is progressively gaining traction as a competitive alternative to other existing energy storage devices. While ILs have the capacity to act independently as electrolytes, their exclusive use is not a viable approach for various practical considerations. One such hindrance is the inherently high viscosity of ILs, which poses a substantial challenge to their widespread commercialization. Moreover, the prohibitive cost of ultra-pure ILs makes them an economically unfeasible choice as the sole constituent of electrolytes. A more rational strategy involves incorporating ILs as part of a composite electrolyte system to facilitate ionic mobility and enhance overall device efficiency [152].

2.6. *Advantages and Disadvantages of Supercapacitors*

Supercapacitors offer several advantages over other energy storage technologies, including:

- **High Power Density:** Supercapacitors have a much higher power density than batteries, allowing for faster charging and discharging rates.
- **Fast Charge–Discharge Rates:** Supercapacitors can be charged and discharged much faster than batteries, making them suitable for applications requiring rapid energy delivery.
- **Long Cycle Life:** Supercapacitors have a much longer cycle life than batteries, with some devices capable of withstanding hundreds of thousands or even millions of charge–discharge cycles.
- **Wide Operating Temperature Range:** Supercapacitors can operate over a wider temperature range than batteries, making them suitable for use in harsh environments.
- **High Efficiency:** Supercapacitors have high energy efficiency, with minimal energy loss during charging and discharging.

However, supercapacitors also have some disadvantages:

- **Low Energy Density:** Supercapacitors have a lower energy density than batteries, meaning they cannot store as much energy for a given size and weight.
- **Voltage Drop During Discharge:** The voltage of a supercapacitor decreases linearly during discharge, which can be a problem for some applications.
- **High Cost:** Supercapacitors can be more expensive than batteries, although the cost is decreasing as the technology matures.

3. Design and Fabrication Strategies

3.1. The Fundamentals of Supercapacitor Electrode Design

The architecture of supercapacitor electrodes significantly influences their overall effectiveness, energy storage capacity, and operational efficiency [153]. By scrutinizing and refining these fundamental design aspects, scientists are equipped to augment the performance of these electrodes, thereby paving the path for the next generation of high-performance energy storage solutions, as outlined in Table 3.

Table 3. Comparison of several electrode design principles for supercapacitors. With permission from ref. [153].

Design Principle	Advantages	Applications	Limitations
Surface Area and Porosity	Higher charge storage capacity, faster ion diffusion, enhanced charge–discharge rates, and improved capacitance retention.	Supercapacitor electrodes, energy storage in portable electronics, power buffering in renewable energy systems, hybrid electric vehicles, and grid stabilization and frequency regulation.	Challenging to achieve ultrahigh porosity and limitations in mass loading due to pore structure.
Electrode Material Selection	High specific capacitance, excellent charge–discharge characteristics, wide range of available materials, and tunable properties for various applications.	Energy storage in electric vehicles, portable electronics, renewable energy storage, and aerospace and defense applications.	Limited voltage windows for some materials may exhibit poor cycling stability and high cost for certain advanced materials.
Electrical Conductivity	Low internal resistance, efficient charge transfer, and enhanced power delivery.	High-power applications, rapid charge–discharge cycles, high-frequency applications, and reduced energy losses.	Some materials may suffer from poor conductivity, compatibility with certain electrolytes, and energy harvesting.
Electrolyte Compatibility	Efficient ion transport, improved charge–discharge rates, lower internal resistance, and a wide range of available electrolytes.	Portable electronics, renewable energy systems, electric vehicles, and grid stabilization and frequency regulation.	Limited operating voltage window for certain electrolytes and potential for electrolyte decomposition.
Binder and Additive Selection	Enhanced electrode stability, improved electrode–electrolyte interface, and better mechanical integrity.	Supercapacitor electrodes, portable electronics, renewable energy storage, and hybrid electric vehicles.	Potential for binder and additive decomposition: binder content may reduce the effective surface area.
Electrode Thickness and Mass Loading	Higher energy density, enhanced ion transport, improved power density, and tailored properties for specific applications.	Supercapacitor electrodes, energy storage in portable electronics, renewable energy systems, and electric vehicles.	Limited mass loading may affect overall capacitance, thicker electrodes may lead to slower charge–discharge rates, and mass loading may affect mechanical stability.
Scalability and Cost-Effectiveness	Cost-effective manufacturing, large-scale production feasibility, commercial viability, and potential for integration with existing processes.	Supercapacitor electrodes, energy storage in portable electronics, renewable energy systems, and electric vehicles.	Scalability may affect some material properties, scalability may introduce fabrication challenges, and cost-effectiveness may compromise performance.

3.1.1. Surface Area and Porosity

A crucial aspect of supercapacitor electrode design is enhancing the surface area and porosity, which significantly affects the performance. This is because an increased surface area provides a greater number of sites for charge storage, which is essential for improving the device's energy density. Moreover, an elevated porosity level facilitates better electrolyte infiltration and ion diffusion, leading to enhanced charge transport within the electrode material. To achieve these goals, scientists often employ strategies like constructing hierarchical nanostructures and utilizing various templating approaches. These methods aim to create a porous architecture that allows for optimal interaction between the electrolyte and the electrode material, thereby enhancing the overall efficiency of the supercapacitor [154].

3.1.2. Selection of Electrode Material

Selecting the appropriate electrode material is essential for optimizing the specific capacitance and energy storage capabilities of supercapacitors. Carbon-based substances, including activated carbon, graphene, and carbon nanotubes, are frequently utilized in these devices, as they exhibit exceptional surface areas, electrical conductivity, and longevity. Additionally, researchers have investigated the potential of transition metal oxides, conductive polymers, and composite materials to further improve the efficiency and adapt the characteristics of supercapacitor electrodes to specific requirements. These alternative options contribute to advancing the field by broadening the range of suitable materials for such applications [155].

3.1.3. Electrical Conductivity

The significance of electrical conductivity in the context of supercapacitor electrode design cannot be overstated. It is a fundamental parameter that influences the efficiency of electron transfer across the electrode material, which in turn affects the device's internal resistance and its capacity to rapidly charge and discharge. In order to optimize this conductivity, electrode architectures often integrate carbon-based substances and conductive additives. These components are meticulously selected and engineered to facilitate superior electron transport, thereby ensuring that the supercapacitor operates with minimal energy loss and achieves high power densities. The strategic incorporation of such materials enhances the electrochemical performance of the device, contributing to the overall effectiveness and longevity of the supercapacitor [156].

3.1.4. Electrolyte Compatibility

Critical to the development of supercapacitor electrodes is the selection of an appropriate electrolyte that is compatible with the chosen material and provides high ionic conductivity. The suitability of the electrolyte is contingent upon the specific requirements of the electrode material and the intended application of the supercapacitor. In this context, a range of aqueous and non-aqueous electrolytes with different salt concentrations are employed to optimize performance. Additionally, the exploration of ionic liquids and gel electrolytes has emerged as a promising avenue to enhance the safety features of these energy storage devices. These alternative electrolyte forms are increasingly being considered for their potential contributions to the overall efficiency and reliability of supercapacitors [157].

3.1.5. Binder and Additive Selection

Essential components known as binders and additives are incorporated within the anode and cathode structures to maintain the integrity of the active material particles and enhance their stability. The careful selection of a suitable binder is crucial as it ensures robust adhesion among particles and significantly reduces the occurrence of detrimental interactions with the electrolyte. Furthermore, the inclusion of conductive additives, such as carbon black or carbon nanotubes, serves to amplify the electrical conductivity of the electrode, thereby contributing to the overall performance of the battery cell. These materials are meticulously chosen to complement the properties of the active electrode substances and foster a cohesive and functional assembly that can efficiently facilitate the flow of electrons and ions, which is pivotal for the battery's energy storage and release mechanisms [158].

3.1.6. Mass Loading and Electrode Thickness

The energy storage capability of a supercapacitor is significantly impacted by the thickness and mass loading of its electrodes. The optimization process for these factors entails a delicate equilibrium between enhancing the content of the active material to obtain a higher capacitance and preserving efficient ion transport and accessibility. To attain an ideal balance between energy and power density, scientists meticulously calibrate the aforementioned parameters, ensuring that the design enhances both performance metrics without compromising either. This approach is underpinned by a thorough understanding of the underlying physical and chemical principles that govern the operation of supercapacitors [159].

3.1.7. Scalability and Cost-Effectiveness

When engineering supercapacitor electrodes, it is essential to prioritize not only performance enhancement but also scalability and economic viability. The production processes must be adaptable to large-scale manufacturing to maintain performance consistency without incurring significant costs or technical barriers. Furthermore, the selection of materials and implementation of fabrication methods should be based on cost-effective principles, ensuring that supercapacitors remain competitive within the broader energy storage market. This approach is crucial for the commercialization and widespread adoption of supercapacitors as a reliable and efficient energy storage solution. The emphasis on scalability and cost efficiency will facilitate the integration of these devices into various industrial and consumer applications, thereby advancing the overall sustainability of energy storage technologies [160].

4. Electrode Fabrication Techniques and Architecture

The effectiveness of supercapacitor electrodes is significantly shaped by the techniques and designs utilized in their production. This discussion has focused on various manufacturing approaches aimed at creating electrodes with augmented surface area, porosity, and electrical conductivity. These enhancements are crucial for achieving superior performance in energy storage and ensuring long-term reliability. By employing such strategies, researchers are able to optimize the electrochemical properties of these devices, leading to substantial advancements in the field. The emphasis on these fabrication methodologies is essential for the continuous development and refinement of supercapacitor technology [161].

4.1. Chemical Vapor Deposition (CVD)

The chemical vapor deposition (CVD) technique is a widely implemented and highly versatile process that allows for the creation of superior quality thin films, nanomaterials, and nanostructures with meticulous control over their chemical makeup, morphological features, and crystalline organization. In the specific domain of supercapacitor electrode manufacturing, CVD has demonstrated substantial benefits, notably in the synthesis of carbon nanotubes (CNTs) and graphene. These carbon-based materials are celebrated for their remarkable properties, which are highly advantageous in the context of advancing energy storage solutions [162].

4.1.1. Carbon Nanotube (CNT) Growth

The advent of chemical vapor deposition (CVD) has significantly transformed the large-scale synthesis of carbon nanotubes (CNTs), paving the way for their integration into supercapacitors. In this sophisticated technique, a precursor gas rich in carbon is injected into a reaction chamber that maintains an elevated temperature. The precursor gas decomposes under these conditions, resulting in the formation of CNTs on an appropriate substrate. The strategic placement of catalytic nanoparticles on the substrate acts as a nucleation site, facilitating the growth of the CNTs. The CVD process grants researchers precise control over several critical variables such as the temperature, gas flow rates, and growth duration, which in turn influences the CNT properties like the diameter, length, and orientation. Notably, CVD enables the synthesis of vertically aligned CNT arrays. These arrays exhibit remarkable features such as an enhanced surface area for charge storage, rapid ion diffusion for swift charge transfer, and efficient alignment for optimal charge transport, making them highly suitable for use in supercapacitor electrodes [163]. In their 2021 study, Gan et al. [164] developed V_2O_5/VACNT composites employing vanadium(III) acetylacetonate (V(acac)$_3$) from Adamas as the vanadium source within a supercritical CO_2 (scCO_2) medium. The process involved introducing different precursor quantities (20, 40, and 60 mg) into a 100 mL high-pressure reactor in a N_2-filled glove box, along with a VACNT specimen. A minimal benzene volume was added as a cosolvent, and the system was then sealed and preheated to 50 °C. Following preheating, the CO_2 was pressurized to 12 MPa using a gas injection pump. The reactor was maintained at 100 °C for 12 h, allowing the V(acac)$_3$ to dissolve and be absorbed by the VACNTs through scCO_2. The adsorbed precursor mass was determined by subtracting the initial mass from the mass of the post-treatment sample. After cooling and depressurization, the V_2O_5/VACNTs composites were formed by annealing in an air atmosphere, converting V(acac)$_3$ to vanadium oxide. These composites were denoted as VN-T, with T representing the annealing temperature. Yin et al. [165], in 2018, prepared TMOs/VACNTs hybrids utilizing a scCO_2-assisted impregnation technique with tris(2,2,6,6-tetramethyl-3,5-heptanedionato)cobalt(III) [Co(tmhd)$_3$] from Alfa Aesar as the cobalt precursor. The experimental setup included a 50 mL stainless-steel high-pressure cell in a N_2 environment. Typically, the researchers placed VACNT arrays with 20–40 mg Co(tmhd)$_3$ inside the cell. The cell was then sealed and connected to a gas supply system, with CO_2 being introduced and pressurized to 11 MPa at 70 °C using a syringe pump. For 6 h, the cell was kept at 200 °C before undergoing a slow depressurization and cooling process to room temperature. To assess the absorbed precursor quantity, the VACNTs' mass was measured both pre- and post-scCO_2 treatment. The Co(tmhd)$_3$-impregnated VACNTs were annealed in an air atmosphere within a rapid thermal annealing furnace at temperatures ranging from 200 to 300 °C. The resulting cobalt oxide-containing composites were denoted as CVN−T, with T signifying the annealing temperature.

4.1.2. Synthesis of Graphene

The significant progress in chemical vapor deposition (CVD) has been instrumental in the large-scale manufacture of graphene tailored for various energy storage applications. Graphene's distinctive architecture, comprised of sp^2-hybridized carbon atoms arranged in a two-dimensional honeycomb lattice, exhibits exceptional electrical conductivity, mechanical resilience, and a substantial specific surface area. The CVD process entails subjecting a metal catalyst substrate, such as copper or nickel, to a carbon precursor gas like methane or ethylene. As a result, carbon species adhere to the metal surface, undergoing reactions that culminate in the creation of graphene layers as they build up. Post-synthesis, these layers are meticulously transferred to targeted substrates using suitable techniques. This approach is highly favorable for producing supercapacitor electrodes, as it allows for precise control over the thickness of high-quality large-area graphene films [166]. Graphene, being a single-layered carbon allotrope, showcases an unparalleled combination of high electrical conductivity, carrier mobility, and superior mechanical and morphological traits, due to its sp2 hybridization and two-dimensional hexagonal lattice structure [167]. Its exceptional properties distinguish it from other carbon materials in that it does not rely on pore distribution within the solid state, making it highly suitable for high-performance supercapacitors [168]. This versatile material can be structured into various forms, such as zero-dimensional (free-standing dots), one-dimensional (fiber or tube structures), two-dimensional (graphene nanosheets and nanocomposite films), and three-dimensional (graphene foams) configurations, each offering distinct advantages in different supercapacitor designs [169]. To optimize the performance of supercapacitor electrodes, graphene can be modified through processes like nitrogen doping or graphene oxide reduction or by employing graphene nanosheets. These modifications can enhance the material's specific surface area and improve its electrochemical properties. For instance, studies have demonstrated that graphene-based electrodes can achieve a substantial theoretical specific surface area of around 2600 m^2/g, with a specific capacitance of 205 F/g, resulting in a supercapacitor with an energy density of 28.5 Wh/kg and a power density of 10 W/kg [170]. Furthermore, researchers such as Guevara et al. [171] have successfully fabricated a reduced graphene oxide/polyaniline composite supercapacitor utilizing an infrared laser from a Light Scribe recorder drive for the deposition of PANI nanofibers. This approach exemplifies the innovative ways in which graphene's properties can be harnessed to create highly efficient energy storage devices (Figure 18).

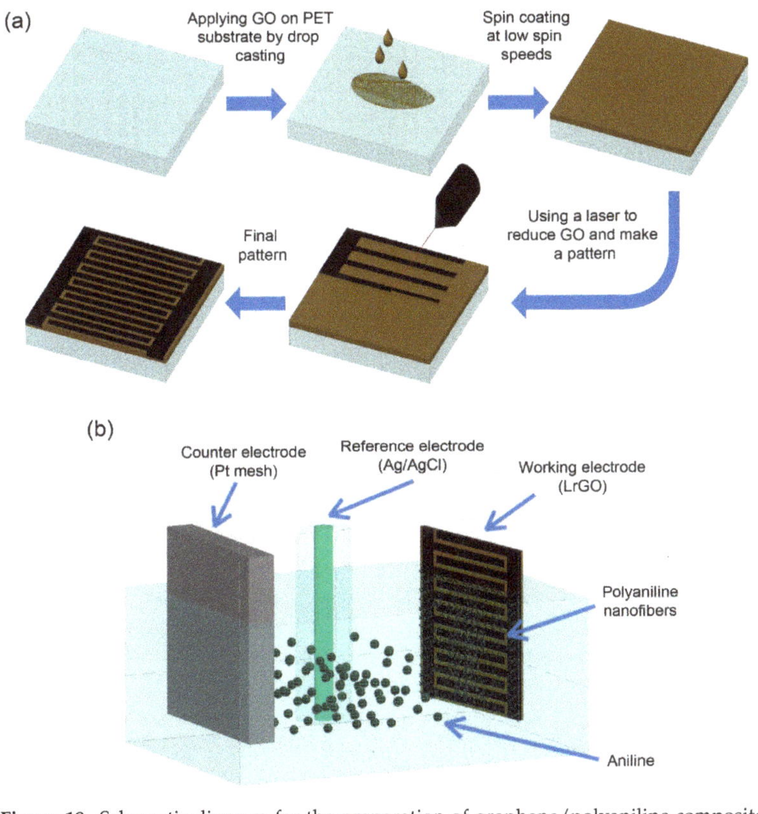

Figure 18. Schematic diagram for the preparation of graphene/polyaniline composite-based SC (**a**) Scheme of the reduction process and (**b**) Electrodeposition of the polyaniline nanofibers. (Reprinted with permission from Elsevier) [171].

The reported composite presented a specific capacitance of 442 farads per gram and preserved 84% of this capacitance over an extensive cycling period of 2000 cycles. In a study conducted by Ogata and colleagues [172], an innovative all-solid-state device was engineered, which manifested both supercapacitor (SC) and battery characteristics. This hybrid system comprised reduced graphene oxide (rGO) and graphene oxide (GO) layers. The SC behavior was observed at operational voltages less than 1.2 V, while the device demonstrated battery-like properties above this threshold. The creation of this rGO/GO/rGO sandwich structure involved the photoreduction of a GO film on both sides, resulting in the formation of the rGO electrodes and an intermediate GO layer that functioned as both an electrolyte and a separator (Figure 19).

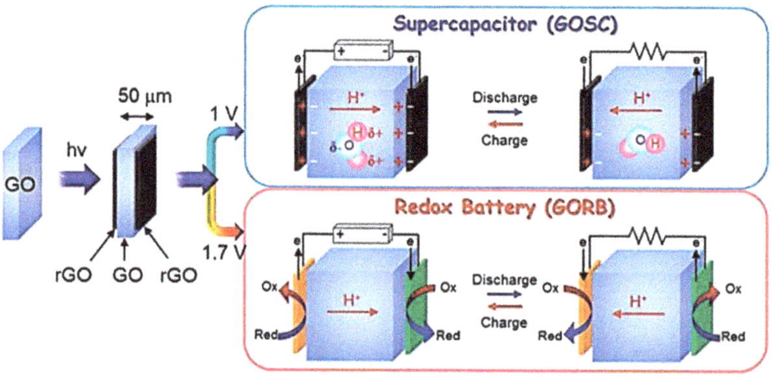

Figure 19. Schematic of the preparation of rGO/GO/rGO device (Reprinted with permission from Royal Society of Chemistry).

In a study by S. Zhongting and colleagues from 2022 [173], $NiCo_2O_4$/rGO composites were synthesized by generating graphene oxide (GO) through a modified Hummers technique, starting with graphite powder. The obtained GO was then diluted in deionized water to form a 0.7 mg mL^{-1} solution for subsequent use. A combination of 20% ethanol and 80% ethylene glycol was utilized to dissolve $Ni(NO_3)_2{\cdot}6H_2O$ and $Co(NO_3)_2{\cdot}6H_2O$ in a 1:2 ratio, resulting in a uniform reserve solution. To achieve different concentrations of GO to metal salt, the researchers added varying amounts of the prepared GO dispersion to the solution. The mixture was sonicated for 30 min to ensure homogeneity. For fabrication, an UC320 ultrasonic spraying system from Siansonic, Beijing, China, was employed. The substrate, Ni foam, was cleaned with ethanol and deionized water before being placed on a hot plate. The precursor solution was fed into the system at 0.6 mL min^{-1} under an ultrasonic power of 1.5 W, and the nozzle was scanned at 2 mm s^{-1} to produce the composites. The mass loading of the active material on the Ni foam, ranging from 0.5 to 1 mg cm^{-2}, was regulated by adjusting the spray duration. The synthesized $NiCo_2O_4$/rGO served as the positive electrode component in the construction of a button-cell asymmetric supercapacitor. The negative electrode, activated carbon (AC), was prepared by combining AC (Super P, YEC-8A), carbon black, and polytetrafluoroethylene (PTFE) in an 8:1:1 weight ratio, followed by grinding with isopropanol and drying at 60 °C for 12 h. The resulting AC sheets were pressed onto the Ni foam to complete the electrode assembly. The electrolyte used for these experiments was 2 M KOH. This approach facilitates the scalable production of such supercapacitors with tailorable properties.

4.1.3. Supercapacitor Electrode Applications

Owing to their distinct characteristics, carbon nanotubes (CNTs) and graphene produced via chemical vapor deposition (CVD) have gained substantial attention in the context of supercapacitor electrode material [174]. The vertical alignment of CNTs offers superior pathways for both charge transfer and ion diffusion, which in turn facilitates high-rate performance and bolsters cycling stability within these devices. Graphene, known for its exceptional specific surface area and electrical conductivity, displays remarkable capacitive properties that significantly enhance the energy storage potential of the electrodes used in supercapacitors. Furthermore, the modulation of CVD growth conditions allows for the optimization of CNT and graphene properties, thereby tailoring their performance to suit a variety of specialized application needs [175].

4.1.4. Future Directions

The continuous development of chemical vapor deposition (CVD) methodologies is expected to lead to significant improvements in the synthesis of carbon nanotubes (CNTs) and graphene, thereby enhancing the performance of supercapacitor electrodes. These advancements will likely result in the fabrication of electrodes with superior characteristics. The exploration of hybrid configurations, like the integration of CNTs and graphene within nanocomposites, presents an attractive opportunity to exploit the synergistic interactions between these materials. This approach is poised to enhance the energy storage capacity of supercapacitors substantially. Additionally, the combination of CVD-produced carbon nanotubes and graphene with diverse nanomaterials, such as metal oxides or conductive polymers, opens up a realm of possibilities for the design of next-generation electrodes. These innovations are expected to introduce tailor-made properties that cater to an extensive array of applications in the energy storage sector [176,177].

4.2. *Template-Assisted Synthesis*

The utilization of template-guided synthesis stands as a flexible and reliable methodology in the creation of nanostructured materials boasting precise morphologies and architectures [173]. This process involves the application of templates that are predefined in their shape and dimensions to direct the deposition or proliferation of materials intended for use in electrodes. The outcome of such synthesis frequently yields materials that display an enhanced surface area, porosity, and notably superior properties in terms of charge storage. These attributes render these materials highly suitable for integration into the design of supercapacitor electrodes.

4.2.1. Using Alumina Membranes as Models

A prevalent template in the realm of template-aided synthesis is porous alumina membranes, alternatively referred to as anodic aluminum oxide (AAO) templates. These membranes are typically fabricated through the electrochemical anodization of aluminum foils, which yield a structure characterized by nanopores that can be precisely controlled in terms of both the diameter and spatial distribution. The AAO template serves as a scaffold for the deposition or growth of active electrode materials, thereby enabling the production of materials with tailored morphologies and properties. This method involves the infusion of precursor substances into the nanopores, followed by specific processing steps to facilitate the synthesis of a variety of materials such as metal oxides, conductive polymers, and carbon-based compounds [178].

4.2.2. Templates Used as Sacrifices

Aside from alumina membranes, template-assisted synthesis for supercapacitor electrodes often employs sacrificial templates, which are materials that can be simply eliminated post-synthesis to reveal a porous architecture. These templates are constructed from substances that are readily removable, and they include organic polymers, such as polystyrene, as well as inorganic materials like silica particles. The process typically involves incorporating these sacrificial materials within the precursor systems, followed by techniques like thermal annealing or etching to extract them and thus create the desired porous morphology [179].

4.2.3. Applications in Supercapacitor Electrodes

The utilization of template-aided synthesis has demonstrated remarkable efficacy in the production of a multitude of nanomaterials tailored for supercapacitor electrodes. One notable example involves the application of alumina membranes as scaffolding structures, which facilitate the creation of metal oxides exhibiting intricate nanoarchitectures, such as hierarchical nanowires or nanotubes [180]. These metal oxides, with their meticulously controlled morphologies, have been found to possess superior electrochemical characteristics, such as a high specific capacitance and remarkable rate capabilities, rendering them highly suitable for integration into supercapacitor technology. Additionally, the employment of template-assisted synthesis has led to the development of porous conducting polymers with enhanced ion permeability and improved capacitive performance [181]. The high specific surface area and porosity inherent to these polymers significantly bolster their effectiveness in charge storage. This methodology remains a vibrant field of study, with persistent research efforts aimed at discovering innovative template materials and refining synthesis procedures. The flexibility of utilizing sacrificial templates allows for the design and fabrication of electrode architectures with a high degree of customization. Furthermore, the integration of template-assisted synthesis with advanced techniques such as inkjet printing or chemical vapor deposition (CVD) presents a promising avenue for the realization of supercapacitor electrodes boasting hierarchical nanostructures tailored for optimized performance [182]. Moreover, the exploration of hybrid electrode materials, which amalgamate the merits of various nanostructures into a single architecture, opens up exciting prospects for the development of supercapacitors that can achieve enhanced energy storage performance. This approach has the potential to significantly improve the capacitive properties and overall efficiency of these energy storage devices.

4.3. Sol–Gel Process

The sol–gel approach represents a highly adaptable and commonly implemented method for producing metal oxides and various other inorganic substances that exhibit high purity and tailorable structures (Figure 20). Within the realm of supercapacitor electrode manufacturing, this process has seen substantial research interest, particularly for the creation of materials based on transition metal oxides, such as manganese oxide and nickel oxide. Additionally, it has been employed to synthesize mixed metal oxides, encompassing binary and ternary systems, which can significantly enhance the performance characteristics of the electrodes [183].

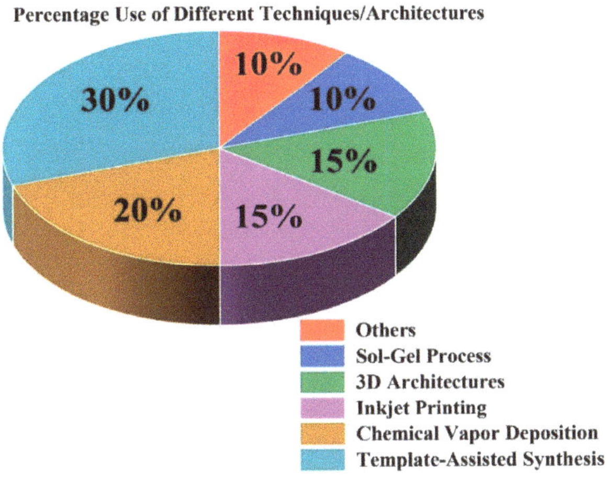

Figure 20. The percentage distribution of different supercapacitor electrode production methods and topologies.

4.3.1. Procedures for the Sol–Gel Method

The sol–gel methodology is a multi-stage process that transitions from a precursor solution, termed as "sol," to a solid–gel matrix. This technique is commonly applied in the fabrication of materials and typically involves the following key phases [184]:

I. Sol Formation: The initial phase involves dissolving metal alkoxides or salts in an appropriate solvent, which can be alcohols or water, resulting in a uniform mixture known as a sol. To adjust the final material's characteristics, dopants or additional chemical agents may be introduced into the sol.

II. Hydrolysis Reaction: Within the sol, the metal–oxygen bonds of the metal alkoxides or salts are broken down by water molecules in a process called hydrolysis. This generates metal hydroxides, which are essential for initiating the polymerization that follows.

III. Polycondensation Process: The metal hydroxides formed during hydrolysis proceed to react with each other, leading to the creation of metal–oxygen–metal (M-O-M) linkages. This step is critical for the formation of a three-dimensional gel framework. The polycondensation reaction can be regulated to achieve the desired porosity and specific surface area in the gel material.

IV. Solvent Removal and Thermal Treatment: Once the gel has formed, the next step is to eliminate the solvent through a drying process. Following this, the gel undergoes calcination at high temperatures, typically ranging from 200 to 800 degrees Celsius. This thermal process serves to remove any residual organic substances and induces a transformation into a crystalline metal oxide structure.

4.3.2. Applications in Supercapacitor Electrodes

The sol–gel technique holds numerous merits in the context of manufacturing supercapacitor electrodes. This approach provides a high degree of precision in the regulation of the metal oxide compounds' composition and structure. Through meticulous manipulation of synthesis conditions and the utilization of a range of metal precursors, it is feasible to create metal oxides that exhibit a variety of nanostructures, including nanoparticles, nanowires, and nanosheets. This flexibility in morphology is crucial as it influences the electrical properties of the resulting material. Additionally, the sol–gel method allows for the seamless introduction of dopants or functional species at the gelation stage. This capability is instrumental in modifying the electrochemical characteristics of the material, such as its specific capacitance and the mechanisms responsible for charge storage. These tailored properties are essential in enhancing the performance of supercapacitors, making the sol–gel process an attractive choice for researchers and engineers in the field [185].

4.3.3. Advantages and Limitations

Sol–gel synthesis provides a number of noteworthy benefits in comparison to traditional solid-state procedures for the fabrication of metal oxides, such as the capability to conduct reactions at lower temperatures, the homogeneous distribution of dopants within the material, and the precise manipulation of the nanostructure [186]. These features are particularly beneficial when aiming to optimize the electrochemical properties of the resulting compounds. Nonetheless, this technique can face certain limitations, such as the stringent requirement for the precise regulation of reaction parameters and potential complexities in scaling up to produce substantial amounts of material efficiently.

4.3.4. Future Directions

Continuous investigation in the sol–gel domain centers on refining the synthesis parameters for metal oxide-based supercapacitor electrodes to elevate their functionality. A noteworthy approach involves the study of composite substances, which combine metal oxides with carbon-containing materials to realize synergistic interactions and superior total performance. Moreover, the adaptation of sol–gel procedures to facilitate large-scale manufacturing is pivotal for the widespread application and integration of supercapacitor technology in practical scenarios [187].

4.4. *Inkjet Printing*

Inkjet printing, a method of depositing materials in a precise and controlled manner, has experienced substantial interest in recent times due to its potential in the realm of additive manufacturing (Figure 20). When applied to the creation of supercapacitor electrodes, this technique presents several key advantages. It enables the production of high-resolution devices, affords great flexibility in design, and boasts scalability, which in turn makes it an attractive choice for the fabrication of customized and high-performance supercapacitor electrodes [188].

4.4.1. Supercapacitor Electrode Advantages

Inkjet printing technology provides significant flexibility in designing electrodes for electrochemical devices. One of its main advantages is the ability to produce customized designs, with precise control over the patterning of these electrodes, including complex geometries and optimized thicknesses. This tailored approach enhances the distribution of current, decreases resistive losses, and ultimately leads to superior electrochemical performance, as noted in past studies [189]. Furthermore, the direct deposition method inherent in inkjet printing eliminates the reliance on intricate lithographic processes and the use of masks. This simplification not only reduces the overall complexity of the manufacturing process but also minimizes the amount of material waste generated during the production of these electrodes. The environmental and economic implications of such a reduction in waste are substantial, contributing to the appeal of inkjet printing in the fabrication of electrochemical devices. Moreover, inkjet printing allows for the sequential deposition of various ink formulations, which is essential for the integration of multilayer structures within the electrode framework. This feature enables the construction of electrodes with hierarchical nanostructures, where different materials can be strategically combined to synergistically improve their electrochemical properties. Such hybrid electrodes can capitalize on the unique advantages of each component material, leading to enhanced overall performance.

4.4.2. Applications in Supercapacitor Electrodes

The application of inkjet printing technology in the creation of supercapacitor electrodes has yielded positive results with a variety of materials, including metal oxides, conductive polymers, and carbon-based nanostructures. This approach has facilitated the production of metal oxide nanoparticle-based electrodes with meticulously regulated porosity and extensive surface areas, which significantly enhances the number of ion-accessible locations and thus the capacitive performance. Moreover, the utilization of conducting polymers in inkjet-printed electrodes has exhibited remarkable charge transfer capabilities and swift ionic mobility, leading to a substantial increase in energy storage capacity. This method's adaptability allows for tailored designs that cater to the unique properties of each material, thereby optimizing these supercapacitors' overall functionality [190].

4.4.3. Future Directions

Continuous studies in the domain of inkjet printing for supercapacitor electrode production are concentrated on improving ink compositions, fine-tuning the parameters of the printing process, and examining innovative substances to elevate the efficiency of these electrodes. There is also a significant emphasis on employing eco-friendly solvents and binder components to advance sustainable and environmentally responsible manufacturing methods. Moreover, the integration of inkjet printing with alternative electrode fabrication techniques, such as template-based synthesis or 3D printing, presents a promising approach for the creation of sophisticated supercapacitor devices with customized characteristics suitable for a broad range of uses. This interdisciplinary combination leverages the precision of inkjet technology and the potential of complementary methods to yield supercapacitors with optimized performance and adaptability to various practical scenarios [191].

4.5. Three-Dimensional (3D) Architectures

Emerging as a notable solution to the limitations of conventional planar designs, three-dimensional (3D) architectures have been found to significantly enhance the energy storage capacity and electrochemical performance of supercapacitor electrodes (Figure 20). These innovative structures boast distinct characteristics that effectively tackle critical issues encountered in traditional two-dimensional systems, thereby rendering them highly suitable for next-generation energy storage devices. The intricate 3D configurations provide a substantial surface area for redox reactions, facilitate ion and electron transport, and contribute to an increased volumetric energy density, which are all vital for achieving superior supercapacitor functionality [192].

4.5.1. Types of 3D Architectures

A plethora of 3D architectures have been examined for the purpose of supercapacitor electrodes, each presenting unique advantages in optimizing performance [193].

I. Hierarchical Nanostructures: These architectures consist of a combination of diverse nanosized components, such as nanowires, nanotubes, and nanoparticles, which are assembled to create a multifaceted interconnected network. The hierarchical arrangement of pores within these structures enhances ionic access and facilitates superior charge storage due to the synergistic effects of the various pore sizes and high surface-to-volume ratio.

II. Mesoporous Materials: Materials characterized by mesoporosity, with pore dimensions ranging typically between 2 and 50 nm, are highly effective in supercapacitor applications. The interconnected nature of these mesopores contributes to an extensive surface area and substantial pore volume, which enable efficient ionic diffusion and bolster the capacitive properties of the electrode.

III. Conductive Frameworks: The utilization of 3D conductive frameworks, like graphene foam or carbon nanotube scaffolds, serves as a supportive skeleton for the active electrode material. These frameworks not only ensure rapid electron transfer but also provide mechanical stability to the electrode, as they are designed to accommodate the strain associated with the charge–discharge cycles.

IV. Additive Manufacturing: Employing 3D printing technology in the fabrication of supercapacitor electrodes allows for the creation of intricate and tailored geometries. This method enables precise control over the deposition of layers of electrode material, thus catering to specific design requirements and potentially leading to superior structural integrity and functional efficiency.

4.5.2. Applications in Supercapacitor Electrodes

Three-dimensional architectures have gained extensive utilization in supercapacitor electrode design, showcasing noteworthy performance enhancements over conventional planar configurations. In particular, metal oxides with hierarchical nanostructures have revealed significant improvements in specific capacitance and rate capability. This is largely attributed to the augmentation of the surface area and the facilitation of the ion diffusion pathways these structures provide. Moreover, mesoporous carbon materials have exhibited remarkable capacitive properties and high energy storage density, establishing them as highly suitable candidates for supercapacitor electrodes. Their distinctive porosity and structural hierarchy contribute to their proficient electrical conductivity and overall capacitive performance [194].

4.5.3. Future Directions

The persistent investigation into state-of-the-art 3D designs for supercapacitor electrodes is anticipated to lead to substantial advancements in energy storage effectiveness. The emphasis of these research endeavors lies in meticulously adjusting the pore dimensions, structural configuration, and material composition of these 3D constructs to optimize their energy storage capacity and power handling characteristics. Furthermore, the integration of various materials within the three-dimensional framework and the utilization of hybrid architectures present promising avenues for realizing enhanced electrochemical performance. This, in turn, opens up new possibilities for their application in a broad spectrum of domains, including cutting-edge energy storage systems, wearable electronic devices, and the burgeoning field of electric vehicles [158].

5. Recent Advancements in Electrode Nanotechnology

Recent years have witnessed significant strides in the customization of nanostructures, the investigation of cutting-edge carbon-based nanomaterials, and the exploitation of the capabilities of two-dimensional (2D) nanomaterials. Furthermore, researchers have focused on creating binder-free and nanocomposite electrodes, which are essential components of supercapacitors. Table 4 details the latest developments in the realm of electrode technology for supercapacitors [195]. Additionally, the advancement of sophisticated characterization methods and an increased commitment to environmental conservation have paved the way for the innovation and production of high-performance and eco-friendly supercapacitor electrodes, expanding the horizons for the design and fabrication of such devices [196].

Table 4. Recent developments in supercapacitor electrode nanotechnology. With permission from ref. [153].

Electrode Material	Specific Capacitance (F/g)	Reference
Graphene oxide	200:300	[160]
Carbon nanotubes	100:300	[160]
Manganese dioxide	350:400	[163]
Nickel hydroxide	2000:3000	[176]

5.1. Tailored Nanostructures

The development of customized nanostructures has presented a highly promising path in the realm of electrode nanotechnology for supercapacitors. Through meticulous manipulation of nanomaterials' dimensions, morphology, and chemical makeup, substantial advancements in both the energy storage capacity and the rapidity of charge–discharge cycles have been realized. This innovative research field, with its pronounced potential to revolutionize supercapacitor electrodes' efficiency, holds great promise across various energy storage technology applications [197].

5.1.1. Advantages in Supercapacitor Technology

Supercapacitors, also known as ultracapacitors, stand out as an emerging technology with significant potential in the realms of energy and data management. Their versatility and adaptability are continually expanding due to the ongoing discovery and development of innovative materials and techniques. These devices are poised to transform numerous sectors and fields of interest because of their distinctive properties [198].

In comparison to traditional capacitors and batteries, supercapacitors boast a multitude of benefits, including the following:

I. Enhanced Energy Density: Through the implementation of tailored nanostructures, the specific surface area is substantially increased, thereby augmenting the number of active sites for ionic adsorption and desorption processes. This leads to a higher specific capacitance, allowing for greater energy storage per unit mass or volume.

II. Rapid Charge and Discharge: The meticulously designed nanostructures facilitate efficient ion diffusion through the electrode material. By reducing the diffusion pathways and enhancing the porosity, the ion transport is significantly accelerated. Consequently, supercapacitors are capable of discharging and charging at high rates, rendering them ideal for applications demanding quick energy release.

III. Enhanced Cyclability: Supercapacitors equipped with tailored nanostructures often demonstrate improved mechanical robustness and structural stability over numerous charge–discharge cycles. This characteristic contributes to their prolonged lifespan and minimized maintenance needs.

IV. Customizable Electrode Architecture: The ability to tailor nanostructures permits the fine-tuning of electrode properties to match the particular requirements of diverse applications. Scientists can optimize supercapacitor performance by adjusting the morphology, composition, and porosity of nanomaterials according to specific energy storage demands.

V. Multifunctional Electrode Integration: The strategic design of nanostructures allows for the combination of various functionalities within a single electrode. By merging nanomaterials with complementary traits, such as high conductivity and pseudocapacitance, it is feasible to create hybrid systems capable of electric double-layer capacitance as well as Faradaic redox reactions. This synergistic approach enhances the total energy storage capacity of the device.

5.1.2. Applications in Supercapacitor Technology

Nanostructured electrodes have become increasingly prevalent in the realm of portable electronics, including smartphones, tablets, and wearable technology. These meticulously designed nanostructures confer significant advantages such as reduced weight and enhanced energy storage performance, which in turn provide users with extended operational time and the convenience of rapid charging. In the domain of electric vehicles, tailored nanostructured supercapacitors have demonstrated their suitability for applications involving energy recovery and regenerative braking systems. The remarkable feature of these electrodes is their capability to maintain rapid charge–discharge rates while delivering high energy density, thereby contributing to the overall efficiency and power output enhancement of the vehicles' energy storage systems. Moreover, the integration of renewable energy sources into the electrical grid significantly benefits from the utilization of supercapacitors with tailored nanostructures. By effectively storing surplus energy produced by intermittent sources like solar panels and wind turbines during peak generation periods and releasing it during periods of low production or high consumption, these devices facilitate the stabilization of the grid and enhance the overall functionality of renewable energy systems. At the microscopic level, nanostructured electrodes are revolutionizing the landscape of microelectronics and the Internet of Things (IoT). Their compact nature and high energy density enable the creation of energy storage units that are well-suited for powering small sensors and IoT devices. This advancement allows for the development of self-sustaining and durable power solutions that are essential for the operation of these technologies over extended periods without the need for frequent charging or replacement. The tailored design of these nanostructures is instrumental in achieving the desired performance characteristics, such as power density and energy storage capacity, that are crucial for the successful implementation of IoT networks and various microelectronic applications.

5.2. Advanced Carbon Nanomaterials

The emergence of sophisticated carbon nanomaterials has marked a substantial advancement in the realm of electrode nanotechnology, particularly in the creation of high-performance supercapacitors. These innovative materials exhibit a range of unique structural, electronic, and electrochemical properties, which render them highly desirable for various energy storage purposes. Extensive research has been dedicated to exploring and exploiting these materials, including carbon quantum dots, graphene with strategic dopants, and carbon nitride, among others. The remarkable electrical conductivity, extensive surface area, and superior capacitive characteristics inherent to these substances enable the design of electrodes that boast enhanced energy storage capabilities. These properties contribute significantly to the performance enhancement of supercapacitors, leading to the creation of more efficient and reliable energy storage solutions [199].

5.2.1. Graphene-Based Electrodes

Graphene, a material consisting of a single plane of carbon atoms organized in a two-dimensional honeycomb configuration, is widely recognized as a prominent example of advanced carbon nanomaterials. Its properties include remarkable electrical conductivity and an extensive surface area that facilitate effective charge transfer and offer copious adsorption sites for ions. Electrodes constructed from graphene display high specific capacitance and excellent performance in terms of rate capability, which render them highly suitable for applications requiring rapid storage and discharge of energy [200].

5.2.2. Carbon Quantum Dots (CQDs)

Carbon quantum dots are nanostructures of zero dimensions, which exhibit exceptional properties arising from quantum confinement. Their diminutive size and adjustable energy gap facilitate effective electron entrapment and charge separation, resulting in noteworthy capacitive performance. Given their lightweight nature, eco-compatibility, and robust stability, these materials are highly appealing for utilization in the electrodes of supercapacitors [201].

5.2.3. Doped Graphene and Its Derivatives

The process of incorporating heteroatoms such as nitrogen, boron, or sulfur into graphene's lattice induces structural imperfections and modifies its electronic characteristics, which can significantly bolster its capacitive performance. Specifically, when graphene is doped with nitrogen, it displays pseudocapacitive behavior as a result of the reversible oxidation–reduction processes that occur at the nitrogen-rich locations. These doped forms of graphene, including reduced graphene oxide (rGO), amalgamate the conductive properties inherent to graphene with an augmented number of redox-active moieties. This synergistic combination contributes to a substantial increase in the energy storage capability of electrodes employed in supercapacitors [202].

5.2.4. Carbon Nitride (g-C_3N_4)

A two-dimensional polymeric material known as carbon nitride possesses a distinctive electronic configuration. Its stratified configuration offers a copious number of active locations that are beneficial for the adsorption of ions. Moreover, the ability to adjust its bandgap enables efficient charge segregation and accumulation. Electrodes constructed from carbon nitride exhibit remarkable specific capacitance and exceptional stability in electrochemical environments. These attributes make carbon nitride a highly promising substance for supercapacitors with prolonged operational lifespans [203].

5.3. 2D Nanomaterials

Electrode nanotechnology has been significantly captivated by the emergence of two-dimensional (2D) nanomaterials, which exhibit remarkable structural and electronic characteristics. Prominent among these materials are graphene, transition metal dichalcogenides (TMDs), and boron nitride, all of which are composed of single- or few-layered atomic arrangements in a 2D lattice. This distinctive configuration confers on them a range of exceptional properties, rendering them highly suitable for utilization in supercapacitors [204].

Benefits

The utilization of two-dimensional (2D) nanomaterials in supercapacitors offers several compelling advantages that enhance their performance and functionality. These materials, due to their high specific surface area, have all their atoms exposed, which provides a multitude of active sites for ion adsorption. This inherent feature is conducive to efficient charge storage and results in a substantially higher specific capacitance, which is a critical factor for energy storage systems. Electrical conductivity is another significant benefit of 2D nanomaterials, particularly in the case of graphene and certain transition metal dichalcogenides (TMDs) such as MoS_2. Their sp2 hybridized carbon atoms and high electron mobility confer remarkable electrical properties, which are essential for rapid electron transfer during charge–discharge cycles. This characteristic minimizes the internal resistance within the electrode material, thereby facilitating swift and efficient energy storage and retrieval processes. Furthermore, the ultrathin nature of these materials leads to quantum confinement effects, which alter the electronic properties of the 2D structures. These effects are manifested as a tunable bandgap and augmented charge separation. As a result, the quantum confined 2D nanomaterials demonstrate superior capacitive behavior, which is crucial for achieving high energy density and power in supercapacitors. This level of electronic control allows for the optimization of the electrochemical performance of the electrode materials. The layered morphology of certain 2D materials, such as TMDs, permits ion intercalation between the layers. This phenomenon not only enhances ionic diffusion dynamics but also expands the availability of active sites for charge storage. Consequently, the layered architecture of these nanomaterials significantly contributes to the enhancement of both capacitance and rate capability. Additionally, some 2D nanomaterials like TMDs and graphene oxide (GO) exhibit redox-active characteristics. This property enables them to support Faradaic charge storage, a mechanism that complements the electric double-layer capacitance. The presence of pseudocapacitance in these materials significantly bolsters the overall energy storage capacity of supercapacitor electrodes, making them increasingly comparable to traditional batteries regarding energy density. Lastly, the mechanical flexibility of 2D nanomaterials is an attribute that makes them particularly suitable for integration into flexible and wearable devices. Their atomically thin nature allows for the preservation of high energy storage efficiency without sacrificing the flexibility and adaptability required for such applications. This flexibility opens new avenues for the development of lightweight and conformable energy storage solutions, which are essential for the burgeoning field of flexible electronics and wearable technology.

5.4. Binder-Free Electrodes

Binder-free electrodes represent a significant advancement in the realm of supercapacitor technology, with the primary objective of overcoming the limitations inherent in their binder-containing counterparts. This innovative design omits the use of binders, which are typically inert substances that serve to cohere the various electrode components. By doing away with these binders, such electrodes present a multitude of benefits, including superior electrochemical properties, elevated energy storage capabilities, and improved mechanical robustness. These enhancements position binder-free electrodes as a highly attractive candidate for the development of next-generation supercapacitor systems [163].

Advantages

The utilization of binder-free electrodes in energy storage systems presents several compelling benefits that distinguish them from conventional binder-containing counterparts. These advantages stem from the maximization of active material content, which enhances specific capacitance due to increased active surface area utilization. The absence of binders in these electrodes also facilitates superior ion diffusion kinetics by eliminating barriers that may impede ion movement, particularly under high-rate charge–discharge conditions. This leads to expedited charge and discharge processes and superior power delivery. Binder-free electrodes typically exhibit reduced internal resistance, which is attributed to the direct contact established between the active materials and the current collectors. This minimizes interfacial resistance, thereby promoting superior charge transport dynamics and overall electrochemical performance. Such electrodes are known for their enhanced electrochemical stability across numerous charge–discharge cycles, as the omission of binders mitigates the likelihood of detrimental side reactions that can degrade the electrode's structure and impair long-term functionality. Moreover, these electrodes often showcase remarkable mechanical resilience and flexibility, which can be attributed to their open and interconnected structural configuration. This property allows for the efficient distribution of mechanical strain during cyclic operations, thereby lowering the risk of cracks or delamination. Consequently, they are highly suitable for integration into flexible and wearable electronics, as well as other applications requiring adaptable energy storage solutions. The production of binder-free electrodes is generally more straightforward and cost-effective due to the streamlined fabrication process. Techniques such as direct deposition and self-assembly are commonly employed, simplifying the manufacture and enhancing scalability. These electrodes find extensive use in various energy storage devices, including supercapacitors in portable technologies, electric vehicles, and grid-scale storage systems, where their high energy storage density and rapid charge–discharge capabilities are particularly advantageous. Furthermore, the flexibility and mechanical stability of binder-free electrodes make them ideal for incorporation into flexible and wearable electronics, such as smart fabrics and MEMS devices. These qualities also extend to miniature and microelectronic applications, where space constraints necessitate compact yet powerful energy storage solutions. In the realm of electrochemical sensors, the improved charge transport kinetics and stability of binder-free electrodes are highly valuable. They enable rapid response times and efficient ion diffusion, which are essential for the accurate and reliable detection of various analytes. Their robust performance underpins their suitability for use in a broad spectrum of sensor types. When assembled into supercapacitor modules, binder-free electrodes contribute to systems that excel in high-power density and stability, making them highly applicable across diverse sectors such as renewable energy, energy management, and power backup solutions. Their integration into these modules enhances the efficiency and reliability of operations within these domains. In essence, the adoption of binder-free electrodes in energy storage technology represents a significant step forward in terms of performance, longevity, and adaptability, opening up a plethora of potential uses and innovations.

5.5. Electrodes Made of Nanocomposite Materials: Improved Performance via Cooperation

The construction of nanocomposite electrodes presents an intriguing and adaptable strategy within supercapacitor technology, capitalizing on the synergistic outcomes that arise from the amalgamation of various nanomaterials to achieve superior electrochemical efficacy. Through the integration of nanomaterials that exhibit complementary characteristics, these electrodes provide a multitude of merits, such as augmented energy storage density, enhanced electrical conductance, and bolstered mechanical resilience. This segment delves into the conceptual framework, positive impacts, and practical implementations of nanocomposite electrodes in the continuous progression of supercapacitors [205].

5.5.1. Design Principles

Creating effective nanocomposite electrodes involves careful consideration of various fundamental aspects, as detailed in prior research studies [206].

I. Material Combination: These electrodes often consist of a blend of two or more nanomaterials with distinct properties. It is essential to select components that complement each other, such as one contributing to high specific capacitance and the other enhancing electrical conductivity. The harmonious interplay of these materials leads to superior composite performance in electrochemical reactions.
II. Interface Optimization: The efficiency of charge transfer within nanocomposite electrodes is significantly influenced by interfacial interactions. Therefore, it is critical to engineer interfaces that promote rapid electron movement and ion diffusion. This synergistic behavior ensures that each material's unique properties are fully utilized to enhance the electrode's overall electrochemical effectiveness.
III. Nanostructure Manipulation: The shape, size, and arrangement of nanomaterials within the electrode are crucial determinants of its electrochemical characteristics.

By controlling the nanostructure, researchers can tailor the properties to optimize the electrochemical performance while simultaneously bolstering the mechanical robustness. This precision in design allows for the maximization of the electrode's potential in various energy storage and conversion applications.

5.5.2. Advantages

Nanocomposite electrodes exhibit several key advantages that significantly impact the performance of supercapacitors. The integration of nanomaterials with distinct capacitive characteristics leads to a synergistic enhancement in specific capacitance, which is greater than what is achievable with single-component systems. This synergistic effect is crucial, as it directly correlates with the energy storage capacity of the device, thereby leading to improved operational efficiency. The electrical conductivity of nanocomposite electrodes is also notably superior due to the incorporation of highly conductive materials such as graphene or carbon nanotubes. This enhancement facilitates efficient electron flow across the electrode's structure, thereby minimizing the internal resistance and contributing to better power output. The combination of these materials allows for an optimized balance between ionic and electronic conduction, which is essential for high-performance supercapacitors. Moreover, the unique interactions between the various nanomaterials within a nanocomposite can give rise to synergistic effects that amplify the overall energy storage capabilities of the electrode. For instance, a combination of a pseudocapacitive material with a conductive nanostructure can engender a hybrid system that benefits from both electric double-layer capacitance and Faradaic redox processes. This synergy can lead to a significant boost in the supercapacitor's energy density and rate capability. Mechanical stability is another critical aspect, where nanocomposite electrodes excel. The interplay between different nanomaterials can effectively redistribute mechanical stress during the charge–discharge cycles, thereby diminishing the likelihood of structural damage or deformation. This robustness translates into longer operational lifetimes and improved reliability for supercapacitors. Furthermore, nanocomposite electrodes offer remarkable flexibility and adaptability in their design. Researchers are able to select specific nanomaterial blends and meticulously adjust their proportions to cater to the precise demands of various supercapacitor applications. This tailored approach allows for the optimization of electrochemical properties, enabling the development of supercapacitors that are well-suited to diverse technological contexts, from wearable electronics to electric vehicles [207].

5.6. *Advanced Characterization Techniques*

The latest developments in the field of electrode nanotechnology involve the utilization of sophisticated analytical methods to probe the intricate electrochemical properties of nanomaterials and the dynamics at electrode–electrolyte interfaces. Techniques such as in situ microscopy, X-ray spectroscopy, and electrochemical impedance spectroscopy are being employed to scrutinize the phenomena of charge accumulation and dispersion within the electrodes throughout charge and discharge cycles. The integration of these approaches facilitates an enhanced comprehension of the underlying mechanisms governing the performance of energy storage devices, thereby allowing for the rational design and optimization of electrode materials and configurations to achieve superior storage capacity and stability [208].

5.7. *Sustainability of the Environment: Adopting Green Energy Storage Technologies*

The importance of environmental sustainability in the realm of supercapacitor technology is undeniable, as the world faces an urgent need to adopt cleaner and more ecologically responsible energy storage solutions. These devices exhibit characteristics that are in harmony with sustainable objectives, which in turn enhances their allure as alternatives to traditional energy storage systems that prioritize eco-friendliness [209].

5.7.1. Environmental Benefits

Significant environmental benefits are associated with the utilization of supercapacitors. These include the following:

I. Minimized Ecological Footprint: The constituent materials of supercapacitors, like activated carbon, carbon nanotubes, and conductive polymers for electrodes, are generally non-toxic and pose a lower environmental risk compared to traditional batteries, which are based on heavy metals and potentially harmful chemicals. This reduces the environmental strain during their entire lifecycle, from manufacturing to disposal.

II. Enhanced Energy Efficiency: Supercapacitors boast high efficiency in energy storage and release, which translates to reduced energy waste during charge and discharge cycles. This efficiency contributes to a decrease in the frequency of replacements, thereby lowering the environmental footprint associated with production and disposal.

III. Exceptional Endurance: With the capacity to undergo countless charge–discharge cycles without substantial performance loss, supercapacitors demonstrate remarkable durability. This longevity diminishes the necessity for frequent replacements, thus cutting down on the total waste produced and the resources consumed over time.

IV. Quick Energy Exchange: The rapid charging and discharging rates of supercapacitors facilitate efficient energy capture and release, which is particularly beneficial for stabilizing power grids. Their responsive nature allows for better integration of renewable energy sources and reduces dependency on fossil fuel-based power plants that are typically less flexible and more carbon-intensive.

V. Sustainable Disposal: Supercapacitors can be recycled effectively due to the lack of hazardous materials in their construction. This feature enables the recovery of valuable components, such as carbon-based electrodes and current collectors, which can be reused. This recyclability aspect further reduces the environmental impact by curtailing the depletion of resources and the accumulation of electronic waste.

5.7.2. Support for Sustainable Energy Systems

The integration of renewable energy sources, such as solar and wind power, is significantly bolstered by the use of supercapacitors within electrical grids. These devices are instrumental in storing excess energy generated during periods of high production and then discharging it during peak demand, which enhances the stability and efficiency of energy distribution systems [210]. Moreover, supercapacitors are ideal for microgrid applications, offering robust energy management capabilities. Their high-power density and rapid response times facilitate the stabilization and reliability of these localized power networks, thereby reducing dependency on fossil fuels and enhancing energy self-sufficiency within communities or small regions. When it comes to electrified transportation, supercapacitors are a promising component in electric vehicles (EVs) and hybrid electric vehicles (HEVs). Their ability to charge rapidly and provide energy over numerous cycles is particularly advantageous. During regenerative braking, supercapacitors can swiftly store energy and later discharge it for acceleration purposes, which is beneficial in lowering greenhouse gas emissions and enhancing the overall environmental performance of these vehicles. Furthermore, supercapacitors serve as a reliable and efficient energy storage solution for off-grid and remote locations. Their capacity to maintain energy supply effectively and their reduced reliance on fossil fuels makes them a suitable alternative to diesel generators, thus contributing to a decrease in carbon footprint and pollution in such areas. Lastly, within the realm of electronics, supercapacitors are gaining ground in the powering of green and IoT devices. Their eco-conscious design and long operational lifespan, coupled with high-power density, make them an attractive choice for achieving energy-efficient and sustainable performance across various electronic gadgets.

6. Features of Electrode Performance: Assessing the Important Metrics

The evaluation of supercapacitor electrode efficiency is based on several key parameters, which significantly influence their energy storage capacity, power delivery, and cyclic endurance. These performance indicators are essential in guiding the enhancement of electrode architecture and material selection to suit particular use cases, as outlined in Table 5. The significance of these characteristics lies in their role in the creation of highly effective and dependable supercapacitors. Through the continuous refinement of electrode design and the judicious choice of materials informed by these metrics, researchers aim to promote the widespread integration of supercapacitors across a spectrum of energy storage systems, thereby facilitating their practical application in various industries [211].

Table 5. Important features used in research applications for supercapacitor electrodes. With permission of ref. [153].

Characteristic	Description	Importance	Measurement Unit	Reference
Specific Capacitance	Energy storage capacity per unit mass or surface area	Fundamental for energy storage efficiency	Farads per gram (F/g) or farads per square centimeter (F/cm^2)	[212]
Energy Density	Total energy storage capacity per unit volume or mass	Determines overall device performance	Watt-hours per kilogram (Wh/kg) or watt-hours per liter (Wh/L)	[213]
Power Density	Rate of energy charge and discharge	Crucial for high-power applications	Watts per kilogram (W/kg) or watts per liter (W/L)	[214]
Cycling Stability	Ability to maintain performance over charge–discharge cycles	Ensures long-term device reliability	N/A	[215]
Rate Capability	Ability to handle rapid charge and discharge rates	Important for dynamic energy storage applications	N/A	[216]
Equivalent Series Resistance (ESR)	Internal resistance affecting energy conversion efficiency	Influences power performance	Ohms	[199]
Cycle Life	Number of charge–discharge cycles before performance degradation	Important for device lifespan and cost-effectiveness	N/A	[166]

7. Advanced Characterization Techniques

7.1. Analytical Characterization Methods

When evaluating the properties of electrode materials, two primary aspects are typically considered: dimensions and morphology. The examination encompasses the size distribution, the tendency towards agglomeration, the presence of surface charges, and the surface chemistry features. These characteristics significantly impact various performance metrics and potential uses. After the synthesis of the electrode materials, their crystalline organization and chemical makeup are meticulously studied, yet a universally accepted protocol remains elusive. Supercapacitors derive their capacitance predominantly from electrode surface interactions. Therefore, understanding the surface properties, including the surface area and pore structures, is critical. Functional groups present on these surfaces are also of great importance. To achieve a comprehensive analysis, researchers often employ techniques such as the Brunauer–Emmett–Teller (BET) method of gas adsorption for examining macropores, mesopores, and micropores within a specified relative pressure range. The density functional theory (DFT)-based approach to interpreting nitrogen adsorption isotherms is particularly useful for elucidating micropore characteristics. Moreover, surface analysis tools like Fourier transform infrared spectroscopy (FTIR) and X-ray photoelectron spectroscopy (XPS) are indispensable for the study of pseudocapacitance electrodes, which rely heavily on the nature of their surface functionalities.

7.1.1. X-Ray Techniques

X-Ray Diffraction (XRD)

X-ray diffraction (XRD) is a commonly employed analytical tool for investigating the properties of electrode materials, offering insights into their crystalline structure, phase identification, lattice parameters, and crystallite dimensions. This technique is particularly suitable for powdered samples, providing a volume-averaged representation of the material [217–219]. To put it succinctly, the phenomenon of peak broadening in X-ray diffraction patterns can be attributed to various microscopic factors such as the presence of microstrains, nanograins, subgrains, and dislocations, as well as the existence of small crystallites within the material's structure. This effect is indicative of the material's microstructure and is influenced by these features that deviate from an ideal perfectly ordered crystalline arrangement [220–223]. Additionally, shifts in peak positions are often associated with internal stress mechanisms, the occurrence of stacking faults, and the homogeneous distribution of vacancies, which can be a result of chemical variations within the material. The specific location or spacing of these peaks serves as a fingerprint for identifying the material's crystallographic group, whereas the intensity and morphology of the peaks offer insights into the constituent elements and their spatial distribution within the crystal lattice.

X-Ray Absorption Spectroscopy (XAS)

This methodology presents a straightforward approach to quantifying the chemical states of singular elements within a material matrix and is particularly applicable for observing the oxidation levels of substances participating in supercapacitor electrochemical processes. Since every element exhibits unique absorption edges that are tied to the distinct binding energies of its electrons, XAS is a highly selective and valuable tool for material analysis. EXAFS is recognized for its high sensitivity, allowing it to determine the oxidation state of materials even when present in low concentrations. However, its practical application necessitates the use of synchrotrons for spectrum recording, which results in higher costs and less widespread availability compared to other methods [224–229]. In research conducted by Kim and their colleagues, the effect of heat treatment on nickel oxide (NiO) structural defects with respect to their use in supercapacitor electrodes was scrutinized using X-ray absorption spectroscopy (XAS) [230]. The findings from both the EXAFS and XANES analyses indicated that the heat treatment process led to the transformation of non-stoichiometric Ni1-xO into a stoichiometric NiO structure. This is attributed to the inherent defect healing capability of NiO, which suggests that defective NiO structures can potentially enhance the supercapacitors' electrochemical performance. In a separate investigation, Kim et al. [231] utilized in situ Mn K-edge fluorescence XAS to explore the electronic and structural changes in MnO_2 that were electrodeposited on carbon paper for electrochemical capacitors. Dong and their team [232] also leveraged XAS at the K-edge of C, O, and Mn to evaluate the electronic configuration of MnO_2/functionalized carbon nanotube (f-CNT) nanocomposites. They discovered that the f-CNT functionalization significantly improved the charge storage capacity of the carbonaceous materials, primarily due to the facilitation of electrolyte ion transport by the surface functional groups. The XAS analysis yielded comparable spectral profiles for all the examined samples, signifying a similar local electronic and atomic environment. However, variations in peak intensities were observed, which were attributed to interfacial effects following MnO_2 deposition on f-CNT. The stability of the graphitic framework in f-CNT post-MnO_2 deposition was evidenced by the A5 peak in the in situ XAS spectra, which remained unaffected by any peak shifts. This indicates that the f-CNT's good conductivity was maintained. Additionally, the A5 peak intensity for MnO_2/f-CNT, which corresponds to the density of unoccupied π^* states, was observed to be lower than that of f-CNT alone. This suggests that there was an electron transfer from MnO_2 to the π^* state of f-CNT, which is consistent with the metallic nature of MnO_2. The higher intensity of the B5 and C5 peaks in MnO_2/f-CNT, as opposed to f-CNT, pointed towards a robust interaction between MnO_2 and carbon, which caused a substantial electron transfer from C=O π^* to C-O σ^*, thus creating a highly unoccupied state at the local level. Dong et al. [233] provided an extensive study on the electronic structure of various MnO_2 composites by integrating in situ XAS with electrochemical measurements. They observed that MnO_2 composites with reduced graphene oxide (rGO) experienced the least alteration in peak intensities during the charging/discharging process, compared to those with CNTs. This minimal change can be explained by the homogenous distribution of MnO_2 on the rGO surface, which effectively decreased the ion diffusion pathway at the interface. This structural configuration allows for more efficient charge storage and transport, highlighting the importance of material composition and architecture in optimizing supercapacitor performance.

X-Ray Photoelectron Spectroscopy (XPS)

Pseudocapacitance in supercapacitors is primarily contingent upon the electrochemical reactions occurring at the electrode material's surface. To thoroughly investigate these processes, X-ray photoelectron spectroscopy (XPS) is frequently utilized due to its proficiency. While standard XPS provides insights into the surface chemistry, angle-resolved XPS allows for the study of even thinner layers. The utility of XPS extends to various aspects of supercapacitor research, such as ligand exchange interactions, surface functional groups, and heteroatom distribution. Researchers like Toupin et al. [234] have employed XPS to examine the chemical state transitions of MnO_2 during charge/discharge cycles, noting shifts from Mn(III) to Mn(IV) due to the redox reactions with electrolyte ions like sodium (Na^+). Tõnisoo et al. [235] further demonstrated the value of in situ XPS in understanding the charge storage mechanisms in molybdenum carbide-based electrodes for electrochemical double-layer capacitors (EDLCs). They implemented a two-electrode setup within the supercapacitor for real-time monitoring of the electrode/electrolyte interface. This approach underscores the importance of combining XPS with other analytical methods to comprehensively study the complex surface interactions and bulk properties of supercapacitor materials.

7.1.2. Electron Microscopy

Transmission Electron Microscopy (TEM)

The technique known as transmission electron microscopy (TEM) explains how a thin object interacts with an electron beam that has a constant current density. The resulting images are constructed from the information obtained from the transmitted electrons, making TEM a widely applied tool for examining nanomaterial dimensions and morphology, due to its high precision and direct imaging capabilities [236,237]. To mitigate these issues and ensure reliable outcomes, TEM is frequently complemented by other analytical methods like UV–Vis spectroscopy and dynamic light scattering (DLS) that are capable of detecting a higher number of nanoparticles with less intrusive sample preparation. These combined approaches offer a more thorough understanding of the homogeneity and structure of various superlattice composites consisting of isostructural to multi-atomic crystals [238].

High-Resolution TEM (HRTEM)

High-resolution transmission electron microscopy (HR-TEM) is a specialized TEM imaging approach that capitalizes on phase-contrast imaging, an advanced technique that integrates both transmitted and scattered electron waves to construct visual representations. It is particularly indispensable when the goal is to observe and study the intricate crystallographic structure of individual nanoparticles, revealing details beyond the reach of standard TEM resolution. Thus, HR-TEM has become the preeminent tool for elucidating the internal architecture of nanoscale substances [239–241].

Electron Diffraction (ED)

Electron diffraction (ED), alternatively referred to as selected area electron diffraction (SAED), is a crucial microscopic tool utilized in the study of crystal structures. This method is typically performed within the confines of a transmission electron microscope (TEM) or scanning electron microscope (SEM), where it leverages the phenomenon of electron backscatter diffraction. However, the technique encounters limitations when dealing with large quantities of particles, as the ensuing diffraction patterns can become complex and challenging to interpret on an individual basis [242,243].

Scanning Electron Microscopy (SEM)

Scanning electron microscopy (SEM) is frequently utilized for the detailed examination of nanomaterials, due to its capacity to produce high-resolution images at the nanoscale [244]. The microscopic analysis of electrode materials, as illustrated in (Figures 21–23), reveals a significant research focus on supercapacitors utilizing 3D printing technology. Despite the substantial interest in this area, the stringent rheological demands for uniform ink formulation pose challenges to the realization of complex true 3D designs that could potentially enhance energy storage. Xue and colleagues [245] have employed a stereolithography technique to create metal composite lattices with octet-truss configurations through electroless plating, followed by the deposition of 3D hierarchical porous graphene on these structures to form quasi-solid supercapacitors. Examining the morphology of the synthesized reduced graphene oxide (rGO) (Figure 21) through low and high magnification field emission scanning electron microscopy (FESEM), (Figure 21a) clearly demonstrates the establishment of a porous architecture on the metallic frameworks. This unique structure is attributed to the generation of H_2 bubbles during the conversion of H^+ to rGO in a hydrochloric acid solution, which are subsequently released. This innovative porosity enables the efficient utilization of the active material, thereby leading to an increase in the capacitance. Further insight into the material's composition and distribution is provided by the scanning electron microscopy–energy dispersive X-ray spectroscopy (SEM-EDX) images, as presented in Figure 21b,c. These images confirm the presence of the constituent elements and their uniform distribution throughout the lattice. The lattice consists of rGO strands with a diameter of 1 mm, which feature consistent pores of 200 μm, as seen in Figure 21d,e. This structural feature is advantageous for enhancing ion diffusion, a critical aspect for improving the supercapacitor's performance. The formation of multicavities with diameters of 5 μm on the rGO sidewalls, as depicted in Figure 21f, along with the stacking of nanopores resulting from the folded graphene nanosheets, shown in Figure 21g,h, contribute to the enhancement of the electrochemical properties. This hierarchical porous architecture, coupled with the wrinkled graphene nanosheets, highlighted in Figure 21i, offers multiple pathways for ion diffusion, which is vital for optimizing the material's performance. These intricate structural characteristics are essential for achieving high capacitance and efficient energy storage in supercapacitors. The combination of lattices and porous rGO facilitates the creation of a supercapacitor with superior electrochemical properties, addressing the limitations imposed by conventional 2D architectures.

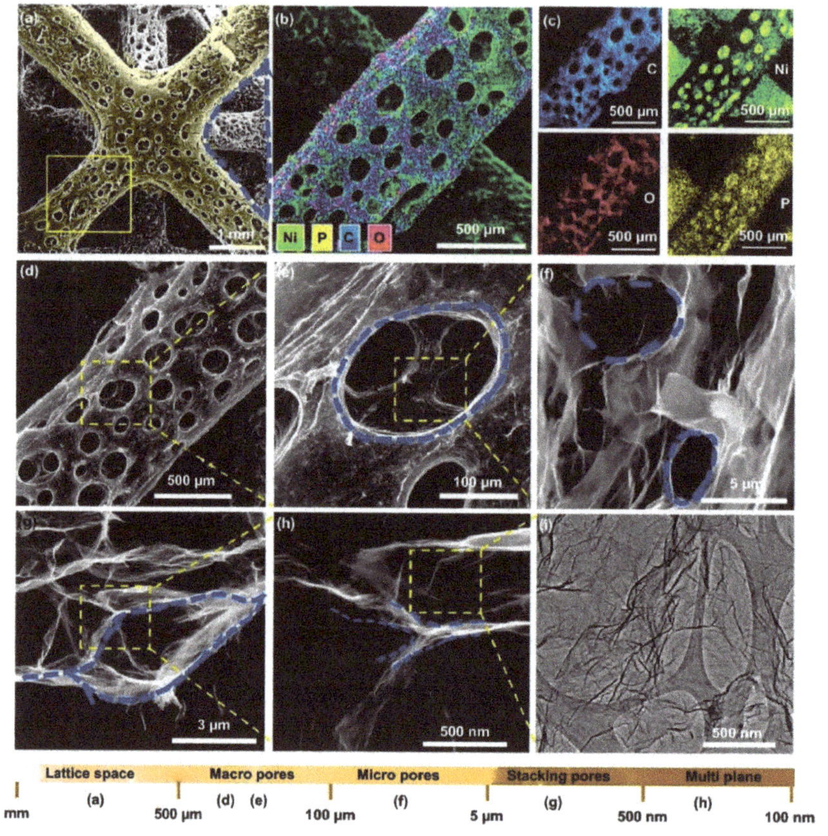

Figure 21. Figure depicting a multi-level porous lattice architecture. (**a**) An individual porous lattice unit is observed under field emission scanning electron microscopy (FESEM). (**b**–**d**) Scanning electron microscopy (SEM) combined with energy-dispersive X-ray spectroscopy (EDX) mapping showcases the intricate detail of a single lattice component. The image reveals a lattice with continuous porosity, highlighting its unique structural characteristics. (**e**) Magnification of the seamless porous framework. (**f**) The sidewalls of the rGO feature microporosity, contributing to the lattice's high surface area. (**g**) The lattice's porosity arises from the layered arrangement of graphene sheets, which creates distinct pore patterns. (**h**) These pores are derived from the multi-layered nature of the graphene sheets, enhancing the material's overall porosity. (**i**) Transmission electron microscopy (TEM) provides insight into the graphene sheets' morphology, displaying a wrinkled and wavy configuration. (Reproduced from Ref. [245], with permission from Springer and the Creative Commons CC BY license).

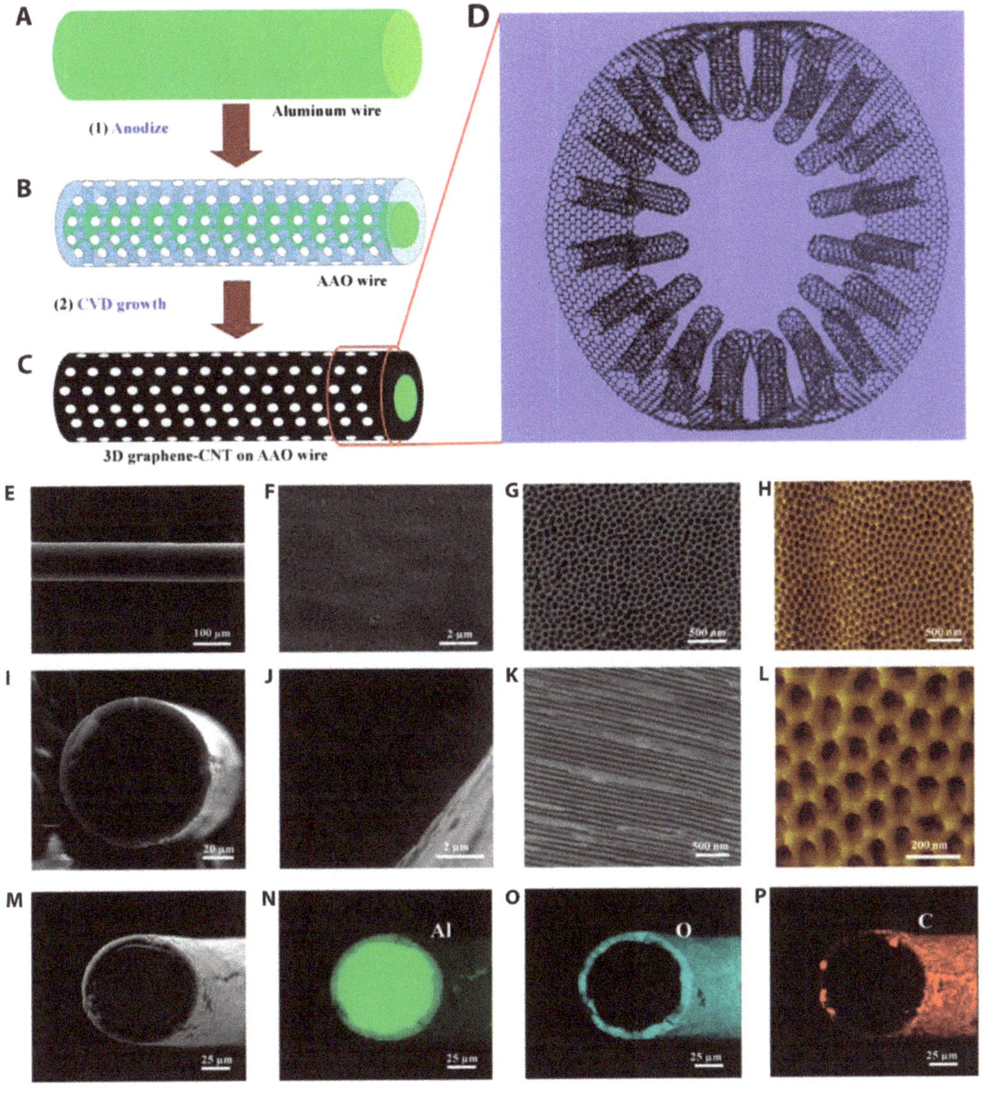

Figure 22. Schematics illustrating the microstructures and synthesis of a three-dimensional graphene–RACNT fiber. (**A**) Wire made of aluminum. (**B**) Aluminum wire that has been surface anodized (AAO wire). (**C**) The AAO wire's three-dimensional graphene–RACNT structure. (**D**) The pure 3D graphene–RACNT structure is shown schematically. Top-view SEM pictures of the 3D graphene–RACNT fiber at various magnifications are shown in (**E–G**). SEM pictures of the 3D graphene–RACNT fiber cross section (**I–K**). AFM pictures of the 3D graphene–RACNT fiber (**H**,**L**). The 3D graphene–RACNT fiber's (**M–P**) SEM image (**M**) and matching EDX elemental mapping of (**N**) aluminum, (**O**) oxygen, and (**P**) carbon. Reproduced from ref. [246].

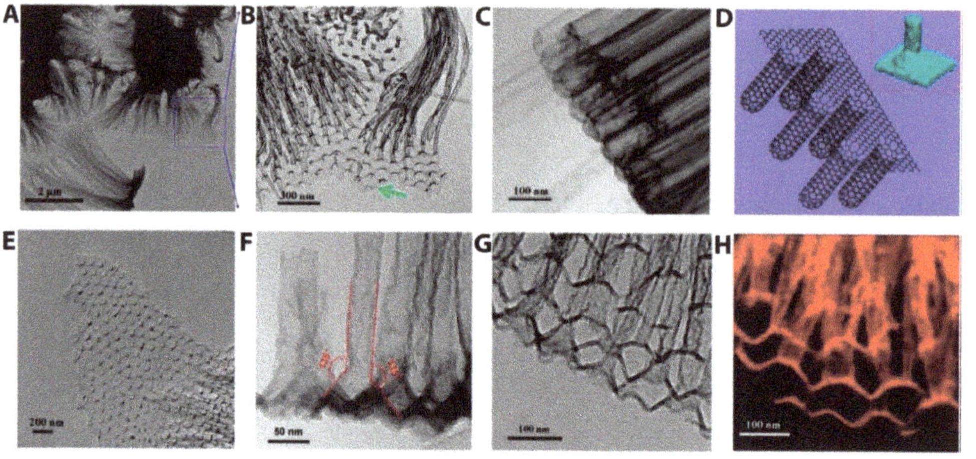

Figure 23. (**A**,**B**) A graphene sheet attached to the RACNTs' exposed tips. (**C**) RACNTs' closed end. (**D**) Three-dimensional graphene–RACNT network schematic; inset displays MD simulations' energy-minimized structure (Supplementary Materials). (**E**) The 3D graphene–RACNT network's broken graphene sheet. (**F**) TEM picture of the 3D graphene–RACNT side view around the graphene–nanotube interface. (**G**,**H**) The 3D graphene–RACNT structure's constituent RACNTs' cross-sectional views (**G**) and matching carbon mapping (**H**). Reproduced from ref. [246].

Xue and colleagues [246] introduced a streamlined single-step CVD process for producing 3D graphene/CNT hollow fibers with radially oriented CNT arrays encapsulated by graphene cylinders for energy conversion and storage purposes. The synthesis of the catalyst is outlined in a sequence of steps depicted in Figure 22, progressing from (A) to (C). The 3D graphene–RACNT fibers' structure was examined through various microscopic methods. An SEM micrograph in Figure 22E exhibits a uniform lateral diameter of 100 μm for these fibers. Upon closer inspection at higher magnifications, the open RACNT tip arrays are visible with a pore diameter of about 50 nm, consistent with the AFM measurements displayed in Figure 22F–H,L. Cross-sectional images of the 3D graphene-RACNT fiber, presented in Figure 22I–K, indicate a shell thickness of approximately 6 μm, encircling an unanodized Al core, which is proportional to the RACNT length determined by anodization duration. The homogeneous CNT distribution is evident in the high-magnification views (Figure 22J,K). Moreover, SEM-EDX X-ray mapping, as shown in Figure 22N–P, confirms the central Al composition of the fibers, surrounded by C and O elements, aligning with the theoretical model presented in Figure 22D. This confirms the effective formation of the intended catalyst structure.

The meticulously structured 3D graphene–RACNT fiber, comprising a substantial graphene tube that envelops the RACNT arrays along its inner cavity, is obtained through an aluminum and alumina core removal process in a 1 M KOH solution. The TEM images of these fibers post-extraction are presented in Figure 23A–H. Upon examination of Figure 23A, it becomes evident that the RACNTs are intimately connected to the graphene sheets, which feature a uniform distribution of accessible pores. This observation aligns with the SEM and AFM data. The green indication in the image marks a fragmented section of the graphene–RACNT structure. The formation of CNT bundles and detached pieces is likely a consequence of surface energy-driven aggregation that occurred during the aluminum and alumina dissolution. The stability of the 3D graphene–RACNT arrays may be primarily attributed to the encasing graphene nanotube closely interlinked with the RACNTs. Further TEM imagery, as shown in Figure 23B, provides a closer look at the

RACNTs attached to the graphene sheets, with the green marker highlighting a dislodged edge piece. The termination points of the RACNTs arrays are revealed in Figure 23C, illustrating that the nanotubes' tips situated away from the graphene sheet are all sealed. This configuration corresponds with the 3D architecture depicted in Figure 23D, which serves as the fundamental building element for the creation of these hollow fibers. The open ends of the RACNTs are seen to be flawlessly connected to the graphene nanosheets. The cross-sectional perspective of the synthesized 3D catalyst network, displayed in Figure 23F, indicates an approximate bonding angle of 135° rather than the typical 90° at the graphene–RACNT interface. Lastly, Figure 23G offers a TEM image of an individual hollow 3D catalyst, while Figure 23H presents an elemental map of carbon within the catalyst, showcasing its distribution and structure.

7.1.3. Scanning Probe Microscopy

Atomic Force Microscopy (AFM)

Atomic force microscopy (AFM) is an advanced imaging method capable of generating high-resolution three-dimensional surface images. Additionally, it can analyze low-density materials that typically exhibit poor contrast in electron microscopy, thus broadening the scope of its analytical capabilities [246–248]. When it comes to supercapacitors, the electrode materials undergo structural and chemical changes during charge/discharge cycles, which can impact their long-term stability. Despite these transformations being crucial to understand in order to improve the device performance, the exact mechanisms underlying these changes remain elusive. AFM provides a valuable tool for examining the intricate details of these electrode materials, potentially shedding light on the phenomena occurring at the nanoscale during the operational lifecycle of supercapacitors. This insight can be instrumental in optimizing their design and enhancing their robustness. Thomas and his colleagues [249] conducted a study focusing on the evolution of two-dimensional (2D) electrode materials during charge/discharge cycles, utilizing a multi-faceted approach. They employed Kelvin probe force microscopy (KPFM) to monitor the surface potential of 2D WS_2 while it underwent periodic charging and discharging, as depicted in Figure 24. The study revealed that a strain developed within the WS_2 layers due to the intercalation and subsequent deintercalation of sodium ions, which in turn caused an increase in the number of accessible electrode active sites and thus an enhancement in the capacity. The AFM topographic images of the synthesized 2D material, presented at various stages of the charge/discharge process in Figure 24a,d,g, showed minor fluctuations in grain size, yet the overall structure remained compact. The KPFM-derived potential difference values, displayed in Figure 24b,e,h, highlight the variations in the estimated surface potential of the electrode, which are attributable to the cyclically induced strains in the 2D material. The intercalation of heteroatoms into the layered structure of the electrode is known to cause strain buildup within the material's domains. The strain is inversely proportional to the Fermi velocity when considered as a function of wave vector "k", and it results in an increase in the surface potential. Before the initiation of the charge/discharge process, the mean surface potential of the WS_2 electrode material was approximately 4.82 eV, as seen in Figure 24c. However, as the material experienced the early stages of the charge/discharge process, the potential increased due to the intercalation of sodium ions. This phenomenon is linked to the expansion of the interlayer spacing in WS_2, which provides additional active sites for ion interaction and thus enhances the capacitance. The surface potential reached approximately 4.95 eV after 2500 cycles, as indicated in Figure 24f, demonstrating significant strain in the 2D material. Interestingly, no further noticeable effects of strain on the electrode domains were observed beyond 2500 cycles, suggesting that the electrochemical surface area stabilized, and the material achieved its maximum capacity improvement. After

10,000 cycles, a slight increase in surface potential to approximately 4.97 eV was noted, as seen in Figure 24i. This suggests that the capacitance remained relatively constant once the WS_2 nanosheets had accommodated their maximum strain.

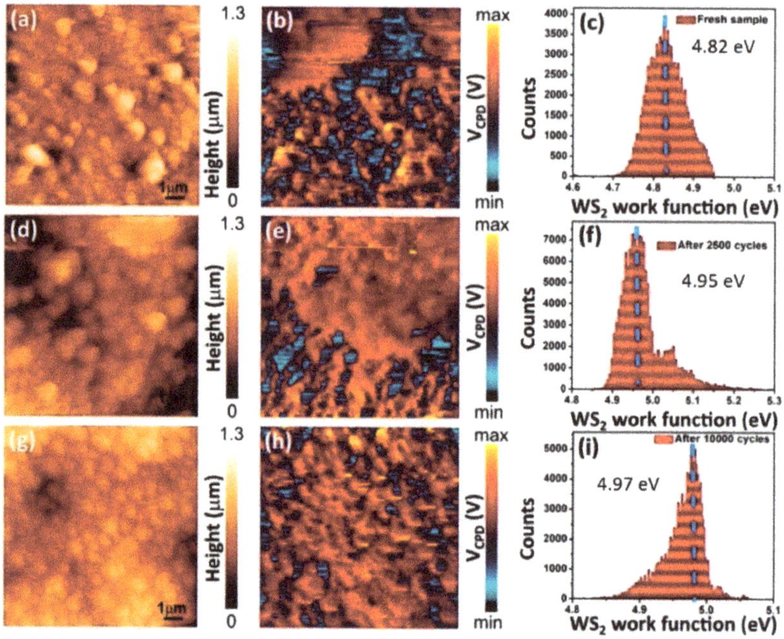

Figure 24. AFM topography pictures of the WS2 electrode at the first, 2500th, and 10,000th cycles are shown in (**a**,**d**,**g**). The WS2 electrode's KPFM signal pictures at the first, 2500th, and 10,000th cycles (**b**,**e**,**h**) display the relative VCPD measured across the electrode. The WS2 electrode's work function histogram (**c**,**f**,**i**) was computed using the KPFM pictures taken at the first, 2500th, and 10,000th cycles. (Reproduced with approval from Ref. [249].

Scanning Electrochemical Microscopy (SECM)

Scanning electrochemical microscopy (SECM) provides high spatial resolution in determining redox properties, making it an instrumental technique in electrocatalytic research, corrosion studies, charge transfer kinetics elucidation, and mechanism dissection. Its utility has extended to energy storage and conversion domains, serving as a sophisticated analytical instrument capable of scrutinizing diverse electrode processes. Unlike cyclic voltammetry, which assesses effective electrode areas rather than geometric, SECM visualizes electrochemical currents directly at the electrode interface, offering insights into the interplay between the in situ structure and surface charge transfer mechanisms [250].

7.1.4. Spectroscopic Techniques

Fourier Transform Infrared (FTIR) Spectroscopy

Fourier transform infrared spectroscopy (FTIR) is an analytical technique that involves examining the absorption of infrared radiation to identify functional groups within materials. FTIR technology boasts several advantages compared to conventional infrared spectroscopy: (i) it enables the acquisition of a complete spectrum in a matter of seconds, as opposed to the minutes required by traditional methods; (ii) it operates at elevated scan rates, which is advantageous when analyzing substances that elute from a column; (iii) enhanced sensitivity is achieved due to the minimal "background noise"; (iv) the use of digital signal processing allows for the subtraction of spectra from solvents or known impurities, thereby improving the data quality; (v) it can produce a spectrum from a minuscule sample by accumulating data from multiple scans to obtain a composite result [251–254].

Raman Spectroscopy (RS)

Raman spectroscopy (RS) is commonly utilized to explore the correlations between structural and electrochemical properties during electrode cycling processes. In the context of in situ applications, RS is particularly beneficial for analyzing the variation in the distribution of small-diameter nanotubes and the specific capacitance of electric double-layer capacitors (EDLCs) that incorporate chemically functionalized single-walled carbon nanotubes (SWCNT). By monitoring the displacement of the G^+ line in relation to non-functionalized SWCNT, researchers can gain simultaneous insights into the structural evolution and electrochemical performance of these materials. This approach provides valuable data for enhancing our understanding of the underlying mechanisms governing the behavior of nanotube-based EDLCs during their operational lifecycle [255,256].

Laser Raman Mapping

Laser Raman mapping, also recognized as Raman imaging, is a sophisticated method that offers insights into the intricate properties of two-dimensional materials such as graphene, graphene oxide, and TMDs [257,258]. This enhancement technique allows for the visualization of molecules with lower intrinsic Raman scattering efficiencies, offering a powerful tool in the study of complex systems.

7.1.5. Magnetic Techniques

In Situ Nuclear Magnetic Resonance (NMR) Spectroscopy

Nuclear magnetic resonance (NMR) is as a highly effective tool capable of scrutinizing local structures and molecular dynamics. Its elemental sensitivity equips it to independently discern ionic samples, making it a suitable candidate for exploring the charging mechanisms within supercapacitors during their charge and discharge cycles. In the context of ex situ NMR analysis, a supercapacitor is maintained at a specific potential, then dismantled, allowing the subsequent recording of the NMR spectra of the electrolyte species contained within the electrode film. Conversely, in situ NMR offers the advantage of analyzing the variations in the local environment of ions within the electrical double layer (EDL) during the operational phase of the supercapacitor cells [259]. This technique has significantly contributed to the qualitative understanding of charge storage mechanisms across various supercapacitor configurations. Brunner and his colleagues [260] utilized solid-state NMR techniques, specifically H^1, B^{11}, and C^{13}, to explore the interaction between electrolyte ions and carbonaceous materials with a well-defined porosity structure at the interface. This method was applied to supercapacitors with AC electrode materials and organic electrolytes to determine the extent of charge storage. Their in situ NMR findings indicated the presence of two distinct charge storage mechanisms: (1) at potentials lower than 0.75 V, the negative electrode's micropores exhibit a notable desorption of electrolyte ions, while the positive electrode's ion adsorption remains relatively unchanged with increasing voltage; (2) above 0.75 V, the positive electrode experiences a significant increase in adsorbed electrolyte ions, whereas desorption persists at the negative electrode. Grey and his research team [261] conducted an exhaustive study where they employed in situ NMR to quantify the anionic and cationic species within supercapacitors fabricated from microporous carbon. Their investigation demonstrated that the charge storage processes at the positive and negative electrodes are fundamentally dissimilar. For the positive electrode in a $TEABF_4$ electrolyte system with an organic solvent, the charging mechanism involves cation exchange with anions, whereas the negative electrode predominantly stores charge through cation adsorption.

7.1.6. Surface Area Measurements

Brunauer–Emmett–Teller (BET) Method

Electrode materials in supercapacitors undergo surface reactions such as the adsorption/desorption of charge carriers and redox processes at the juncture where they meet the electrolyte, exhibiting either electrical double-layer capacitance (EDLC) or pseudocapacitance. The performance of these devices is heavily contingent upon the surface area of the active materials constituting the electrodes. Although electrolytes are crucial for facilitating the storage and transfer of charge, their diffusion is inherently limited within the porous structure of the electrode materials. The extent to which these pores can be effectively utilized is determined by the specific pore dimensions. This technique allows for a precise quantification of the surface properties essential for optimizing the supercapacitor efficiency and performance [262,263].

8. Applications of Supercapacitors

8.1. DC Microgrids

DC microgrids are typically supported by a combination of renewable energy sources and the utility grid to manage their power supply. However, when dealing with fluctuating DC loads, this can lead to variations in voltage or current, which in turn puts transient stress on the DC bus. Some power sources like fuel cells (FCs) and batteries might struggle to accommodate these rapid changes, as they are designed for steady-state operations. The efficiency and longevity of these components can be significantly compromised by such dynamic load conditions. Supercapacitors (SCs), on the other hand, are recognized for their superior power density (PD) in comparison to FCs and batteries. This characteristic allows them to be highly effective in handling transient power needs. In light of this, a transient power supply system based on supercapacitors has been tailored specifically for use in DC microgrids [264]. The multi-bus DC microgrid configuration shown in (Figure 25) incorporates a supercapacitor bank within a dedicated DC bus to effectively address the issue of DC load fluctuations. This setup allows the microgrid to maintain a stable voltage profile and ensures reliable power delivery, even under conditions of varying demand. The integration of SCs enhances the overall performance of the system by providing instantaneous power support, thereby reducing the stress on the FCs and batteries and ultimately extending their operational lifetimes.

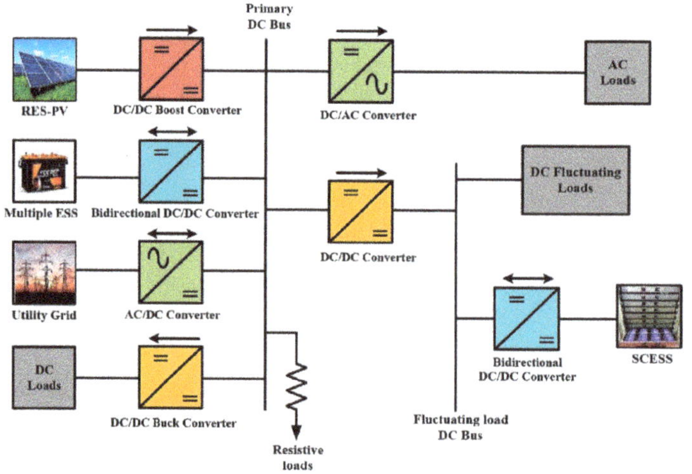

Figure 25. SCESS-based DC microgrid with DC fluctuating loads, with permission from ref. [264].

The methodology presented involves crafting and implementing a control strategy that employs a rate-limited low pass filter to effectively handle substantial transient high-power loads. The central role of the rate limiter is to regulate the velocity of the current variations, thus preserving the system's safe operating environment. The architecture of the load converter control system is comprised of an external voltage loop and an internal current loop. The voltage loop provides a voltage or current reference signal to a low pass filter-rate limiter controller (LPFRLC). The LPFRLC's output is partitioned into a low-frequency segment and a high-frequency segment, which are independently supplied to the corresponding current control loops of the load converter and the solid-state capacitor (SC) converter. Notably, during transient conditions, the high-frequency component acts as the current reference for the SC converter's regulator. In steady-state operation, the SC's current contribution, known as the SCESS current, is maintained at zero. The design process for the PI controllers within the voltage and current loops is predicated on the system's small-signal model. It is essential to ensure that the bandwidth of the inner current loops surpasses that of the outer voltage loop, which contributes to the systemic stability and bolsters the system's robustness against disturbances emanating from either the power supply or the load. This theoretical framework is corroborated by a synthesis of simulation models and empirical verifications. Additionally, an extensive review of the literature was undertaken by various scholars, scrutinizing the applications of battery-solid-state capacitor hybrid energy storage systems (HESSs). This scrutiny delves into the nuanced workings of their control systems and provides a comprehensive understanding of their overall performance characteristics [265].

8.2. Electric Vehicles

Supercapacitors (SCs) have become an essential component in electric vehicle (EV) systems, primarily due to their ability to recover regenerative energy efficiently during the braking process. In EVs equipped with a combination of fuel cell (FC) stacks and lithium-ion (Li-ion) batteries, the FC stack generally requires a higher power rating than the Li-ion battery pack. This issue is mitigated through the integration of SC modules, which exhibit superior characteristics in managing surge currents in contrast to both batteries and FCs, as indicated by previous studies [266,267]. The various configurations of EVs, with or without the inclusion of SC modules, are illustrated in Figure 26a–c. These configurations demonstrate the flexibility and enhancement that SCs provide to the overall performance of the vehicle. Specifically, SCs are adept at addressing the dynamic peak power demands, offering high charge/discharge rates. They are capable of supplying additional power during acceleration phases and effectively absorbing excess energy during braking. The bidirectional converter plays a crucial role in connecting the SC modules to the direct current (DC) link within the EV's power system. It operates in two primary modes: buck mode, which allows for voltage step-down, and boost mode, which steps the voltage up. Under steady-state conditions, the Li-ion battery remains responsible for supplying the necessary energy to power the vehicle. However, the integration of SC modules significantly enhances the system's capability to manage transient power requirements, thereby extending the lifespan of both the battery and the FCs. This is achieved by reducing the load on the battery during peak power situations and preventing the FC from frequently cycling between high and low power outputs. The utilization of SCs in EVs, therefore, not only optimizes energy storage and retrieval but also ensures that the critical components, namely the battery and FCs, are protected from the potential degradation caused by frequent and intense power fluctuations. This synergistic approach leverages the strengths of each component, leading to an overall more robust and efficient propulsion system

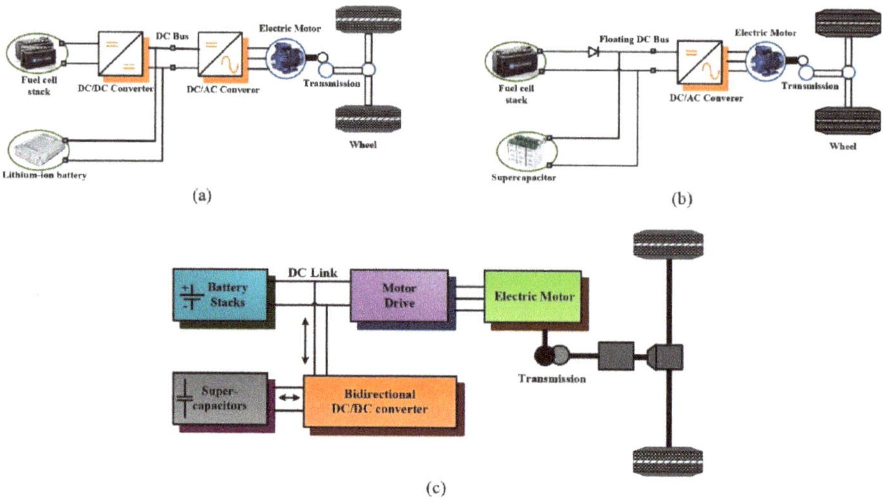

Figure 26. Electric vehicles (**a**) without the SC (**b**) and (**c**) with the SC, with permission from ref. [264].

A number of research endeavors have underscored the significance of assessing the performance of supercapacitors (SCs) within the framework of regenerative braking (RB) systems, akin to the configuration outlined in [268]. This experimental apparatus encompasses a flywheel, an AC induction motor, an inverter, a bidirectional DC–DC converter, and SC modules. The flywheel is preliminarily accelerated to 3000 rpm with the aid of an external DC supply, which is subsequently disconnected before the RB test begins. The RB process unfolds in five separate sets of commands. It is of considerable note that the state of charge of the SCs has a profound effect on the current and power uptake during RB, largely due to the current constraints imposed by the power electronic converters. The study scrutinizes the system's overall efficiency under these circumstances. Another investigation, as mentioned in [269], grapples with the power discrepancy between the SC energy storage system (ESS) and the power grid by introducing a multi-parameter collaborative power prediction control system. This cutting-edge approach amalgamates multi-parameter power forecasting with an optimization controller and incorporates under voltage and overvoltage protections. This strategy effectively precludes the grid from receiving energy during RB, thereby slashing unnecessary energy conversion losses. When juxtaposed with conventional double-closed loop control and braking unit energy consumption control methods, this novel control strategy exhibits substantial energy efficiency enhancements of 20% and 45%, respectively. For electric vehicles (EVs), researchers have put forth a modular reconfigurable multi-source inverter (MSI) concept, as presented in [270]. This innovative design allows for the seamless integration of SCs and battery packs without relying on a magnetic structure. The MSI employs space vector modulation (SVM) for precise governance over switching actions, showcasing a lower weight, reduced costs, and streamlined control. Furthermore, a deterministic state of charge (SOC) regulator is developed to ensure the consistent functioning of the SC bank. In the context of EVs, a C-rate control-based battery/SC hybrid energy storage system (HESS) is proposed in [271]. This system is particularly adept at managing the fluctuating power requirements during acceleration and deceleration. Through the utilization of a PWM control technique in tandem with a current-sensing approach, the system reduces the number of load and SC current sensors, thereby simplifying its architecture. The DC–DC converter's current command is computed using the C-rate control method, which takes into account the SC voltage and the vehicle's operational states. This ingenious solution not only elongates the battery's lifespan by 1.5 years but also optimizes the utilization of regenerated energy. Moreover, the effectiveness of an on-board energy storage device (ESD) is scrutinized within the context of electric railway transportation, as detailed in [272]. The researchers constructed an all-encompassing mathematical model of a train equipped with an ESD by employing the Modeltrack simulation tool, integrating factors such as the velocity constraints, track geometry, and train attributes. Additionally, a detailed timetable was factored into the analysis. The crux of the study lay in identifying the optimal ESD size, which in turn engendered cost savings, enhanced the energy utilization, and decreased the CO_2 emissions. This research is pivotal for guiding the design and practical application of optimized ESDs within DC light transportation systems, thereby contributing significantly to the advancement of energy-efficient mobility solutions.

8.3. Smart Phones and Note Computers

The diminished capacity adaptor of portable computing devices offers notable advantages, such as lower cost and compact lightweight designs. Nonetheless, it is associated with significant battery degradation, especially when continuously connected to wall power despite the increased power demands. To address this undesirable issue, the integration of supercapacitors (SCs) with batteries in a hybrid architecture is crucial. The SC–battery hybrid setup, as depicted in the schematic diagram (Figure 27) for smartphones and notebook computers, employs a structured approach to optimize system configuration [273].

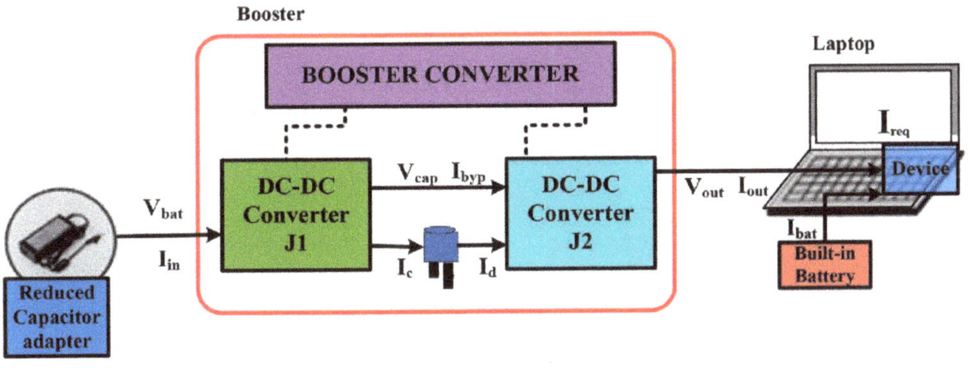

Figure 27. Schematic illustration of an SC–battery hybrid energy storage system.

This methodology entails the measurement and analysis of the insufficient current profile, followed by the establishment of an upper threshold for compensatory energy. This boundary is set to strike a balance between the hybrid structure's cost and the battery's longevity. The design process proceeds with the exploration of the buffer voltage swing and capacitance parameters. Finally, the optimal booster configuration is verified through rigorous analysis. The diagram illustrates J1 and J2 as the charging and discharging DC-DC converters, with J1 charging a supercapacitor array up to a specified voltage limit. The discharging converter, J2, outputs a voltage that matches the fully charged native battery's nominal voltage. An evaluation of the battery lifespan under idle conditions and with or without the booster reveals that with an ideally designed booster, the battery's operational life can be enhanced by 49.6%, surpassing the previously reported 46.8% improvement [274].

8.4. Industrial Drives

An open-winding electric vehicle motor with a power rating of 110 kilowatts has been coupled with a supercapacitor–battery hybrid configuration, as presented in the literature [275]. The system's architecture, shown in (Figure 28), employs a dual inverter drive that interfaces with two distinct energy storage devices. The intricate design allows for a sophisticated power distribution mechanism, which is managed through vector-based algorithms.

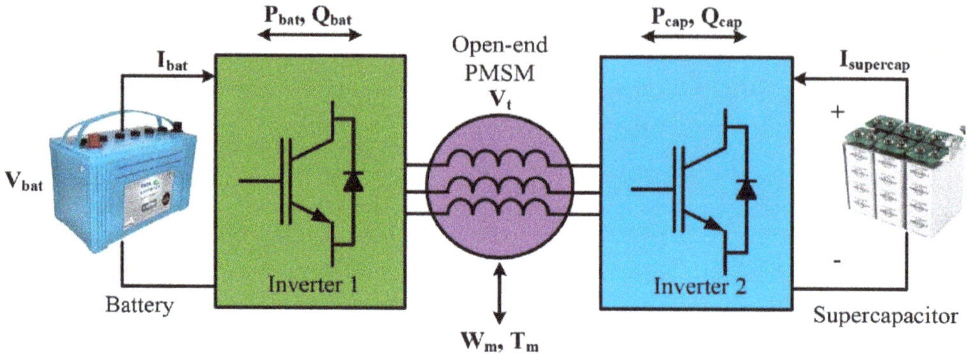

Figure 28. SC–battery hybrid energy storage system for industrial drive applications.

Given the substantial disparity in the electrical dynamics of supercapacitors and batteries, the energy management component of this setup assumes precedence to maintain the supercapacitor's operational voltage within acceptable limits throughout the vehicle's operational cycle. This approach is essential to optimize the overall performance and lifespan of the hybrid drive system. Supercapacitors (SCs) exhibit remarkable efficiency in both active and reactive power support due to their distinct configuration. This particular setup obviates the requirement for an additional DC/DC converter, which is typically used to either inject or retrieve energy from the SC. This enhancement contributes to a notable increase in light-load efficiency, as the SC inverter is bypassed. Moreover, the elimination of the separate converter reduces the overall switching losses, which further improves the system's efficiency. This design also enables the utilization of high-voltage electric vehicle (EV) motors, surpassing the capabilities of conventional configurations. To substantiate these claims, comprehensive experimental evaluations were conducted. These tests were primarily focused on validating the rule-based power management system, ensuring its long-term effectiveness in managing SC energy, facilitating stand-still charging capabilities, and safeguarding the SC against potential operational hazards. The integration of advanced intelligent controllers, founded on intricate AI/optimization algorithms, allows for seamless operation and adaptive decision-making, thereby enhancing the overall performance of the power management system. Another significant advantage of this setup is that the power management system can be decoupled from the low-level controls governing the dual inverter drive. This separation streamlines the development process by reducing the dependency on software engineers with specialized expertise in electric drivetrain operations. Consequently, it simplifies the integration and facilitates the adoption of this technology in various electric vehicle applications.

8.5. Wind Power

Supercapacitors (SCs) exhibit remarkable proficiency in countering the volatility inherent in wind power [276]. A typical wind energy conversion system (WECS), as illustrated in (Figure 29), comprises a mimicking converter, supercapacitor modules, a charge regulator, and an energy storage system founded on batteries. The mimicking converter is responsible for transmitting wind energy to the supercapacitors through a DC–DC conversion process. A charge controller facilitates the transfer of stored energy from the supercapacitors to the battery, which can be configured to operate in either a constant voltage or constant current mode. This design ensures that the power electronic converters are safeguarded against current stress during the battery charging phase and significantly enhances the battery's lifespan by mitigating power fluctuations caused by abrupt load changes and varying wind speeds. Further research indicates that ultracapacitor energy storage systems (ESS) can effectively manage wind speed variations [277]. By employing predictive control, the efficiency of harnessing energy can be notably increased. This control mechanism compensates for the fluctuations in the rotational speed of the induction generator, which is crucial for maintaining a consistent power supply. Comprehensive MATLAB/SIMULINK simulations demonstrate the effectiveness of this approach, revealing an energy increase of 296 J relative to systems lacking predictive control. The extent of this enhancement is contingent upon the turbine's size, with larger-scale wind farms benefiting from greater energy yields. Consequently, the integration of predictive control within wind energy systems not only amplifies the amount of energy produced but also has the potential to reduce the network costs significantly.

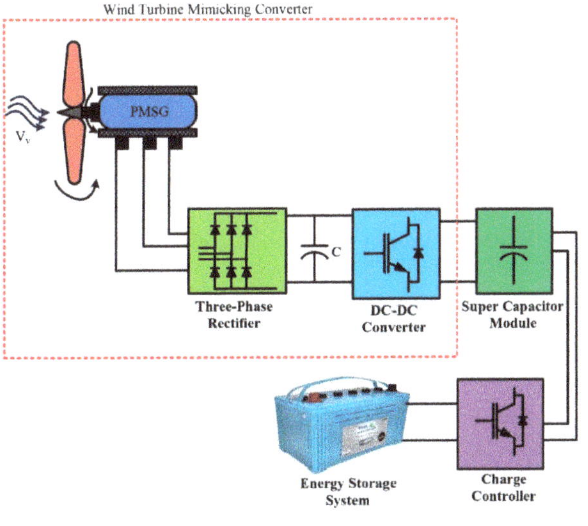

Figure 29. Schematic representation of an SC–battery hybrid energy storage system integrated with a wind energy system.

8.6. Multi-Level Inverter with ESSs

The depiction in Figure 30 illustrates a configuration where multilevel inverters (MLIs) are driven by a combination of solar photovoltaic (PV) systems and fuel cell (FC) stacks within hybrid energy storage systems (HESSs) [278]. Each H-bridge module within these MLIs is linked to the direct current (DC) link via a full bridge DC–DC converter, which integrates the distinct energy sources. These power sources are specifically designed to interface with the solar PV and FC stacks. The MLI's output is then distributed to single-phase microgrid loads. Fluctuations in power delivery from the solar PV systems are a significant concern for grid stability. However, supercapacitors (SCs) play a crucial role in mitigating this issue by either generating or absorbing active power as needed, thereby maintaining a consistent power supply to the MLI during periods of solar PV variation. Furthermore, the SCs have the capability to provide reactive power support, enhancing the overall performance and resilience of the MLI against power fluctuations at the load interface. Under steady and stable solar PV power conditions, the supercapacitors operate in a dormant state, only activating when power fluctuations occur to ensure smooth operation of the microgrid. This interplay between SCs and MLI effectively manages the inherent variability of solar PV power, contributing to the overall stability and efficiency of the HESS-integrated microgrid setup.

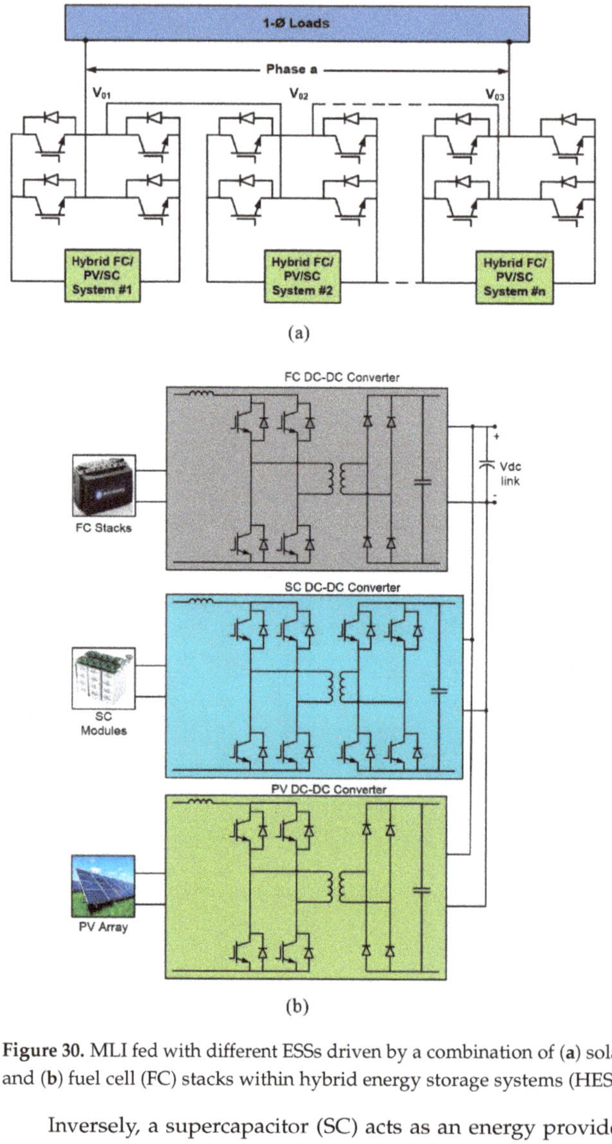

Figure 30. MLI fed with different ESSs driven by a combination of (**a**) solar photovoltaic (PV) systems and (**b**) fuel cell (FC) stacks within hybrid energy storage systems (HESSs).

Inversely, a supercapacitor (SC) acts as an energy provider for an asymmetrically constructed e-shaped clamped x-type multilevel inverter (AMECXMI)-based dynamic voltage restorer (DVR). The AMECXMI is engineered to produce a voltage waveform that closely resembles a sine wave but with a staircase configuration, requiring a minimal count of switching components. This unique design enables it to generate adjustable staircase sinusoidal voltages, which are vital for addressing a variety of voltage disturbances within the DVR framework. A graphical illustration of the DVR equipped with an SC-integrated

AMECXMI is presented in (Figure 31). This setup demonstrates substantial promise for enhancing the fault ride-through (FRT) and power quality (PQ) capabilities of a doubly-fed induction generator (DFIG)-driven wind turbine [279]. The fractional-order super-twisting sliding mode control (FOSTSMC) is employed to regulate the energy supply from the SC, aiming to produce the specified staircase sinusoidal output by the AMECXMI. To evaluate its performance under fault conditions, the DFIG rotor is fitted with a series resistive limiter. Moreover, the DC link is combined with a superconducting DC chopper. This setup addresses various PQ concerns, including voltage fluctuations, voltage flicker, voltage swell, and voltage sag. The effectiveness of the FOSTSMC is then systematically compared with alternative control methods such as fuzzy control, standard sliding mode control (SMC), and proportional–integral–derivative (PID) control to determine its superiority in managing these challenges. The comparison underscores the proficiency of the FOSTSMC in optimizing the FRT and PQ functions within the DVR system.

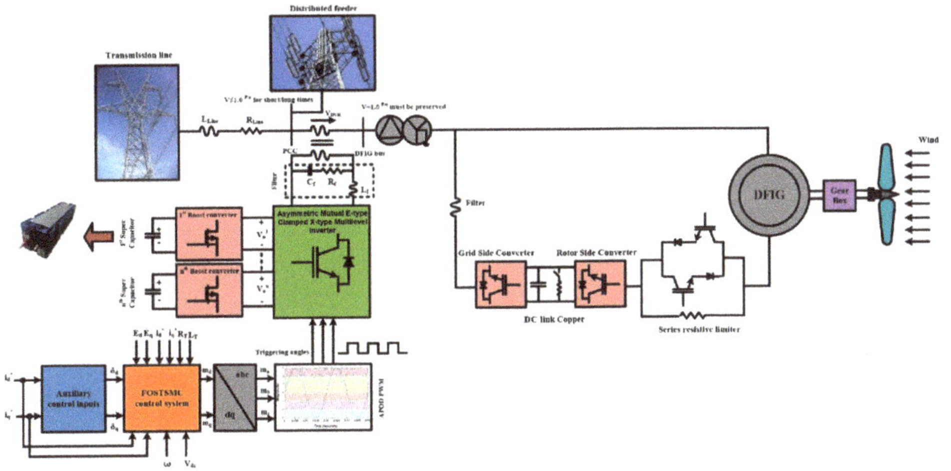

Figure 31. Schematic of DVR based on the SC storage equipped AMECXMI, with permission from ref. [279].

8.7. Wave Energy Converters

One research endeavor focused on incorporating a supercapacitor/battery hybrid setup within the framework of vented oscillating water column (VOWC) wave energy converters [280]. The study devised a conceptual model of a VOWC array situated in a nearshore detached breakwater environment (VOWCDBW), as illustrated in (Figure 32). The investigation encompassed four primary scenarios: isolated VOWCs, a VOWCDBW configuration with two devices spaced at a threefold separation, a trio of individual VOWCs in an array, and three separate VOWCDBW arrays. The objective was to explore the implications of these different arrangements on the energy storage sizing and power quality. The power management system (PMS) operates by monitoring the state of charge (SoC) of both the supercapacitors (SCs) and batteries. In instances where the SoC of the SCs falls below an upper limit, excessive power is directed to charge them during power excesses. In contrast, during periods of power deficiency, the PMS draws from the SCs until they hit a lower threshold, thus averting detrimental overcharging and overdischarging and thereby enhancing the lifespan of the entire system. The study's findings suggest that an array configuration of VOWCs is notably efficient in mitigating power fluctuations and diminishing the necessity for energy storage system (ESS) capacity when compared to other arrangements. This leads to an improved power quality delivery to the grid.

Another investigation underscored the utilization of supercapacitors as an energy storage solution for wave energy converters (WECs) [281]. The Hamiltonian surface shaping and power flow control (HSSPFC) methodology was implemented to efficiently regulate and optimize the electrical energy transmission from a WEC to the shore-based power grid. This study was predicated upon preliminary wave tank experiments of the mechanical system, which served to establish simulation models to assess the performance and operational conditions. Various SC configurations were examined with respect to the performance metrics, mass, volume, and practical storage capacity. The preliminary analysis favored passive SC energy storage systems due to their inherent simplicity and reduced weight. Through comprehensive simulations, the Skeleton 53 F configuration was identified as the most suitable option, offering superior system efficiency.

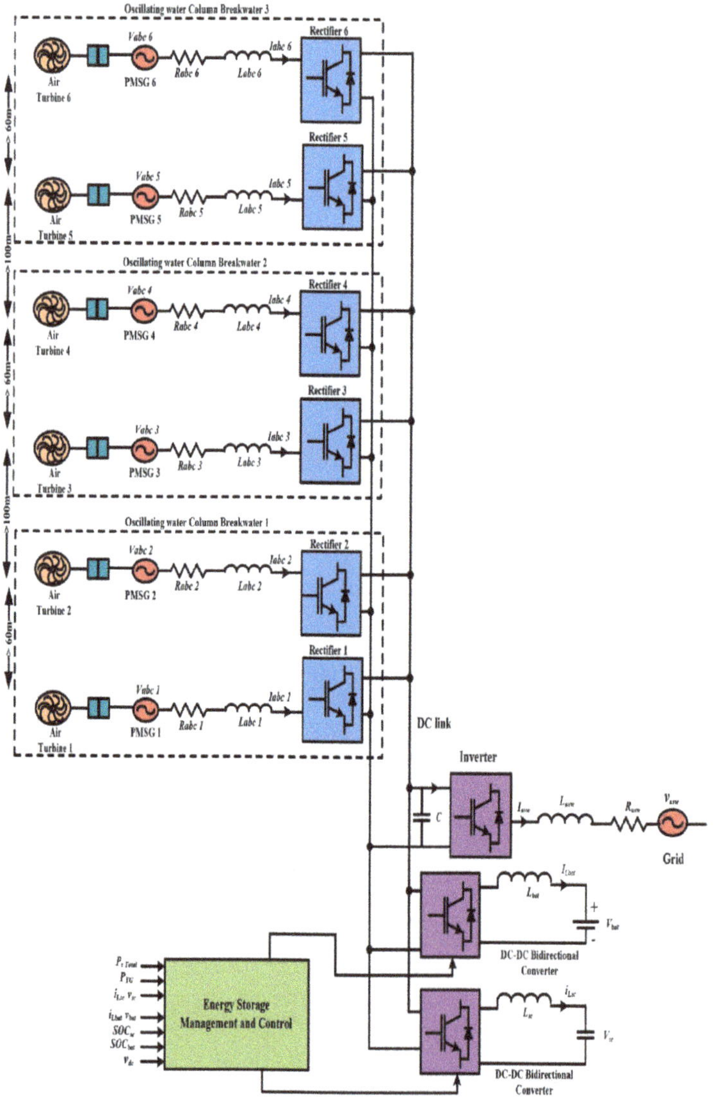

Figure 32. Schematic diagram of hybrid array of VOWCs in the VOWCDBW.

8.8. Hand-Held Applications

The supercapacitor (SC) configuration presented is tailored to function as a power supply for portable devices, eliminating the need for conventional external batteries [282]. This design surpasses standard buck converters due to its flexibility in managing diverse input voltages, load currents, operational amplifier supply voltages, and the correlation between the comparator's supply voltage and the switching pulse amplitude. The system employs a pair of SCs to provide a declining input voltage to the buck converter, which is adept at handling load currents spanning from 0.3 to 1.2 amperes. The output voltage is meticulously maintained at a constant level through an advanced feed-forward closed-loop PWM control system that dynamically adjusts the duty cycle of the switching pulses in accordance with the fluctuating input voltages of the supercapacitors. This unique approach enables the system to deliver a steady voltage for over 4.73 h. Moreover, the operational switching frequency is designed within an eco-friendly and secure range of 1 to 2.5 kilohertz. In the context of similar research, advancements in supercapacitor module technology have been proposed, including a DC–DC converter topology that operates independently of batteries and integrates both feedback and feedforward control systems. These studies underscore the importance of the leakage resistance with respect to the charging current and introduce an innovative charging method based on the gamma function to enhance the efficiency and decrease the charging duration [283].

8.9. Wireless Charging

The SC-based EV wireless charging utilizes resonant inductive coupling, as detailed in a previous study [284]. The research employed MATLAB (MATLAB R2025a) and ANSYS MAXWELL (Maxwell 2025 R1) softwares to perform simulations, with ANSYS MAXWELL focusing on parameter estimation for both the transmitter and receiver coils. MATLAB was then used to calculate the system's efficiency and output power. This wireless charger boasts a power efficiency exceeding 75% and a high power transfer density and operates with a reduced coil size. It can maintain this performance across a larger distance range of 4–24 cm and has a maximum operating frequency of 100 kHz, delivering an output power of 100 W. In a separate study, an innovative WPT-integrated hybrid ESS (WPT-HESS) was presented [285]. This system combines a battery and SC through a DC–DC converter to interface with the WPT system. The WPT employs an LCC-S compensation network to ensure stable direct current voltage. The power flow between the WPT and HESS is managed by an integral terminal sliding mode controller (ITSMC), which optimizes the system's efficiency while enabling the SC to charge to full capacity. The system's stability was scrutinized under the Lyapunov theory framework, and its performance was benchmarked against a standard sliding mode controller (SMC) and proportional–integral–derivative (PID) controller to demonstrate robustness. Furthermore, there was a study on seamlessly integrated wireless charging micro-supercapacitors (IWC-MSCs) that showcased the high performance metrics [286]. These IWC-MSCs are characterized by an impressive energy density (ED) of 463.1 μWhcm^{-2}, along with an ultrahigh areal capacitance of 454.1 mFcm^{-2}. To evaluate the practical application, these components were integrated into a toy car model. The wireless charging system exhibited a power transfer efficiency of approximately 52.8%.

8.10. Comparative Study of EES Systems

The comparative study of different types of EES systems are depicted in Table 6 [287–290] and Table 7 [290–293]. The efficiency, discharge time, cost, and environmental impacts of EES systems are considered for this study.

Table 6. Comparison of EES systems in terms of efficiency, discharge time, and cost [290].

System	Max. Power Rating (MW)	Efficiency (%)	Discharge Time	Cost/KW (USD)	Cost/KWh (USD)	Energy Density (Wh/L)
PHS	3000	70–85	4 h–16 h	600–2000	5–100	0.2–2
CAES	1000	40–70	2 h–30 h	400–800	2–50	2–6
FES	20	70–95	sec–mins	250–350	1000–5000	20–80
Lead–acid	100	80–90	1 min–8 h	300–600	200–400	50–80
NiCd/NiMH	40		sec–hours	500–1500	800–1500	60–150
Li-ion	100	85–95	1 min–8 h	1200–4000	600–2500	200–400
Metal–air	0.01	50	secs–day	100–250	10–60	500–10,000
Sodium–sulfur	0.05–8	75–90	sec–hours	1000–3000	300–500	150–250
RFB/HFB	100	60–85	hours	700–2500	150–1000	20–70
H2	100	25–45	min–week		10	600
Fuel Cell	50	60–80	secs–day	10,000		500–3000
SMES	10 MW	95	millisec–secs	200–300	1000–10,000	0.2–2.5
Thermal	150	80–90	hours	200–300	30–60	70–210

Table 7. Comparison of EES systems in terms of environmental impact [290].

System	Lifetime/Cycles	Environmental Impact	Description of Impact
PHS	30–60 years	-ve	Cutting trees and landscapes for reservoirs
CAES	20–40 years	-ve	Remains from fossil fuel
FES	20,000–100,000	Negligible	
Lead–acid	6–40 years	-ve	Toxic residues
NiCd/NiMH	10–20 years	-ve	Toxic residues
Li-ion	1000–10,000	-ve	Toxic residues
Metal–air	100–300	Very small	Slight residues
Sodium–sulfur	10–15 years	-ve	Toxic residues
RFB/HFB	12,000–14,000	-ve	Toxic residues
H_2	5–30 years	Yes	Emission of hydrogen in atmosphere can create disturbance in the distribution of methane and ozone, thereby causing imbalance
Fuel Cell	5–15 years	-ve	Remains from fossil fuel
SMES	20 years	-ve	High magnetic field
Thermal	30 years	Small	Releasing charge into atmosphere

The Ragone diagram is a graphical representation employed to evaluate the performance of diverse energy storage devices (ESDs) in terms of their power and energy capabilities [294]. It plots energy density on one axis, measured in watt-hours per kilogram (Wh/kg), and power density on the other axis, measured in watts per kilogram (W/kg). This approach, which often utilizes a logarithmic scale for both axes, enables a comprehensive comparison of various ESDs that may exhibit significantly different characteristics. For instance, batteries, capacitors, supercapacitors, and flywheels each occupy distinct regions within the Ragone plot, reflecting their unique energy storage and power delivery properties. The decision on which specific ESD to select is largely influenced by the power density demands of the intended application, as illustrated in the Figure 33, which showcases the relative performances of battery, supercapacitor, and traditional capacitor technologies [295].

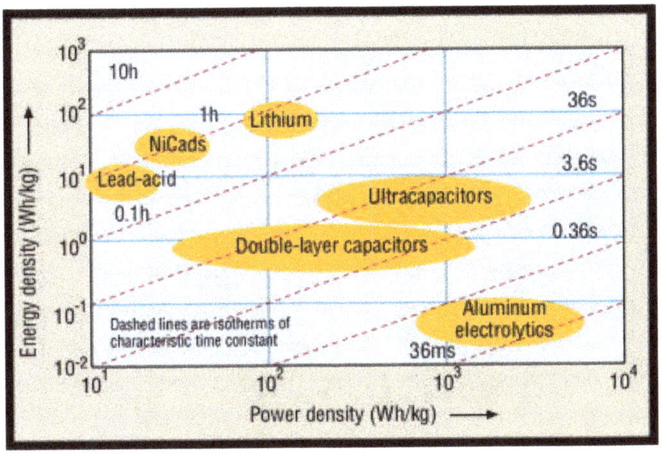

Figure 33. The Ragone plot showing energy density vs. power density, with permission from ref. [296].

The inclined curves (representing characteristic time constants) on a Ragone plot illustrate the differential rates at which devices charge and discharge. On one end of the spectrum, capacitors are capable of transferring power in microseconds, showcasing exceptional dynamic behavior. Conversely, devices exhibiting less impressive performance may need several hours to produce and supply energy. Supercapacitors offer a balanced middle ground between these extremes, providing a practical compromise in terms of power and energy delivery. Portable technology encompasses a diverse set of energy storage requirements. Some devices demand substantial energy reserves to support sustained high-power operations over extended periods, necessitating high-energy storage solutions. In contrast, other applications may only require brief bursts of power before replenishment, focusing on high power density. The Ragone plot effectively visualizes these distinctions, positioning batteries towards the high-energy density and low-power density segment. This makes them ideal for scenarios where prolonged operation without recharge is essential. Batteries, while suitable for long-duration power supply, are not without their limitations. These include a limited lifespan and the need for costly maintenance. Conversely, capacitors boast high power densities and extended lifetimes, with the additional benefit of a substantial number of charge cycles. Despite having lower energy densities, capacitors are advantageous in scenarios requiring quick charge and discharge cycles [297].

9. Future Trends

The illustration labeled as Figure 34 underscores the emerging patterns and hurdles pertinent to solid-state batteries (SSBs). This encompasses the exploration and development of innovative materials for electrodes and electrolytes and efforts to enhance energy density, addressing the issue of cell voltage disparities, the refinement of modeling techniques specific to SSBs, and the establishment of comprehensive industrial standards to govern their production and use. These areas represent the cutting-edge advancements and challenges within the SSB domain.

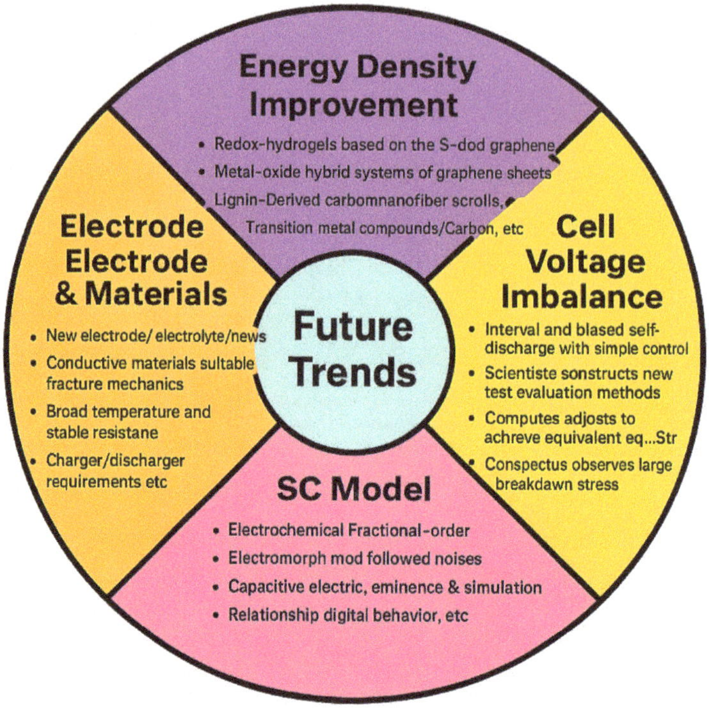

Figure 34. Schematic diagram representing the future trends of supercapacitors.

In recent years, the field of energy storage has undergone rapid evolution, driven by emerging demands in flexible electronics, electric vehicles, and large-scale grid integration. Hybrid supercapacitor systems have attracted growing attention due to their ability to bridge the performance gap between traditional electrochemical capacitors and lithium-ion batteries, offering both high power density and enhanced energy density. At the same time, the global emphasis on sustainability has accelerated research into recyclable electrode materials, green synthesis methods, and environmentally benign electrolytes. These trends underscore the urgent need for a comprehensive and timely review that consolidates recent breakthroughs in material innovation, architectural design, and process scalability. By addressing these developments, this review not only captures the current state of the art but also serves as a strategic guide for future advancements in next-generation energy storage technologies.

9.1. Electrode and Electrolyte Materials

To improve the functionality of supercapacitors (SCs), the advancement of innovative SC materials is crucial. This necessitates an integration of various experimental and computational studies to guide the selection of appropriate electrode and electrolyte components [298–302]. It is imperative to develop fabrication techniques that can produce highly porous and hollow-structured materials, as these architectures enable a greater amount of charge storage and thus enhance energy density (ED). For optimal performance, electrolytes must possess a stable and extensive potential window, as well as excellent ionic conductivity. A smaller equivalent series resistance (ESR) is also essential to achieve both high ED and power density (PD). Additionally, there should be a harmonious relationship between the morphology of the electrode material, the behavior of the electrolyte ions, and the charge storage mechanisms involved. Hybrid supercapacitors (HSCs) have the potential to provide increased PD and ED simultaneously. However, the development of HSCs faces challenges concerning cost-effective and scalable manufacturing processes and the establishment of precise procedures to manage the surface properties of constituent materials. Furthermore, in the context of liquid-based ionic capacitors (LICs), research must concentrate on overcoming the challenges related to novel material innovation, establishing comprehensive characterization benchmarks, and understanding the effects of cycling frequency on longevity. It is also vital to model LICs in the high-frequency domain to assess their suitability for high-frequency applications. Beyond the material aspects, several other factors require investigation to ensure the broader applicability of supercapacitors. These include the lifespan of the devices, the rate of self-discharge, the operational temperature range, and the degradation of separators and current collectors. Moreover, the design and implementation of efficient thermal energy management systems are critical considerations for LICs. Addressing these challenges is essential for advancing the technology and expanding the practical use of supercapacitors. In particular, research efforts should focus on minimizing the self-discharge rate to prolong the operational lifespan of these energy storage systems.

9.2. Energy Density Improvement

Supercapacitors (SCs) are recognized for their remarkable power density (PD) and lower energy density (ED) in comparison to batteries. To overcome the gap between the EDs of SCs and batteries, numerous strategies have been proposed. One promising approach involves the utilization of redox hydrogels constructed on the foundation of sulfur-doped graphene, which has demonstrated a notable enhancement in ED, reaching 21.3 Wh/kg [303]. Additionally, researchers have suggested mass-balancing techniques within graphene composites to further boost the ED [304]. The performance of supercapacitors has also been substantially improved by the integration of thiol-functionalized graphene oxide scrolls, resulting in an impressive ED of 206 Wh/kg and maintaining this performance over an extensive lifespan of 20,000 cycles [305]. Another avenue for advancement lies in the realm of volumetric ED enhancement, which has been achieved through the application of lignin-derived carbon nanofiber in SCs [306]. This material has displayed significant improvements in both volumetric ED and cyclic stability. To gain a deeper understanding of these improvements, a comprehensive review study was conducted, focusing on the enhancement methods applied in hybrid solid-state supercapacitors [307]. The evolution of TMCs/carbon hybrid electrodes is also under scrutiny, with attention being given to structural design strategies such as the electronic structure, interface optimization, and the use of conductive carbon skeletons to enhance the overall performance [308]. These hybrid systems aim to reconcile the high power density and energy density characteristics. However, they face critical challenges concerning the efficient utilization of carbon as a conductive support and the delicate balance required between achieving high mass loadings of nanoscale TMCs and preventing their aggregation. Addressing these issues is essential for the continued development and practical application of these advanced energy storage solutions.

9.3. Cell Voltage Imbalance

The production of a consistent voltage across supercapacitor (SC) cells is essential due to variations inherent in the manufacturing process. Typically, an individual SC cell yields approximately 2.7 volts, necessitating their serial connection to achieve the desired voltage levels for modules. However, this can lead to cell voltage imbalances, which are exacerbated by repeated charging and discharging cycles. To address this issue, advanced voltage balancing systems are essential for reliable and efficient operation. One approach to mitigate this problem involves the implementation of a multiwinding transformer-based equalization circuit for SC modules [309]. This design allows for both inter-cell and inter-module voltage balancing, offering a simple control mechanism, increased speed of equalization, and enhanced reliability. Alternatively, a series LC resonant converter serves as a voltage equalizer to circumvent the intricacies associated with multi-winding transformers [310]. This alternative provides the additional benefit of estimating the time required for voltage equalization. Another notable method involves the use of switched-capacitors as voltage equalizers within the SC framework [311]. This technique leverages the SCs themselves as components of the equalizer system, thereby reducing the size and costs. It also eliminates the need for excessive energy transfer components and offers high efficiency, along with the automatic minimization of the equivalent resistance during the equalization process. Furthermore, a consensus-based voltage equalizer circuit has been proposed for use in reconfigurable supercapacitors [312]. This innovative approach is designed to ensure voltage uniformity across the modules by utilizing a strategy that mimics the decision-making process in a group, thereby providing a more adaptable and robust solution to the voltage imbalance challenge.

9.4. SC Model

Pivotal concerns such as the parameter identification and model selection significantly influence the system design process, as highlighted in the literature [313–316]. Supercapacitors (SCs) are categorized into various models including thermal, self-discharge, fractional-order, intelligent, equivalent circuit, and electrochemical. These models' parameters are susceptible to an array of influences stemming from the load characteristics, system stability, environmental conditions, and fluctuating loads. The challenge of accurately estimating the parameters in real time emerges as a crucial aspect. For electric vehicles (EVs), precise model performance is vital for effectively estimating the state of health (SOH) and state of energy (SOE) [317]. In the context of military applications, particularly in satellites and spacecraft power systems, the presence of non-ideal parameters can lead to substantial risks [298]. Ensuring system reliability is paramount when considering both internal and external environmental factors. To address this, researchers have introduced a capacitance prognostic model that depends on power. This approach aims to understand and predict the second-life capacitance dynamics of SC cells once they are no longer used in their primary applications, thereby facilitating their potential reuse [318]. The robustness and sensitivity of the proposed model have been rigorously assessed, employing both quantitative and qualitative evaluation methods to affirm the efficacy.

9.5. Industrial Standard

The performance of SCs can be assessed through several parameters including the cycle life, energy efficiency, power density, energy density, capacitance, and capacity [319]. In contrast, evaluation techniques involve assessments like round-trip efficiency, current interruption, current initiation, power pulse, constant power, current pulse, and constant current. These assessments are influenced by experimental conditions such as the testing temperature, state of charge (SOC), power and voltage operation ranges, hold/rest periods, rebound time, and current density. To ensure uniformity and safety, establishing certain standards is crucial. These encompass a standardized model naming convention, detailed charger specifications, comprehensive safety protocols, and precise methods for gauging the electrical performance. Moreover, production guidelines and material specifications, particularly for electrolytes and electrodes, are essential [320–322]. For instance, IEC 62,576 and IEC 62,391-2 serve as benchmarks for SCs used in HEVs and various electronic devices [322]. These standards are complemented by UL 810A, a proprietary standard developed by Underwriters Laboratories for SCs. Additionally, the European Union's REACH regulation focuses on the production and utilization of chemical substances within the SC industry. There is a significant need for these universal standards to foster the SC industry's healthy growth and the adherence to safety and quality.

10. Conclusions

Supercapacitors have emerged as indispensable energy storage devices, uniquely bridging the gap between conventional capacitors and batteries. Their hallmark advantages—such as their high-power density, rapid charge–discharge cycles, excellent cycle life, and robust temperature tolerance—have positioned them at the forefront of next-generation energy solutions. This review has highlighted the multifaceted progress in supercapacitor technologies, spanning the development of advanced electrode materials—including carbon-based nanostructures, transition metal oxides, conductive polymers, and composite systems—as well as breakthroughs in electrolyte design, from aqueous and organic systems to ionic liquids. Innovations in device architecture, particularly hybrid configurations and flexible platforms, have further propelled the capabilities of supercapacitors across diverse applications, from wearable electronics to electric vehicles and renewable energy storage. Despite this impressive progress, several critical challenges remain. Chief among them is the need to elevate the energy density without compromising the power performance or longevity. The integration of high-capacitance pseudocapacitive and hybrid materials holds promise, yet issues such as poor long-term stability, complex fabrication methods, and limited scalability persist. Furthermore, ensuring compatibility with sustainable and environmentally benign electrolytes is crucial for real-world adoption. Future research should prioritize the development of low-cost, eco-friendly, and scalable synthesis methods for high-performance materials. A deeper understanding of ion transport mechanisms at the electrode–electrolyte interface, guided by advanced characterization and modeling techniques, will be essential. Additionally, engineering device configurations that support miniaturization, mechanical flexibility, and integration with wearable and IoT technologies will unlock new commercial frontiers.

Supplementary Materials: The following supporting information can be downloaded at https://www.mdpi.com/article/10.3390/batteries11060232/s1, Figure S1: (a) XRD patterns of ZnO nanocrystals (circles) and the corresponding simulated patterns (dotted line) obtained for different sizes of the particles, (b) Powder XRD spectra of a series of nanocrystal sizes. From the width of the reflections, the crystalline domain size is calculated. (c) Powder X-ray diffraction of a series of nanocrystal sizes. The stick spectrum gives the bulk reflections with relative intensities.

Author Contributions: Conceptualization, M.S. and A.M.A.-D.; methodology, M.S. and A.M.A.-D.; validation, M.S., V.Z. and A.Z.; formal analysis, M.S.; investigation, M.S. and A.Z.; resources, J.G., J.M.B., V.Z. and A.Z.; data curation, M.S., V.Z., T.N.A.E., M.F. and A.M.A.-D.; writing—original draft preparation, M.S., A.M.A.-D., T.N.A.E. and M.F.; writing—review and editing, M.S. and A.M.A.-D.; visualization, M.S., A.M.A.-D., J.M.B., J.G. and V.Z.; supervision, A.M.A.-D., V.Z. and A.Z.; project administration, J.G., V.Z. and A.Z.; funding acquisition, J.G., V.Z. and A.Z. All authors have read and agreed to the published version of the manuscript.

Funding: This research was funded by Spanish MICIN, EU (Horizon Europe), Government of the Basque Country and the research, development, and innovation authority (RADIA) with grant number PID2022-141373NB-I00, HORIZON-CL4-2023-RESILIENCE-01, IT1670-22.

Data Availability Statement: The original contributions presented in this study are included in the article. Further inquiries can be directed to the corresponding authors.

Acknowledgments: The authors are thankful for the technical and human support provided by SGIker of UPV/EHU (Medidas Magn' eticas Gipuzkoa) and European funding (ERDF and ESF). The authors wish to acknowledge the funding by the Spanish MICIN, under PID2022-141373NB-I00 project, by EU (Horizon Europe) under "INFINITE" (HORIZON-CL5-2021-D5-01-06) and "HARMONY" (HORIZON-CL4-2023-RESILIENCE-01) and by the Government of the Basque Country, under Elkartek (MOSINCO and ATLANTIS) projects and by under the scheme of "Ayuda a Grupos Consolidados" (Ref.: IT1670-22). In addition, MS wishes to acknowledge the funding within the Maria Zambrano contract by the Spanish Ministerio de Universidades and European Union—Next Generation EU ("Financiado por la Uni' on Europea-Next Generation EU").

Conflicts of Interest: The authors declare no conflicts of interest.

References

1. Ali, Z.M.; Martin, P.; Aleem, S.; Jurado, F.; Gandoman, F.H. Applications of Energy Storage Systems in Enhancing Energy Management and Access in Microgrids: A Review. *Energies* **2023**, *16*, 5930. [CrossRef]
2. Faisal, M.; Hannan, M.A.; Ker, P.J.; Hussain, A.; Mansor, M.B.; Blaabjerg, F. Review of Energy Storage System Technologies in Microgrid Applications: Issues and Challenges. *IEEE Access* **2018**, *6*, 35143–35164. [CrossRef]
3. Katsanevakis, M.; Stewart, R.A.; Lu, J. Aggregated applications and benefits of energy storage systems with application-specific control methods: A review. *Renew. Sustain. Energy Rev.* **2017**, *75*, 719–741. [CrossRef]
4. Hayat, A.; Ali, H.; Ajmal, Z.; Alshammari, A.; Alghamdi, M.M.; El-Zahhar, A.A.; Almuqati, N.; Sohail, M.; Abu-Dief, A.M.; Khan, S.; et al. Emerging breakthroughs in covalent triazine frameworks: From fundamentals towards photocatalytic water splitting and challenges. *Prog. Mater. Sci.* **2025**, *147*, 101352. [CrossRef]
5. Ali, H.; Iqbal, O.; Sadiq, M.; Hassan, J.U.; Al Alwan, B.; El Jery, A.; Abu-Dief, A.M.; El-Kasaby, R.A.; Hayat, A.; Yue, D.; et al. Graphdiyne-based metal-free catalysts: Innovations in synthesis, properties, functionalization, morphology and applications. *Renew. Sustain. Energy Rev.* **2025**, *215*, 115570. [CrossRef]
6. Kandari, R.; Neeraj, N.; Micallef, A. Review on Recent Strategies for Integrating Energy Storage Systems in Microgrids. *Energies* **2022**, *16*, 317. [CrossRef]
7. Zhou, X.; Guo, T.; Ma, Y. An overview on microgrid technology. In Proceedings of the 2015 IEEE International Conference on Mechatronics and Automation (ICMA), Beijing, China, 2–5 August 2015; pp. 76–81.
8. Aghmadi, A.; Mohammed, O.A. Energy Storage Systems: Technologies and High-Power Applications. *Batteries* **2024**, *10*, 141. [CrossRef]
9. Rekioua, D. Energy Storage Systems for Photovoltaic and Wind Systems: A Review. *Energies* **2023**, *16*, 3893. [CrossRef]
10. Czagany, M.; Hompoth, S.; Keshri, A.K.; Pandit, N.; Galambos, I.; Gacsi, Z.; Baumli, P. Supercapacitors: An Efficient Way for Energy Storage Application. *Materials* **2024**, *17*, 702. [CrossRef]
11. Wustoni, S.; Ohayon, D.; Hermawan, A.; Nuruddin, A.; Inal, S.; Indartono, Y.S.; Yuliarto, B. Material Design and Characterization of Conducting Polymer-Based Supercapacitors. *Polym. Rev.* **2023**, *64*, 192–250. [CrossRef]
12. Shin, S.-J.; Gittins, J.W.; Balhatchet, C.J.; Walsh, A.; Forse, A.C. MetalOrganic Framework Supercapacitors: Challenges and Opportunities. *Adv. Funct. Mater.* **2024**, *34*, 2308497. [CrossRef]
13. Saidon, M.S.; Romli, M.I.F.; Othman, S.M.; Mustafa, W.A.; Manan, M.R.; Aihsan, M.Z. A Review on BLDC Motor Application in Electric Vehicle (EV) using Battery, Supercapacitor and Hybrid Energy Storage System: Efficiency and Future Prospects. *J. Adv. Res. Appl. Sci. Eng. Technol.* **2023**, *30*, 41–59.
14. Katkar, P.K.; Sheikh, Z.A.; Chavan, V.D.; Lee, S.W. In-situ growth of 3D amorphous Ni-Co-Mn phosphate on 2D $Ti_3C_2T_x$ nanocomposite for commercial-level hybrid energy storage application. *J. Mater. Sci. Technol.* **2025**, *206*, 282–296. [CrossRef]
15. Dong, Z.; Zhang, Z.; Li, Z.; Li, X.; Qin, J.; Liang, C.; Han, M.; Yin, Y.; Bai, J.; Wang, C.; et al. A Survey of Battery-Supercapacitor Hybrid Energy Storage Systems: Concept, Topology, Control and Application. *Symmetry* **2022**, *14*, 1085. [CrossRef]
16. Parsa, S.M.; Norozpour, F.; Shoeibi, S.; Shahsavar, A.; Aberoumand, S.; Afrand, M.; Said, Z.; Karimi, N. Lithium-ion battery thermal management via advanced cooling parameters: State-of-the-art review on application of machine learning with exergy, economic and environmental analysis. *J. Taiwan Inst. Chem. Eng.* **2023**, *148*, 104854. [CrossRef]
17. Xu, J.; Cai, X.; Cai, S.; Shao, Y.; Hu, C.; Lu, S.; Ding, S. High-Energy Lithium-Ion Batteries: Recent Progress and a Promising Future in Applications. *Energy Environ. Mater.* **2023**, *6*, e12450. [CrossRef]
18. Rouholamini, M.; Wang, C.; Nehrir, H.; Hu, X.; Hu, Z.; Aki, H.; Zhao, B.; Miao, Z.; Strunz, K. A Review of Modeling, Management, and Applications of Grid-Connected Li-Ion Battery Storage Systems. *IEEE Trans. Smart Grid* **2022**, *13*, 4505–4524. [CrossRef]
19. Bindra, A. Electric Vehicle Batteries Eye Solid-State Technology: Prototypes Promise Lower Cost, Faster Charging, and Greater Safety. *IEEE Power Electron. Mag.* **2020**, *7*, 16–19. [CrossRef]

20. Mekonnen, Y.; Sundararajan, A.; Sarwat, A.I. A review of cathode and anode materials for lithium-ion batteries. In Proceedings of the SoutheastCon 2016, Norfolk, VA, USA, 30 March–3 April 2016; IEEE: New York, NY, USA, 2016; pp. 1–6.
21. Belekar, K.G.; Patil, S.S.; Bhosale, S.B.; Kumbhar, S.S.; Jadhav, G.D.; Parale, V.G.; Lokhande, C.D.; Park, H.H.; Katkar, P.K.; Patil, U.M. Amorphous, binder-free cobalt manganese phosphate cathodes prepared by SILAR method for asymmetric supercapacitors: Harnessing cationic synergy. *Synth. Met.* **2025**, *311*, 117800. [CrossRef]
22. Xiaopeng, C.; Weixiang, S.; Tu, V.T.; Zhenwei, C.; Ajay, K. An overview of lithium-ion batteries for electric vehicles. In Proceedings of the 2012 10th International Power & Energy Conference (IPEC), Ho Chi Minh City, Vietnam, 12–14 December 2012; IEEE: New York, NY, USA, 2012; pp. 230–235.
23. Diouf, B.; Pode, R. Potential of lithium-ion batteries in renewable energy. *Renew. Energy* **2015**, *76*, 375–380. [CrossRef]
24. Salaheldeen, M.; Abu-Dief, A.M.; El-Dabea, T. Functionalization of Nanomaterials for Energy Storage and Hydrogen Production Applications. *Materials* **2025**, *18*, 768. [CrossRef] [PubMed]
25. Pranav, K.K.; Sang-Wha, L. An Interfacial Engineering Approach of Flower-like Li+ Preintercalated Co–Cu Phosphate for Solid-State Hybrid Energy Storage Device. *ACS Sustain. Chem. Eng.* **2024**, *12*, 5927–5942.
26. Maletić, F.; Deur, J.; Erceg, I. A Multitimescale Kalman Filter-Based Estimator of Li-Ion Battery Parameters Including Adaptive Coupling of State-of-Charge and Capacity Estimation. *IEEE Trans. Control. Syst. Technol.* **2022**, *31*, 692–706. [CrossRef]
27. Pranav, K.K.; Abhijit, N.K.; Sahng-Kyoon, J.; Seung-Hyun, C.; Sang-Wha, L. Rational design of redox active amorphous Ni-Mn phosphate anchored on vertical graphene nanohills (VGNHs) for solid-state energy storage device. *J. Alloys Compd.* **2023**, *968*, 171935.
28. Görtz, J.; Aouad, M.; Wieprecht, S.; Terheiden, K. Assessment of pumped hydropower energy storage potential along rivers and shorelines. *Renew. Sustain. Energy Rev.* **2022**, *165*, 112027. [CrossRef]
29. Kocaman, A.S.; Modi, V. Value of pumped hydro storage in a hybrid energy generation and allocation system. *Appl. Energy* **2017**, *205*, 1202–1215. [CrossRef]
30. Ichimura, S. Utilization of cross-regional interconnector and pumped hydro energy storage for further introduction of solar PV in Japan. *Glob. Energy Interconnect.* **2020**, *3*, 68–75. [CrossRef]
31. Olabi, A.; Wilberforce, T.; Ramadan, M.; Abdelkareem, M.A.; Alami, A.H. Compressed air energy storage systems: Components and operating parameters—A review. *J. Energy Storage* **2020**, *34*, 102000. [CrossRef]
32. Bazdar, E.; Sameti, M.; Nasiri, F.; Haghighat, F. Compressed air energy storage in integrated energy systems: A review. *Renew. Sustain. Energy Rev.* **2022**, *167*, 112701. [CrossRef]
33. Nozari, M.H.; Yaghoubi, M.; Jafarpur, K.; Mansoori, G.A. Development of dynamic energy storage hub concept: A comprehensive literature review of multi storage systems. *J. Energy Storage* **2022**, *48*, 103972. [CrossRef]
34. Naeem, S.; Patil, A.V.; Shaikh, A.V.; Shinde, U.P.; Husain, D.; Alam, T.; Sharma, M.; Tewari, K.; Ahmad, S.; Shah, A.A.; et al. A Review of Cobalt-Based Metal Hydroxide Electrode for Applications in Supercapacitors. *Adv. Mater. Sci. Eng.* **2023**, *2023*, 1133559. [CrossRef]
35. Kim, K.; Park, J.; Lee, J.; Suh, S.; Kim, W. Ultrafast PEDOT: PSS/H_2SO_4 Electrical Double Layer Capacitors: Comparison with PANI Pseudocapacitors. *ChemSusChem* **2023**, *16*, e202202057. [CrossRef] [PubMed]
36. Adedoja, O.; Sadiku, E.; Hamam, Y. An Overview of the Emerging Technologies and Composite Materials for Supercapacitors in Energy Storage Applications. *Polymers* **2023**, *15*, 2272. [CrossRef] [PubMed]
37. Hamidah, I.; Ramdhani, R.; Wiyono, A.; Mulyanti, B.; Pawinanto, R.; Hasanah, L.; Diantoro, M.; Yuliarto, B.; Yunas, J.; Rusydi, A. Biomass-Based Supercapacitors Electrodes for Electrical Energy Storage Systems Activated Using Chemical Activation Method: A Literature Review and Bibliometric Analysis. *Indones. J. Sci. Technol.* **2023**, *8*, 439–468. [CrossRef]
38. Goodenough, J.B.; Park, K.S. The Li-ion rechargeable battery: A perspective. *J. Am. Chem. Soc.* **2013**, *135*, 1167–1176. [CrossRef]
39. Zhou, G.; Li, F.; Cheng, H.-M. Progress in flexible lithium batteries and future prospects. *Energy Environ. Sci.* **2014**, *7*, 1307–1338. [CrossRef]
40. Reddy, M.V.; Rao, G.V.S.; Chowdari, B.V.R. Metal oxides and oxysalts as anode materials for Li-ion batteries. *Chem. Rev.* **2013**, *113*, 5364–5457.
41. González, A.; Goikolea, E.; Barrena, J.A.; Mysyk, R. Review on supercapacitors: Technologies and materials. *Renew. Sustain. Energy Rev.* **2016**, *58*, 1189–1206. [CrossRef]
42. Ali, H.; Iqbal, O.; Al Alwan, B.; Abdulrahman, A.; Orooji, Y.; Al-Farraj, E.S.; Sadiq, M.; Imran, S.; Abu-Dief, A.M.; Yue, D.; et al. Morphology-controlled metal-organic frameworks for enhanced photocatalytic performance. *Coord. Chem. Rev.* **2025**, *541*, 216822. [CrossRef]
43. Sun, H.; Li, J.; Liang, W.; Gong, X.; Jing, A.; Yang, W.; Liu, H.; Ren, S. Porous Organic Polymers as Active Electrode Materials for Energy Storage Applications. *Small Methods* **2024**, *8*, e2301335. [CrossRef]
44. Swain, N.; Saravanakumar, B.; Kundu, M.; Schmidt-Mende, L.; Ramadoss, A. Recent trends in template assisted 3D porous materials for electrochemical supercapacitors. *J. Mater. Chem. A* **2021**, *9*, 25286–25324. [CrossRef]

45. Khan, M.; Shakir, S.; Mohammad, B.; Azooz, S.; Ali, R.; Syed, K.G.; Waleed, A.; Hassan, K. Recent advancements in supercapacitors and their charge storage mechanism and progress in transition metal sulfide-based electrodes. *Phys. Scr.* **2024**, *99*, 062001. [CrossRef]
46. Mensah-Darkwa, K.; Nframah Ampong, D.; Agyekum, E.; de Souza, F.M.; Gupta, R.K. Recent Advancements in Chalcogenides for Electrochemical Energy Storage Applications. *Energies* **2022**, *15*, 4052. [CrossRef]
47. Joseph, A.; Thomas, T. Pseudocapacitance: An Introduction. In *Pseudocapacitors: Fundamentals to High Performance Energy Storage Devices*; Gupta, R.K., Ed.; Springer: Cham, Germany, 2024.
48. Kavishka, D.; Dulsha, K.-A. A review of supercapacitors: Materials, technology, challenges, and renewable energy applications. *J. Energy Storage* **2024**, *96*, 112563.
49. Shakil, R.; Shaikh, M.N.; Shah, S.S.; Reaz, A.H.; Roy, C.K.; Chowdhury, A.N.; Aziz, M.A. Development of a Novel Bio-based Redox Electrolyte using Pivalic Acid and Ascorbic Acid for the Activated Carbon-based Supercapacitor Fabrication. Asian. *J. Org. Chem.* **2021**, *10*, 2220–2230. [CrossRef]
50. Salaheldeen, M.; Zhukova, V.; Blanco, J.M.; Gonzalez, J.; Zhukov, A. The Impact of High-Temperature Annealing on Magnetic Properties, Structure and Martensitic Transformation of Ni_2MnGa-based Glass-Coated Microwires. *Ceram. Int.* **2025**, *51*, 4378–4387. [CrossRef]
51. Salaheldeen, M.; Vega, V.; Caballero-Flores, R.; Prida, V.M.; Fernández, A. Influence of nanoholes array geometrical parameters on magnetic properties of Dy–Fe antidot thin films. *Nanotechnology* **2019**, *30*, 455703. [CrossRef]
52. Tiwari, A.P.; Mukhiya, T.; Muthurasu, A.; Chhetri, K.; Lee, M.; Dahal, B.; Lohani, P.C.; Kim, H.-Y. A review of electrospun carbon nanofiber-based negative electrode materials for supercapacitors. *Electrochem* **2021**, *2*, 236–250. [CrossRef]
53. Sheng, L.; Wei, T.; Liang, Y.; Jiang, L.; Qu, L.; Fan, Z. Vertically Oriented Graphene Nanoribbon Fibers for High-Volumetric Energy Density All-Solid-State Asymmetric Supercapacitors. *Small* **2017**, *13*, 1700371. [CrossRef]
54. Jebakumar, T.N.I.E.; Atchudan, R.; Lee, Y.R. Facile synthesis of carbon encapsulated RuO2 nanorods for supercapacitor and electrocatalytic hydrogen evolution reaction. *Int. J. Hydrog. Energy* **2019**, *44*, 2323–2329. [CrossRef]
55. Rudge, A.; Raistrick, I.; Gottesfeld, S.; Ferraris, J.P. A study of the electrochemical properties of conducting polymers for application in electrochemical capacitors. *Electrochim. Acta* **1994**, *39*, 273–287. [CrossRef]
56. Jurewicz, K.; Guterl, C.V.; Frackowiak, E.; Saadallah, S.; Reda, M.; Parmentier, J.; Patarin, J.; Beguin, F. Capacitance properties of ordered porous carbon materials prepared by a templating procedure. *J. Phys. Chem. Solids* **2004**, *65*, 287–293. [CrossRef]
57. Wu, M.; Snook, G.A.; Gupta, V.; Shaffer, M.; Fray, D.J.; Chen, G.Z. Electrochemical fabrication and capacitance of composite films of carbon nanotubes and polyaniline. *J. Mater. Chem.* **2005**, *15*, 2297–2303. [CrossRef]
58. Sivakkumar, S.R.; Kim, W.J.; Choi, J.A.; MacFarlane, D.R.; Forsyth, M.; Kim, D.W. Electrochemical performance of polyaniline nanofibres and polyaniline/multi-walled carbon nanotube composite as an electrode material for aqueous redox supercapacitors. *J. Power Sources* **2007**, *171*, 1062–1068. [CrossRef]
59. Li, H.; Wang, J.; Chu, Q.; Wang, Z.; Zhang, F.; Wang, S. Theoretical and experimental specific capacitance of polyaniline in sulfuric acid. *J. Power Sources* **2009**, *190*, 578–586. [CrossRef]
60. Yang, Q.; Hou, Z.; Huang, T. Self-assembled polypyrrole film by interfacial polymerization for supercapacitor applications. *J. Appl. Polym. Sci.* **2015**, *132*, 41615. [CrossRef]
61. Rajesh, M.; Raj, C.J.; Kim, B.C.; Cho, B.B.; Ko, J.M.; Yu, K.H. Supercapacitive studies on electropolymerized natural organic phosphate doped polypyrrole thin films. *Electrochim. Acta* **2016**, *220*, 373–383. [CrossRef]
62. Snook, G.A.; Kao, P.; Best, A.S. Conducting-polymer-based supercapacitor devices and electrodes. *J. Power Sources* **2011**, *196*, 1–12. [CrossRef]
63. Laforgue, A.; Simon, P.; Sarrazin, C.; Fauvarque, J.F. Polythiophene-based supercapacitors. *J. Power Sources* **1999**, *80*, 142–148. [CrossRef]
64. Patil, B.H.; Jagadale, A.D.; Lokhande, C.D. Synthesis of polythiophene thin films by simple successive ionic layer adsorption and reaction (SILAR) method for supercapacitor application. *Synth. Met.* **2012**, *162*, 1400–1405. [CrossRef]
65. Yogita, D.; Muruganandham, H.; Manoj, K.; Ankur, J.; Debasish, S. Modified transition metal chalcogenides for high performance supercapacitors: Current trends and emerging opportunities. *Coord. Chem. Rev.* **2022**, *451*, 214265.
66. Salaheldeen, M.; Vega, V.; Fernández, A.; Prida, V. Anomalous in-plane coercivity behaviour in hexagonal arrangements of ferromagnetic antidot thin films. *J. Magn. Magn. Mater.* **2019**, *491*, 165572. [CrossRef]
67. Zheng, C.; Zhou, X.; Cao, H.; Wang, G.; Liu, Z. Synthesis of porous graphene/activated carbon composite with high packing density and large specific surface area for supercapacitor electrode material. *J. Power Sources* **2014**, *258*, 290–296. [CrossRef]
68. Xia, H.; Wang, Y.; Lin, J.; Lu, L. Hydrothermal synthesis of MnO_2/CNT nanocomposite with a CNT core/porous MnO_2 sheath hierarchy architecture for supercapacitors. *Nanoscale Res. Lett.* **2011**, *6*, 595. [CrossRef]
69. Salaheldeen, M.; Wederni, A.; Ipatov, M.; Zhukova, V.; Lopez Anton, R.; Zhukov, A. Enhancing the Squareness and Bi-Phase Magnetic Switching of Co_2FeSi Microwires for Sensing Application. *Sensors* **2023**, *23*, 5109. [CrossRef]

70. Xiong, G.; He, P.; Liu, L.; Chen, T.; Fisher, T.S. Plasma-grown graphene petals templating Ni-Co-Mn hydroxide nanoneedles for high-rate and long-cycle-life pseudocapacitive electrodes. *J. Mater. Chem. A* **2015**, *3*, 22940–22948. [CrossRef]
71. Liangliang, T.; Chunyang, J. Conducting polymers as electrode materials for supercapacitors. *Prog. Chem.* **2010**, *22*, 1610–1618.
72. Yang, Y.; Xi, Y.; Li, J.; Wei, G.; Klyui, N.I.; Han, W. Flexible supercapacitors based on polyaniline arrays coated graphene aerogel electrodes. *Nanoscale Res. Lett.* **2017**, *12*, 394. [CrossRef]
73. Yan, J. Preparation of graphene nanosheet/carbon nanotube/polyaniline composite as electrode material for supercapacitors. *J. Power Sources* **2010**, *195*, 3041–3045. [CrossRef]
74. Salaheldeen, M.; Wederni, A.; Ipatov, M.; Zhukova, V.; Zhukov, A. Preparation and Magneto-Structural Investigation of High-Ordered ($L2_1$ Structure) Co_2MnGe Microwires. *Processes* **2023**, *11*, 1138. [CrossRef]
75. Liu, Y.; Zhang, B.; Yang, Y.; Chang, Z.; Wen, Z.; Wu, Y. Polypyrrole-coated α-MoO_3 nanobelts with good electrochemical performance as anode materials for aqueous supercapacitors. *J. Mater. Chem. A* **2013**, *1*, 13582–13587. [CrossRef]
76. Adhikari, S.; Selvaraj, S.; Ji, S.H.; Kim, D.H. Encapsulation of Co_3O_4 nanocone arrays via ultrathin NiO for superior performance asymmetric supercapacitors. *Small* **2020**, *16*, e2005414. [CrossRef] [PubMed]
77. Lin, H. Self-assembled graphene/polyaniline/Co_3O_4 ternary hybrid aerogels for supercapacitors. *Electrochim. Acta* **2016**, *191*, 444–451. [CrossRef]
78. Yang, X. Low-temperature synthesis of sea urchin-like Co-Ni oxide on graphene oxide for supercapacitor electrodes. *J. Mater. Sci. Technol.* **2020**, *55*, 223–230. [CrossRef]
79. Namsheer, K.; Rout, C.S. Conducting polymers: A comprehensive review on recent advances in synthesis, properties and applications. *RSC Adv.* **2021**, *11*, 5659–5697.
80. Azman, N.H.N.; Mamat Mat Nazir, M.S.; Ngee, L.H.; Sulaiman, Y. Graphene-based ternary composites for supercapacitors. *Int. J. Energy Res.* **2018**, *42*, 2104–2116. [CrossRef]
81. Hao, M. Coherent polyaniline/graphene oxides/multi-walled carbon nanotubes ternary composites for asymmetric supercapacitors. *Electrochim. Acta* **2016**, *191*, 165–172. [CrossRef]
82. Xu, L.; Jia, M.; Li, Y.; Zhang, S.; Jin, X. Design and synthesis of graphene/activated carbon/polypyrrole flexible supercapacitor electrodes. *RSC Adv.* **2017**, *7*, 31342–31351. [CrossRef]
83. Li, J.; Jiao, Y. Polyaniline-polypyrrole nanocomposites using a green and porous wood as support for supercapacitors. *Front. Agric. Sci. Eng.* **2019**, *6*, 137–143. [CrossRef]
84. Hussain, I.; Amna, R.; Kalidasan, B.; Rani, G.M.; Bandi, H.; Venkateswarlu, S.; Khan, S.A.; Mohapatra, D.; Zhang, K. MXenes for Various Applications: Recent Trends and Future Aspects. *SmartMat* **2025**, *6*, e70012. [CrossRef]
85. Borah, A.J.; Natu, V.; Biswas, A.; Srivastava, A. A review on recent progress in synthesis, properties, and applications of MXenes. *Oxf. Open Mater. Sci.* **2025**, *5*, itae017. [CrossRef]
86. Pathaare, Y.; Reddy, A.M.; Sangrulkar, P.; Kandasubramanian, B.; Satapathy, A. Carbon hybrid nano-architectures as an efficient electrode material for supercapacitor applications. *Hybrid. Adv.* **2023**, *3*, 100041. [CrossRef]
87. Lakshmi, K.C.S.; Vedhanarayanan, B. High-performance supercapacitors: A comprehensive review on paradigm shift of conventional energy storage devices. *Batteries* **2023**, *98*, 202. [CrossRef]
88. Arnold, A.M.; Singh, J.; Sydlik, S.A. The Role and Future of Functional Graphenic Materials in Biomedical and Human Health Applications. *Biomacromolecules* **2025**, *26*, 2015–2042. [CrossRef]
89. Al-dolaimy, F.; Saraswat, S.K.; Abed Hussein, B.; Abdul-Reda Hussein, U.; Saeed, S.M.; Kareem, A.T.; Abdulwahid, A.S.; Mizal, T.L.; Muzammil, K.; Alawadi, A.H.; et al. A review of recent advancement in covalent organic framework (COFs) synthesis and characterization with a focus on their applications in antibacterial activity. *Micron* **2024**, *179*, 103595. [CrossRef]
90. Argirusis, C.; Katsanou, M.-E.; Alizadeh, N.; Argirusis, N.; Sourkouni, G. Recent Advances in the Application of MOFs in Supercapacitors. *Batteries* **2025**, *11*, 181. [CrossRef]
91. Ramavath, J.N.; Raja, M.; Kumar, S.; Kothandaraman, R. Mild acidic mixed electrolyte for high-performance electrical double layer capacitor. *Appl. Surf. Sci.* **2019**, *489*, 867–874. [CrossRef]
92. Salaheldeen, M.; Zhukova, V.; Ipatov, M.; Zhukov, A. Unveiling the Magnetic and Structural Properties of (X_2YZ; X = Co and Ni, Y = Fe and Mn, and Z = Si) Full-Heusler Alloy Microwires with Fixed Geometrical Parameters. *Crystals* **2023**, *13*, 1550. [CrossRef]
93. Ibukun, O.; Jeong, H.K. Effects of aqueous electrolytes in supercapacitors. *New Phys. Sae Mulli* **2019**, *69*, 154–158. [CrossRef]
94. Bakhshandeh, M.B.; Kowsari, E. Functionalization of partially reduced graphene oxide by metal complex as electrode material in supercapacitor. *Res. Chem. Intermed.* **2020**, *46*, 2595–2612. [CrossRef]
95. Macchi, S.; Siraj, N.; Watanabe, F.; Viswanathan, T. Renewable-resource-based waste materials for supercapacitor application. *ChemistrySelect* **2019**, *4*, 492–501. [CrossRef]
96. Wen, Y.; Zhang, L.; Liu, J. Hierarchical porous carbon sheets derived on a MgO template for high-performance supercapacitor applications. *Nanotechnology* **2019**, *30*, 295703. [CrossRef] [PubMed]
97. Jaidev, J.R.I.; Ak, M. Polyaniline–MnO_2 nanotube hybrid nanocomposite as supercapacitor electrode material in acidic electrolyte. *J. Mater. Chem.* **2011**, *21*, 1760. [CrossRef]

98. Kang, J.; Jayaram, S.H.; Rawlins, J.; Wen, J. Characterization of thermal behaviors of electrochemical double layer capacitors (EDLCs) with aqueous and organic electrolytes. *Electrochim. Acta* **2014**, *144*, 200–210. [CrossRef]
99. Oyedotun, K.O.; Masikhwa, T.M.; Lindberg, S. Comparison of ionic liquid electrolyte to aqueous electrolytes on carbon nanofibres supercapacitor electrode derived from oxygen-functionalized graphene. *Chem. Eng. J.* **2019**, *375*, 121906. [CrossRef]
100. Balaji, S.S.; Karnan, M.; Anandhaganesh, P. Performance evaluation of B-doped graphene prepared via two different methods in symmetric supercapacitor using various electrolytes. *Appl. Surf. Sci.* **2019**, *491*, 560–569. [CrossRef]
101. Chen, J.; Chang, B.; Liu, F. Modification of porous carbon with nitrogen elements to enhance the capacitance of supercapacitors. *J. Mater. Sci.* **2019**, *54*, 11959–11971. [CrossRef]
102. Salaheldeen, M.; Nafady, A.; Abu-Dief, A.M.; Díaz Crespo, R.; Fernández-García, M.P.; Andrés, J.P.; López Antón, R.; Blanco, J.A.; Álvarez-Alonso, P. Enhancement of Exchange Bias and Perpendicular Magnetic Anisotropy in CoO/Co Multilayer Thin Films by Tuning the Alumina Template Nanohole Size. *Nanomaterials* **2022**, *12*, 2544. [CrossRef]
103. Lai, C.; Sun, Y.; Lin, B. Synthesis of sandwich-like porous nanostructure of Co_3O_4-rGO for flexible all-solid-state high-performance asymmetric supercapacitors. *Mater. Today Energy.* **2019**, *13*, 342–352. [CrossRef]
104. Mahieddine, A.; Adnane-Amara, L.; Gabouze, N.; Khenfer, K.; Bahdaouia, A.; Belkhettab, I. Synthesis, characterization, and effect of electrolyte concentration on the electrochemical performance of $Li_2Cu(WO_4)_2$ as a high-performance positive electrode for hybrid supercapacitors. *J. Energy Storage* **2022**, *56*, 106011. [CrossRef]
105. Dawoud, H.D.; Tahtamouni, A.; Bensalah, T. Sputtered manganese oxide thin film on carbon nanotubes sheet as a flexible and binder-free electrode for supercapacitors. *Int. J. Energy Res.* **2019**, *43*, 1245–1254. [CrossRef]
106. Sopčić, S.; Šešelj, N.; Kraljić Roković, M. Influence of supporting electrolyte on the pseudocapacitive properties of MnO_2/carbon nanotubes. *J. Solid State Electrochem.* **2019**, *23*, 205–214. [CrossRef]
107. Ghaly, H.A.; El-Deen, A.G.; Souaya, E.R.; Allam, N.K. Asymmetric supercapacitors based on 3D graphene-wrapped V_2O_5 nanospheres and Fe_3O_4@3D graphene electrodes with high power and energy densities. *Electrochim. Acta* **2019**, *310*, 58–69. [CrossRef]
108. Wang, P.; Wen, Y.; Lu, J. Manganese oxide(III)/carbon hybrids with interesting morphologies as improved active materials for supercapacitors. *Int. J. Hydrog. Energy* **2019**, *44*, 13623–13631. [CrossRef]
109. Peng, L.; Liang, Y.; Huang, J. Mixed-biomass wastes derived hierarchically porous carbons for high-performance electrochemical energy storage. *ACS Sustain. Chem. Eng.* **2019**, *7*, 10393–10402. [CrossRef]
110. Mary, A.J.C.; Bose, A.C. Incorporating $Mn^{2+}/Ni^{2+}/Cu^{2+}/Zn^{2+}$ in the Co_3O_4 nanorod: To investigate the effect of structural modification in the Co_3O_4 nanorod and its electrochemical performance. *ChemistrySelect* **2019**, *4*, 160–170. [CrossRef]
111. Durai, G.; Kuppusami, P.; Maiyalagan, T. Supercapacitive properties of manganese nitride thin film electrodes prepared by reactive magnetron sputtering: Effect of different electrolytes. *Ceram. Int.* **2019**, *45*, 17120–17127. [CrossRef]
112. Fenoy, G.E.; Schueren, B.; Scotto, J. Layer-by-layer assembly of iron oxide-decorated few-layer graphene/PANI: PSS composite films for high performance supercapacitors operating in neutral aqueous electrolytes. *Electrochim. Acta* **2018**, *283*, 1178–1187. [CrossRef]
113. Xu, C.; Wei, C.; Li, B. Charge storage mechanism of manganese dioxide for capacitor application: Effect of the mild electrolytes containing alkaline and alkaline-earth metal cations. *J. Power Sources* **2011**, *196*, 7854–7859. [CrossRef]
114. Nguyen, H.V.T.; Kwak, K.; Lee, K.-K. 1,1-Dimethylpyrrolidinium tetrafluoroborate as novel salt for high-voltage electric double-layer capacitors. *Electrochim. Acta* **2019**, *299*, 98–106. [CrossRef]
115. Yang, P.-Y.; Ju, S.-P.; Hsieh, H.-S. Electrolytic molecule in-pore structure and capacitance of supercapacitors with nanoporous carbon electrodes: A coarse-grained molecular dynamics study. *Comput. Mater. Sci.* **2019**, *166*, 293–302. [CrossRef]
116. Lw, L.F. Cell optimisation of supercapacitors using a quasi-reference electrode and potentiostatic analysis. *J. Power Sources* **2019**, *424*, 52–60.
117. Lu, X.; Zhang, Y.; Zhong, H. Molten-salt strategy for fabrication of hierarchical porous N-doped carbon nanosheets towards high-performance supercapacitors. *Mater. Chem. Phys.* **2019**, *230*, 178–186. [CrossRef]
118. Ou, J.; Tu, W.; Zeng, S.-Z.; Yao, Y.; Zhang, Q.; Wu, H.; Lan, T.; Liu, S.; Zeng, X. High-performance supercapacitors based on hierarchically porous carbons with a three-dimensional conductive network structure. *Dalton Trans.* **2019**, *48*, 5271–5284.
119. Ghosh, S.; Sahoo, G.; Polaki, S.R. Enhanced supercapacitance of activated vertical graphene nanosheets in hybrid electrolyte. *J. Appl. Phys.* **2017**, *122*, 214902. [CrossRef]
120. Sun, J.; Iakunkov, A.; Rebrikova, A.T.; Talyzin, A.V. Exactly matched pore size for the intercalation of electrolyte ions determined using the tunable swelling of graphite oxide in supercapacitor electrodes. *Nanoscale* **2018**, *10*, 21386–21395. [CrossRef]
121. Vijayakumar, M.; Rohita, D.S.; Rao, T.N.; Karthik, M. Electrode mass ratio impact on electrochemical capacitor performance. *Electrochim. Acta* **2019**, *298*, 347–359. [CrossRef]
122. Centeno, T.A.; Sereda, O.; Stoeckli, F. Capacitance in carbon pores of 0.7 to 15 nm: A regular pattern. *Phys. Chem. Chem. Phys.* **2011**, *13*, 12403–12406. [CrossRef]

123. Shim, Y.; Jung, Y.; Kim, H.J. Graphene-based supercapacitors: A computer simulation study. *J. Phys. Chem. C* **2011**, *115*, 23574–23583. [CrossRef]
124. Varanasi, S.R. Complementary effects of pore accessibility and decoordination on the capacitance of nanoporous carbon electrochemical supercapacitors. *J. Phys. Chem. C* **2015**, *119*, 28809–28818. [CrossRef]
125. Li, J.; Tang, J.; Yuan, J. Interactions between graphene and ionic liquid electrolyte in supercapacitors. *Electrochim. Acta* **2016**, *197*, 84–91. [CrossRef]
126. Ue, M. Chemical capacitors and quaternary ammonium salts. *Electrochemistry* **2007**, *75*, 565–572. [CrossRef]
127. Kim, D.; Lee, G.; Kim, D.; Yun, J.; Lee, S.S.; Ha, J.S. High performance flexible double-sided micro-supercapacitors with an organic gel electrolyte containing a redox-active additive. *Nanoscale* **2016**, *8*, 15611–15620. [CrossRef]
128. Ahmed, S.; Ahmed, A.; Rafat, M. Impact of aqueous and organic electrolytes on the supercapacitive performance of activated carbon derived from pea skin. *Surf. Coat. Technol.* **2018**, *349*, 242–250. [CrossRef]
129. Chaudhari, S.; Sharma, Y.; Archana, P.S.; Jose, R.; Ramakrishna, S.; Mhaisalkar, S.; Srinivasan, M. Electrospun polyaniline nanofibers web electrodes for supercapacitors. *J. Appl. Polym. Sci.* **2013**, *129*, 1660–1668. [CrossRef]
130. Wu, F.; Chen, R.; Wu, F. Binary room-temperature complex electrolytes based on $LiClO_4$ and organic compounds with acylamino group and its characterization for electric double layer capacitors. *J. Power Sources* **2008**, *184*, 402–407. [CrossRef]
131. Wee, G.; Soh, H.Z.; Cheah, Y.L. Synthesis and electrochemical properties of electrospun V_2O_5 nanofibers as supercapacitor electrodes. *J. Mater. Chem.* **2010**, *20*, 6720–6725. [CrossRef]
132. Chen, H.-Y.; Wee, G.; Al-Oweini, R. A polyoxovanadate as an advanced electrode material for supercapacitors. *ChemPhysChem* **2014**, *15*, 2162–2169. [CrossRef]
133. Azam, M.A.; Dorah, N.; Seman, R.N.A.R. Electrochemical performance of activated carbon and graphene-based supercapacitor. *Mater. Technol.* **2014**, *30*, A14–A17. [CrossRef]
134. Azam, M.A.; Manaf, M.S.A.; Ahsan, Q. Lithium-Ion supercapacitor using vertically-aligned carbon nanotubes from direct growth technique, and its electrochemical characteristics. *Port. Electrochim. Acta* **2019**, *37*, 167–178. [CrossRef]
135. Que, H.; Chang, C.; Yang, X.; Jiang, F.; Li, M. Heteroatoms doped hierarchical porous carbon materials for high performance supercapacitor electrodes. *Int. J. Electrochem. Sci.* **2019**, *14*, 3477–3493. [CrossRef]
136. Xia, H.; Hu, J.; Li, J.; Wang, K. Electrochemical performance of graphene-coated activated mesocarbon microbeads as a supercapacitor electrode. *RSC Adv.* **2019**, *9*, 7004–7014. [CrossRef] [PubMed]
137. Zhang, Y.; Dyatkin, B.; Cummings, P.T. Molecular investigation of oxidized graphene: Anatomy of the double-layer structure and ion dynamics. *J. Phys. Chem. C* **2019**, *123*, 12583–12591. [CrossRef]
138. Ray, A.; Saruhan, B. Application of ionic liquids for batteries and supercapacitors. *Materials* **2021**, *14*, 2942. [CrossRef]
139. Kongsawatvoragul, K.; Kalasina, S.; Kidkhunthod, P.; Sawangphruk, M. Charge storage mechanisms of cobalt hydroxide thin film in ionic liquid and KOH electrolytes for asymmetric supercapacitors with graphene aerogel. *Electrochim. Acta* **2019**, *324*, 134854. [CrossRef]
140. Adusei, P.K.; Gbordzoe, S.; Kanakaraj, S.N.; Hsieh, Y.Y.; Alvarez, N.T.; Fang, Y.; Johnson, K.; McConnell, C.; Shanov, V. Fabrication and study of supercapacitor electrodes based on oxygen plasma functionalized carbon nanotube fibers. *J. Energy Chem.* **2020**, *40*, 120–131. [CrossRef]
141. Mendhe, A.; Panda, H.S. A review on electrolytes for supercapacitor device. *Discov. Mater.* **2023**, *3*, 29. [CrossRef]
142. Fic, K.; Gorska, B.; Bujewska, P.; Béguin, F.; Frackowiak, E. Selenocyanate-based ionic liquid as redox-active electrolyte for hybrid electrochemical capacitors. *Electrochim. Acta* **2019**, *314*, 1–8. [CrossRef]
143. Chen, Z.; Li, Z.; Ma, X.; Wang, Y.; Zhou, Q.; Zhang, S.A. New DMF-derived ionic liquid with ultra-high conductivity for high-capacitance electrolyte in electric double-layer capacitor. *Electrochim. Acta* **2019**, *319*, 843–848. [CrossRef]
144. Liu, T.; Wang, K.; Chen, Y.; Zhao, S.; Han, Y. Dominant role of wettability in improving the specific capacitance. *Green Energy Environ.* **2019**, *4*, 171–179. [CrossRef]
145. Momodu, D.; Sylla, N.F.; Mutuma, B.; Bello, A.; Masikhwa, T.; Lindberg, S.; Matic, A.; Manyala, N. Stable ionic-liquid-based symmetric supercapacitors from Capsicum seed-porous carbons. *J. Electroanal. Chem.* **2019**, *838*, 119–128. [CrossRef]
146. Gong, X.; Li, S.; Lee, P.S. A fiber asymmetric supercapacitor based on FeOOH/PPy on carbon fibers as an anode electrode with high volumetric energy density for wearable applications. *Nanoscale.* **2017**, *9*, 10794–10801. [CrossRef] [PubMed]
147. Jeon, H.; Han, J.H.; Yu, D.M.; Lee, J.Y.; Kim, T.H.; Hong, Y.T. Synthesis of mesoporous reduced graphene oxide by Zn particles for electrodes of supercapacitor in ionic liquid electrolyte. *J. Ind. Eng. Chem.* **2017**, *45*, 105–110. [CrossRef]
148. Kim, D.; Keum, K.; Lee, G.; Kim, D.; Lee, S.S.; Ha, J.S. Flexible, water-proof, wire-type supercapacitors integrated with wire-type UV/NO_2 sensors on textiles. *Nano Energy* **2017**, *35*, 199–206. [CrossRef]
149. Urbain, F.; Smirnov, V.; Becker, J.P.; Lambertz, A.; Yang, F.; Ziegler, J.; Kaiser, B.; Jaegermann, W.; Rau, U.; Finger, F. Multijunction Si photocathodes with tunable photovoltages from 2.0 V to 2.8 V for light induced water splitting. *Energy Environ. Sci.* **2016**, *9*, 145–154. [CrossRef]

150. Hu, H.; Wang, J.; Cui, C.; Qian, W. Highly electroconductive mesoporous activated carbon fibers and their performance in the ionic liquid-based electrical double-layer capacitors. *Carbon* **2019**, *154*, 1–6.
151. Yu, L.; Chen, G.Z. Ionic liquid-based electrolytes for supercapacitor and supercapattery. *Front. Chem.* **2019**, *7*, 272. [CrossRef]
152. Eftekhari, A. Supercapacitors utilising ionic liquids. *Energy Storage Mater.* **2017**, *9*, 47–69. [CrossRef]
153. Dar, M.A.; Majid, S.R.; Satgunam, M.; Siva, C.; Ansari, S.; Arularasan, P.; Ahamed, S.R. Advancements in Supercapacitor electrodes and perspectives for future energy storage technologies. *Int. J. Hydrog. Energy* **2024**, *70*, 10–28. [CrossRef]
154. Naikwade, M.B.; Katkar, P.K.; Lee, S.W. Understanding the impact of porosity on Li-ion diffusion enhancement in micro-sized silicon particles for advanced batteries. *Ceram. Int.* **2024**, *50*, 54778–54790. [CrossRef]
155. Hayat, A.; Sohail, M.; Alzahrani, A.Y.A.; Ali, H.; Abu-Dief, A.M.; Amin, M.S.; Alenad, A.M.; Al-Mhyawi, S.R.; Al-Hadeethi, Y.; Ajmal, Z.; et al. Recent advances in heteroatom-doped/hierarchically porous carbon materials: Synthesis, design and potential applications. *Prog. Mater. Sci.* **2025**, *150*, 101408. [CrossRef]
156. Wu, Z.; Li, L.; Yan, J.-M.; Zhang, X.-B. Materials design and system construction for conventional and new-Concept supercapacitors. *Adv. Sci.* **2017**, *4*, 1600382. [CrossRef] [PubMed]
157. Wang, H.; Casalongue, H.; Liang, Y.; Dai, H. $Ni(OH)_2$ nanoplates grown on graphene as advanced electrochemical pseudocapacitor materials. *J. Am. Chem. Soc.* **2010**, *132*, 7472–7477. [CrossRef] [PubMed]
158. Miller, J.R.; Simon, P. Electrochemical capacitors for energy management. *Science* **2008**, *321*, 651–652. [CrossRef]
159. Li, Y.; Kinloch, I.A.; Windle, A.H. Direct spinning of carbon nanotube fibers from chemical vapor deposition synthesis. *Science* **2004**, *304*, 276–278. [CrossRef]
160. Simon, P.; Gogotsi, Y.; Dunn, B. Where do batteries end and supercapacitors begin? *Science* **2014**, *343*, 1210–1211. [CrossRef]
161. Singh, V.; Kuthe, S.; Skorodumova, N.V. Electrode Fabrication Techniques for Li Ion Based Energy Storage System: A Review. *Batteries* **2023**, *9*, 184. [CrossRef]
162. Iijima, S. Helical microtubules of graphitic carbon. *Nature* **1991**, *354*, 56–58. [CrossRef]
163. Liu, C.; Yu, Z.; Neff, D.; Zhamu, A.; Jang, B.Z. Graphene-based supercapacitor with an ultrahigh energy density. *Nano Lett.* **2010**, *10*, 4863–4868. [CrossRef]
164. Sun, G.; Ren, H.; Shi, Z.; Zhang, L.; Wang, Z.; Zhan, K.; Yan, Y.; Yang, J.; Zhao, B. V_2O_5/vertically-aligned carbon nanotubes as negative electrode for asymmetric supercapacitor in neutral aqueous electrolyte. *J. Colloid Interface Sci.* **2021**, *588*, 847–856. [CrossRef]
165. Yin, Y.; Xu, Y.; Zhou, Y.; Yan, Y.; Zhan, K.; Yang, J.; Li, Y.; Zhao, B. Millimeter-Long Vertically Aligned Carbon-Nanotube-Supported Co3O4 Composite Electrode for High-Performance Asymmetric Supercapacitor. *ChemElectroChem* **2018**, *5*, 1394. [CrossRef]
166. Liu, Z.; Zhang, X.; Nishihara, H.; Kyotani, T. Carbon materials with a well-developed but interconnected porous texture for supercapacitors. *Carbon* **2010**, *48*, 4526–4634.
167. Cheng, Q.; Tang, J.; Ma, J.; Zhang, H.; Shinya, N.; Qin, L.C. Graphene and carbon nanotube composite electrodes for supercapacitors with ultra-high energy density. *Phys. Chem. Chem. Phys.* **2011**, *13*, 17615–17624. [CrossRef] [PubMed]
168. Karthika, P.; Rajalakshmi, N.; Dhathathreyan, K.S. Functionalized exfoliated graphene oxide as supercapacitor electrodes. *Soft Nanosci. Lett.* **2012**, *2*, 59–66. [CrossRef]
169. Kim, T.; Jung, G.; Yoo, S.; Suh, K.S.; Ruoff, R.S. Activated graphene-based carbons as supercapacitor electrodes with macro- and mesopores. *ACS Nano* **2013**, *7*, 6899–6905. [CrossRef]
170. Wang, Y.; Shi, Z.; Huang, L.; Ma, Y.; Wang, C.; Chen, M.; Chen, Y. Supercapacitor devices based on graphene materials. *J. Phys. Chem. C* **2009**, *113*, 13103–13107. [CrossRef]
171. Ladrón-de-Guevara, A. Reduced graphene oxide/polyaniline electrochemical supercapacitors fabricated by laser. *Appl. Surf. Sci.* **2019**, *467–468*, 691–697. [CrossRef]
172. Ogata, C.; Kurogi, R.; Hatakeyama, K.; Taniguchi, T.; Koinuma, M.; Matsumoto, Y. All-graphene oxide device with tunable supercapacitor and battery behaviour by the working voltage. *Chem. Commun.* **2016**, *52*, 3919–3922. [CrossRef]
173. Shi, Z.; Sun, G.; Yuan, R.; Chen, W.; Wang, Z.; Zhang, L.; Zhan, K.; Zhu, M.; Yang, J.; Zhao, B. Scalable fabrication of $NiCo_2O_4$/reduced graphene oxide composites by ultrasonic spray as binder-free electrodes for supercapacitors with ultralong lifetime. *J. Mater. Sci. Technol.* **2022**, *99*, 260–269. [CrossRef]
174. Baughman, R.H.; Zakhidov, A.A.; de Heer, W.A. Carbon nanotubes–the route toward applications. *Science* **2002**, *297*, 787–792. [CrossRef]
175. Elrouby, M.; Abu-Dief, A.M.; Mohamed, I.M.A. Facile synthesis and electrochemical characterization of erbium oxide and hydroxide for supercapacitor applications. *Ionics* **2024**, *30*, 5699–5711. [CrossRef]
176. Ali, H.; Orooji, Y.; Ajmal, Z.; Abboud, M.; Abu-Dief, A.M.; Al-Ola, K.A.A.; Hassan, H.M.; Yue, D.; Guo, S.R.; Hayat, A. A comprehensive Review based on the synthesis, properties, morphology, functionalization, and potential applications of transition metals nitrides. *Coord. Chem. Rev.* **2025**, *526*, 216353. [CrossRef]

177. Kumar, A.; Kumar, R.; Singh, R.K. Recent trends in template assisted 3D porous materials for supercapacitor applications. *J. Mater. Chem. A* **2021**, *9*, 22422–22482.
178. Kumar, A.; Kumar, R.; Singh, R.K. Template-free synthesis of Se-nanorods-rGO nanocomposite for application in supercapacitors. *Nanotechnol. Rev.* **2019**, *8*, 661–670.
179. Chen, Z.; Ren, W.; Gao, L.; Liu, B.; Pei, S.; Cheng, H.M. Three-dimensional flexible and conductive interconnected graphene networks grown by chemical vapour deposition. *Nat. Mater.* **2011**, *10*, 424–428. [CrossRef]
180. Sivakumar, S.; Sivakumar, P. Electrochemical performance evaluation of template- assisted and template-free synthesized NiO nanomaterials for supercapacitor applications. *J. Solid State Electrochem.* **2021**, *25*, 2505–2515.
181. Li, J.; Wang, X.; Huang, Q.; Gamboa, S.; Sebastian, P.J. Sol–gel derived manganese oxide as electrode material for electrochemical capacitors. *J. Power Sources* **2006**, *160*, 1505–1509.
182. Kumar, A. Sol gel synthesis of zinc oxide nanoparticles and their application as nano-composite electrode material for supercapacitor. *J. Mol. Struct.* **2020**, *1220*, 128654. [CrossRef]
183. Wang, Y.; Xia, Y. Bottom-up versus top-down approaches to the synthesis of monodispersed colloidal nanocrystals of complex materials. *Nano Res.* **2006**, *6*, 672–687.
184. Zhang, H.; Yu, L. Recent advances in sol–gel derived materials for supercapacitor applications: A mini review. *J. Mater. Sci. Mater. Electron.* **2017**, *28*, 13383–13393.
185. Salaheldeen, M.; Abu-Dief, A.M.; Martínez-Goyeneche, L.; Alzahrani, S.O.; Alkhatib, F.; Álvarez-Alonso, P.; Blanco, J.Á. Dependence of the Magnetization Process on the Thickness of $Fe_{70}Pd_{30}$ Nanostructured Thin Film. *Materials* **2020**, *13*, 5788. [CrossRef] [PubMed]
186. Liang, J.; Tong, K.; Pei, Q. A high-performance supercapacitor-battery hybrid energy storage device based on graphene-enhanced electrode materials with ultrahigh energy density. *Energy Environ. Sci.* **2010**, *3*, 442–450.
187. Zhang, Y.; Zhang, L.; Zhou, C. Review of chemical vapor deposition of graphene and related applications. *Acc. Chem. Res.* **2013**, *46*, 2329–2339. [CrossRef] [PubMed]
188. Zhang, R.; Sun, T. Ink-based additive manufacturing for electrochemical applications. *Heliyon* **2024**, *10*, e33023. [CrossRef]
189. Niu, Z.; Zhang, L.; Liu, L.; Zhu, B.; Dong, H.; Chen, X. All-solid-state flexible ultrathin micro-supercapacitors based on graphene. *Adv. Mater.* **2012**, *24*, 5088–5092. [CrossRef]
190. Kohlmeyer, R.R.; Chen, H.; Liu, J.; Chen, Z. Three-dimensional carbon nanotube–polyaniline hybrid materials for pseudocapacitive electrodes. *J. Mater. Chem.* **2011**, *21*, 9491–9496.
191. Lewandowski, A.; Galinski, M. Ionic liquids as electrolytes for Li-ion batteries—An overview of electrochemical studies. *J. Power Sources* **2010**, *194*, 601–609. [CrossRef]
192. Xu, Q.; Lv, Y.; Dong, C.; Sreeprased, T.S.; Tian, A.; Zhang, H.; Tang, Y.; Yu, Z.; Li, N. Three-dimensional micro/nanoscale architectures: Fabrication and applications. *Nanoscale* **2015**, *7*, 10883–10895. [CrossRef]
193. Wang, G.; Zhang, L.; Zhang, J. A review of electrode material for electrochemical supercapacitors. *Chem. Soc. Rev.* **2012**, *41*, 797–828. [CrossRef]
194. Zhang, L.L.; Zhou, R.; Zhao, X.S. Graphene-based materials as supercapacitor electrodes. *J. Mater. Chem.* **2010**, *20*, 5983–5992. [CrossRef]
195. Meher, S.K.; Gupta, R.K.; Singh, A.K. Graphene oxide and its derivatives as electrode material for supercapacitor. *J. Mater. Sci. Mater. Electron.* **2017**, *28*, 12633–12649.
196. Salaheldeen, M.; Méndez, M.; Vega, V.; Fernández, A.; Prida, V.M. Tuning Nanohole Sizes in Ni Hexagonal Antidot Arrays: Large Perpendicular Magnetic Anisotropy for Spintronic Applications. *ACS Appl. Nano Mater.* **2019**, *2*, 1866–1875. [CrossRef]
197. Wang, Y.-G. Nickel hydroxide/carbon nanotube composite electrodes for supercapacitors. *J. Power Sources Jan.* **2009**, *189*, 1270–1276.
198. Li, H.; Kang, Z.; Liu, Y.; Lee, S.T. Carbon nanodots: Synthesis, properties and applications. *J. Mater. Chem.* **2012**, *22*, 24230–24253. [CrossRef]
199. Kumar, A.; Kumar, R.; Singh, R.K. Advances in supercapacitor development: Materials, processes, and applications. *J. Electron. Mater.* **2022**, *52*, 96–129.
200. Wang, H.; Maiyalagan, T.; Wang, X. Review on recent progress in nitrogen-doped graphene: Synthesis, characterization, and its potential applications. *ACS Catal.* **2012**, *2*, 781–794. [CrossRef]
201. Zhang, J.; Chen, X.; Takanabe, K.; Maeda, K.; Domen, K.; Epping, J.D.; Fu, X. Synthesis of a carbon nitride structure for visible-light catalysis by copolymerization. *Angew. Chem. Int. Ed.* **2010**, *49*, 441–444. [CrossRef]
202. Chhowalla, M.; Shin, H.S.; Eda, G.; Li, L.J.; Loh, K.P.; Zhang, H. The chemistry of two- dimensional layered transition metal dichalcogenide nanosheets. *Nat. Chem.* **2013**, *5*, 263–275. [CrossRef]
203. Li, Y.; Zhang, Y.; Wang, C. Binder-free electrodes and their application for Li-ion batteries. *Nanoscale Res. Lett.* **2020**, *15*, 112.
204. Wang, J.; Xu, Z. Design and synthesis of MnO_2/carbon nanotube/graphene ternary composite for high-performance supercapacitors. *J. Power Sources* **2014**, *271*, 291–299.

205. Zhang, L.; Zhao, X.; Ji, H.; Li, Y.; Bai, X.; Fan, L.; Ruoff, R.S. Nitrogen-doped holey graphene as an anode material for high-performance lithium-ion batteries. *Adv. Energy Mater.* **2013**, *3*, 1215–1220.
206. B'eguin, F.; Frąckowiak, E. (Eds.) *Supercapacitors: Materials, Systems and Applications*; John Wiley & Sons: Hoboken, NJ, USA, 2013.
207. Phor, L.; Kumar, A.; Chahal, S. Electrode materials for supercapacitors: A comprehensive review of advancements and performance. *J. Energy Storage* **2024**, *84*, 110698.
208. Augustyn, V.; Come, J.; Lowe, M.A.; Kim, J.W.; Taberna, P.L.; Tolbert, S.H.; Dunn, B. High- rate electrochemical energy storage through Li^+ intercalation pseudocapacitance. *Nat. Mater.* **2013**, *12*, 518–522. [CrossRef] [PubMed]
209. Salaheldeen, M.; Martínez-Goyeneche, L.; Álvarez-Alonso, P.; Fernandez, A. Enhancement the perpendicular magnetic anisotropy of nanopatterned hard/soft bilayer magnetic antidot arrays for spintronic application. *Nanotechnology* **2020**, *31*, 485708. [CrossRef]
210. Shah, S.S.; Niaz, F.; Ehsan, M.A.; Das, H.T.; Younas, M.; Khan, A.S.; Aziz, M.A. Advanced strategies in electrode engineering and nanomaterial modifications for supercapacitor performance enhancement: A comprehensive review. *J. Energy Storage* **2024**, *79*, 110152. [CrossRef]
211. Sahoo, P.K.; Kumar, N.; Jena, A.; Mishra, S.; Lee, C.P.; Lee, S.Y.; Park, S.J. Recent progress in graphene and its derived hybrid materials for high-performance supercapacitor electrode applications. *RSC Adv.* **2024**, *14*, 1284–1303. [CrossRef]
212. Kulkarni, P.; Kim, K.-H. Mathematical Modelling and Simulation of supercapacitors. In *Supercapacitors: Theoretical and Practical Devices*; Kim, K.-H., Mitra, S., Eds.; Springer: Berlin/Heidelberg, Germany, 2016; pp. 323–348.
213. Stoller, M.D.; Ruoff, R.S. Best practice methods for determining an electrode material's performance for ultracapacitors. *Energy Environ. Sci.* **2010**, *3*, 1294–1301. [CrossRef]
214. Salaheldeen, M.; Zhukova, V.; Rosero, J.; Salazar, D.; Ipatov, M.; Zhukov, A. Comparison of the magnetic and structural properties of MnFePSi microwires and MnFePSi bulk alloy. *Materials* **2024**, *17*, 1874. [CrossRef]
215. Pandey, D.; Kumar, K.S.; Thomas, J. Supercapacitor electrode energetics and mechanism of operation: Uncovering the voltage window. *Prog. Mater. Sci.* **2023**, *141*, 101219. [CrossRef]
216. Burke, A. Ultracapacitors: Why, how, and where is the technology. *J. Power Sources* **2000**, *91*, 37–50. [CrossRef]
217. Bowen, D.K.; Tanner, B.K. *High Resolution X-Ray Diffractometry and Topography*; CRC Press: Boca Raton, FL, USA, 1998.
218. Cullity, B.D.; Stock, S.R. *Diffractometer and Spectrometer Measurements. Elements of X-Ray Diffraction, 2nd ed*; Addison-Wesley: Boston, MA, USA, 1956; pp. 188–226.
219. Baig, M.M.; Pervaiz, E.; Afzal, M.J. Catalytic activity and kinetic studies of core@ shell nanostructure $NiFe_2O_4$ @TiO_2 for photocatalytic degradation of methyl orange dye. *J. Chem. Soc. Pak.* **2020**, *42*, 531–541.
220. Treacy, M.; Newsam, J.; Deem, M. A general recursion method for calculating diffracted intensities from crystals containing planar faults. *Proc. R. Soc. A* **1991**, *411*, 499–520.
221. Sivasankaran, S.; Sivaprasad, K.; Narayanasamy, R.; Satyanarayana, P. X-ray peak broadening analysis of AA $6061_{100-x} - x$ wt.% Al_2O_3 nanocomposite prepared by mechanical alloying. *Mater. Charact.* **2011**, *62*, 661–672. [CrossRef]
222. Ungar, T. Microstructural parameters from X-ray diffraction peak broadening. *Scr. Mater.* **2004**, *51*, 777–781. [CrossRef]
223. Ali, S.; Karunanithi, R.; Prashanth, M.; Rahman, M.A. X-ray peak broadening on microstructure, and structural properties of titanium and Ti-6Al-4V alloys. *Mater. Today Proc.* **2020**, *27*, 2390–2393. [CrossRef]
224. Wolf, M.; May, B.M.; Cabana, J. Visualization of electrochemical reactions in battery materials with X-ray microscopy and mapping. *Chem. Mater.* **2017**, *29*, 3347–3362.
225. Gu, Q.; Kimpton, J.A.; Brand, H.E.; Wang, Z.; Chou, S. Solving key challenges in battery research using in situ synchrotron and neutron techniques. *Adv. Energy Mater.* **2017**, *7*, 1602831. [CrossRef]
226. De Groot, F. High-resolution X-ray emission and X-ray absorption spectroscopy. *Chem. Rev.* **2001**, *101*, 1779–1808. [CrossRef]
227. Koningsberger, D.; Prins, R. *X-Ray Absorption: Principles, Applications, Techniques of EXAFS, SEXAFS, and XANES*; Wiley-Interscience: Hoboken, NJ, USA, 1988.
228. Kelly, S.; Hesterberg, D.; Ravel, B. Analysis of soils and minerals using X-ray absorption spectroscopy. In *Methods of Soil Analysis Part 5—Mineralogical Methods*; John Wiley & Sons: Hoboken, NJ, USA, 2008; pp. 387–463.
229. Thole, B.; Van der Laan, G. Branching ratio in x-ray absorption spectroscopy. *Phys. Rev. B* **1988**, *38*, 3158. [CrossRef]
230. Nam, K.W.; Yoon, W.S.; Kim, K.B. X-ray absorption spectroscopy studies of nickel oxide thin film electrodes for supercapacitors. *Electrochim. Acta* **2002**, *47*, 3201–3209. [CrossRef]
231. Nam, K.W.; Kim, M.G.; Kim, K.B. In Situ Mn K-edge X-ray absorption spectroscopy studies of electrodeposited manganese oxide films for electrochemical capacitors. *J. Phys. Chem. C* **2007**, *111*, 749–758. [CrossRef]
232. Chang, H.W.; Lu, Y.R.; Chen, J.L.; Chen, C.L.; Lee, J.F.; Chen, J.M.; Tsai, Y.C.; Chang, C.M.; Yeh, P.H.; Chou, W.C.; et al. Nanoflaky MnO_2/functionalized carbon nanotubes for supercapacitors: An in situ X-ray absorption spectroscopic investigation. *Nanoscale* **2015**, *7*, 1725–1735. [CrossRef] [PubMed]
233. Chang, H.W.; Lu, Y.R.; Chen, J.L.; Chen, C.L.; Lee, J.F.; Chen, J.M.; Tsai, Y.C.; Yeh, P.H.; Choue, W.C.; Dong, C.L. Electrochemical and in situ X-ray spectroscopic studies of MnO_2 /reduced graphene oxide nanocomposites as a supercapacitor. *Phys. Chem. Chem. Phys.* **2016**, *18*, 18705–18718. [CrossRef] [PubMed]

234. Toupin, M.; Brousse, T.; B'elanger, D. Charge storage mechanism of MnO_2 electrode used in aqueous electrochemical capacitor. *Chem. Mater.* **2004**, *16*, 3184–3190. [CrossRef]
235. Tõnisoo, A.; Kruusma, J.; Pärna, R.; Kikas, A.; Hirsimäki, M.; Nõmmiste, E.; Lust, E. In situ XPS studies of electrochemically negatively polarized molybdenum carbide derived carbon double layer capacitor electrode. *J. Electrochem. Soc.* **2013**, *160*, A1084–A1093. [CrossRef]
236. Zhang, L.; Shi, W.; Zhang, B. A review of electrocatalyst characterization by transmission electron microscopy. *J. Energy Chem.* **2017**, *26*, 1117–1135. [CrossRef]
237. Klein, T.; Buhr, E.; Frase, C.G. TSEM: A review of scanning electron microscopy in transmission mode and its applications. In *Advances in Imaging and Electron Physics*; Elsevier: Amsterdam, The Netherlands, 2012; pp. 297–356.
238. Song, Z.; Xie, Z.H. A literature review of in situ transmission electron microscopy technique in corrosion studies. *Micron* **2018**, *112*, 69–83. [CrossRef]
239. Zhou, W.; Thomas, J.M. HRTEM surface profile imaging of solids. *Curr. Opin. Solid State Mater. Sci.* **2001**, *5*, 75–83. [CrossRef]
240. Greer, H.F.; Zhou, W. Electron diffraction and HRTEM imaging of beam-sensitive materials. *Crystallogr. Rev.* **2011**, *17*, 163–185. [CrossRef]
241. Rouzaud, J.N.; Clinard, C. Quantitative high-resolution transmission electron microscopy: A promising tool for carbon materials characterization. *Fuel Process. Technol.* **2002**, *77*, 229–235. [CrossRef]
242. Salaheldeen, M.; Vega, V.; Ibabe, A.; Jaafar, M.; Asenjo, A.; Fernandez, A.; Prida, V.M. Tailoring of Perpendicular Magnetic Anisotropy in Dy13Fe87 Thin Films with Hexagonal Antidot Lattice Nanostructure. *Nanomaterials* **2018**, *8*, 227. [CrossRef] [PubMed]
243. Allen, C.; Zhang, C.; Burnell, G.; Brown, A.; Robertson, J.; Hickey, B. A review of methods for the accurate determination of the chiral indices of carbon nanotubes from electron diffraction patterns. *Carbon* **2011**, *49*, 4961–4971. [CrossRef]
244. Polcari, D.; Dauphin-Ducharme, P.; Mauzeroll, J. Scanning electrochemical microscopy: A comprehensive review of experimental parameters from 1989 to 2015. *Chem. Rev.* **2016**, *116*, 13234–13278. [CrossRef] [PubMed]
245. Xue, J.; Gao, L.; Hu, X.; Cao, K.; Zhou, W.; Wang, W.; Lu, Y. Stereolithographic 3D printing-based hierarchically cellular lattices for high-performance quasi-solid supercapacitor. *Nano-Micro Lett.* **2019**, *11*, 46. [CrossRef]
246. Xue, Y.; Ding, Y.; Niu, J.; Xia, Z.; Roy, A.; Chen, H.; Qu, J.; Wang, Z.L.; Dai, L. Rationally designed graphene-nanotube 3D architectures with a seamless nodal junction for efficient energy conversion and storage. *Sci. Adv.* **2015**, *1*, e1400198. [CrossRef]
247. Salaheldeen, M.; Garcia-Gomez, A.; Ipatov, M.; Corte-Leon, P.; Zhukova, V.; Blanco, J.M.; Zhukov, A. Fabrication and Magneto-Structural Properties of Co_2-Based Heusler Alloy Glass-Coated Microwires with High Curie Temperature. *Chemosensors* **2022**, *10*, 225. [CrossRef]
248. Cleveland, J.; Anczykowski, B.; Schmid, A.; Elings, V. Energy dissipation in tapping- mode atomic force microscopy. *Appl. Phys. Lett.* **1998**, *72*, 2613–2615. [CrossRef]
249. Kumar, K.S.; Choudhary, N.; Pandey, D.; Ding, Y.; Hurtado, L.; Chung, H.S.; Tetard, L.; Jung, Y.; Thomas, J. Investigating 2D WS 2 supercapacitor electrode performance by Kelvin probe force microscopy. *J. Mater. Chem. A* **2020**, *8*, 12699–12704. [CrossRef]
250. Ventosa, E.; Schuhmann, W. Scanning electrochemical microscopy of Li-ion batteries. *Phys. Chem. Chem. Phys.* **2015**, *17*, 28441–28450. [CrossRef]
251. Dendisová, M.; Jeništová, A.; Parchaňská-Kokaislová, A.; Matějka, P.; Prokopec, V.; Švecová, M. The use of infrared spectroscopic techniques to characterize nanomaterials and nanostructures: A review. *Anal. Chim. Acta* **2018**, *1031*, 1–14. [CrossRef]
252. Petit, T.; Puskar, L. FTIR spectroscopy of nanodiamonds: Methods and interpretation. *Diam. Relat. Mater.* **2018**, *89*, 52–66. [CrossRef]
253. Khan, M.Z.; Gul, I.H.; Baig, M.M.; Khan, A.N. Comprehensive study on structural, electrical, magnetic and photocatalytic degradation properties of Al 3+ ions substituted nickel ferrites nanoparticles. *J. Alloy. Compd.* **2020**, *848*, 155795. [CrossRef]
254. Coates, J. Interpretation of infrared spectra, a practical approach. In *Encyclopedia of Analytical Chemistry: Applications, Theory and Instrumentation*; John Wiley & Sons: Hoboken, NJ, USA, 2006.
255. Gouadec, G.; Colomban, P. Raman Spectroscopy of nanomaterials: How spectra relate to disorder, particle size and mechanical properties. *Prog. Cryst. Growth Charact. Mater.* **2007**, *53*, 1–56. [CrossRef]
256. Kumar, C.S. *Raman Spectroscopy for Nanomaterials Characterization*; Springer Science & Business Media: Heidelberg, Germany, 2012.
257. Krishnamoorthy, K.; Pazhamalai, P.; Kim, S.J. Two-dimensional siloxene nanosheets: Novel high-performance supercapacitor electrode materials. *Energy Environ. Sci.* **2018**, *11*, 1595–1602. [CrossRef]
258. Krishnamoorthy, K.; Mariappan, V.K.; Pazhamalai, P.; Sahoo, S.; Kim, S.J. Mechanical energy harvesting properties of free-standing carbyne enriched carbon film derived from dehydrohalogenation of polyvinylidene fluoride. *Nano Energy* **2019**, *59*, 453–463. [CrossRef]
259. Blanc, F.; Leskes, M.; Grey, C.P. In situ solid-state NMR spectroscopy of electrochemical cells: Batteries, supercapacitors, and fuel cells. *Acc. Chem. Res.* **2013**, *46*, 1952–1963. [CrossRef]

260. Borchardt, L.; Oschatz, M.; Paasch, S.; Kaskel, S.; Brunner, E. Interaction of electrolyte molecules with carbon materials of well-defined porosity: Characterization by solid-state NMR spectroscopy. *Phys. Chem. Chem. Phys.* **2013**, *15*, 15177–15184. [CrossRef]
261. Griffin, J.M.; Forse, A.C.; Tsai, W.Y.; Taberna, P.L.; Simon, P.; Grey, C.P. In situ NMR and electrochemical quartz crystal microbalance techniques reveal the structure of the electrical double layer in supercapacitors. *Nat. Mater.* **2015**, *14*, 812–819. [CrossRef]
262. Sinha, P.; Datar, A.; Jeong, C.; Deng, X.; Chung, Y.G.; Lin, L.C. Surface Area Determination of porous materials using the Brunauer–Emmett–Teller (BET) method: Limitations and improvements. *J. Phys. Chem. C* **2019**, *123*, 20195–20209. [CrossRef]
263. Mohan, V.B.; Jayaraman, K.; Bhattacharyya, D. Brunauer–Emmett–Teller (BET) specific surface area analysis of different graphene materials: A comparison to their structural regularity and electrical properties. *Solid State Commun.* **2020**, *320*, 114004. [CrossRef]
264. Salaheldeen, M.; Zhukova, V.; Gonzalez, J.; Zhukov, A. The Effect of High-Temperature Annealing on the Magnetic and Structural Properties of (MnFePSi)-Based Glass-Coated Microwires. *Crystals* **2025**, *15*, 311. [CrossRef]
265. Muhammad, K. A review on the selected applications of battery-supercapacitor hybrid energy storage systems for microgrids. *Energies* **2019**, *12*, 4559. [CrossRef]
266. Xun, Q.; Liu, Y.; Zhao, J.; Grunditz, E.A. Modelling and simulation of fuel cell/ supercapacitor passive hybrid vehicle system. In Proceedings of the 2019 IEEE Energy Conversion Congress and Exposition (ECCE), Baltimore, MD, USA, 29 September–3 October 2019; pp. 2690–2696.
267. Ping, W.; Zhao, C.; Zhang, Y.; Jing, L.; Yongping, G. A bidirectional three-level DC-DC converter with a wide voltage conversion range for hybrid energy source electric vehicles. *J. Power Electron.* **2017**, *17*, 334–345.
268. Julius, P.; Dina, I.A. The role of supercapacitors in regenerative braking systems. *Energies* **2019**, *12*, 2683. [CrossRef]
269. Zhang, H.; Zhang, F.; Yang, L.; Gao, Y.; Jin, B. Multi-parameter collaborative power prediction to improve the efficiency of supercapacitor-based regenerative braking system, IEEE Trans. *Energy Convers* **2021**, *36*, 2612–2622. [CrossRef]
270. Salari, O.; Zaad, K.H.; Bakhshai, A.; Jain, P. Reconfigurable hybrid energy storage system for an electric vehicle DC–AC inverter, IEEE Trans. *Power Electron.* **2020**, *35*, 12846–12860.
271. Hsieh, M.-F.; Chen, P.-H.; Pai, F.-S.; Weng, R.-Y. Development of supercapacitor- aided hybrid energy storage system to enhance battery life cycle of electric vehicles. *Sustainability* **2021**, *13*, 7682. [CrossRef]
272. Petru, V.R.; Adam, S.; Marcin, S. On-board energy storage devices with supercapacitors for metro trains—Case study analysis of application effectiveness. *Energies* **2019**, *12*, 1291.
273. Jaemin, K.; Donghwa, S.; Donkyu, B.; Jaehyun, P. Design and optimization of supercapacitor hybrid architecture for power supply-connected batteries lifetime enhancement. *Electronics* **2019**, *8*, 41. [CrossRef]
274. Kim, J.; Pröbstl, A.; Chakraborty, S.; Chang, N. Aging mitigation of power supply-connected batteries. In Proceedings of the 2014 IEEE/ACM International Symposium on Low Power Electronics and Design (ISLPED), La Jolla, CA, USA, 11–13 August 2014.
275. Shi, R.; Semsar, S.; Lehn, P.W. Single-stage hybrid energy storage integration in electric vehicles using vector-controlled power sharing, IEEE Trans. *Ind. Electron.* **2021**, *68*, 10623–10633. [CrossRef]
276. Panhwar, H. Mitigating power fluctuations for energy storage in wind energy conversion system using supercapacitors. *IEEE Access* **2020**, *8*, 189747–189760. [CrossRef]
277. Aranizadeh, A.; Zaboli, A.; Gashteroodkhani, O.A.; Vahidi, B. Wind turbine and ultra-capacitor harvested energy increasing in microgrid using wind speed forecasting, Engineering Science and Technology. *Int. J.* **2019**, *22*, 1161–1167.
278. Yadlapalli, R.T.; Kotapati, A.; Rajani, K. Advancements in power conditioning units for electric vehicle applications: A review. *Int. J. Electr. Hybrid Veh.* **2021**, *13*, 81–115. [CrossRef]
279. Ali, D.F.; Hossein, T. Promoted supercapacitor control scheme based on robust fractional-order super-twisting sliding mode control for dynamic voltage restorer to enhance FRT and PQ capabilities of DFIG-based wind turbine. *J. Energy Storage* **2021**, *42*, 102983.
280. Salaheldeen, M.; Garcia-Gomez, A.; Corte-Leon, P.; Gonzalez, A.; Ipatov, M.; Zhukova, V.; Lopez Anton, R.; Zhukov, A. Manipulation of magnetic and structure properties of Ni_2FeSi glass-coated microwires by annealing. *J. Alloys Compd.* **2023**, *942*, 169026. [CrossRef]
281. Weaver, W.W. Super capacitor energy storage system design for wave energy converter demonstration. In Proceedings of the 2020 International Symposium on Power Electronics, Electrical Drives, Automation and Motion (SPEEDAM), Sorrento, Italy, 24–26 June 2020; pp. 564–570.
282. Aditya, S.S.; Chakraborty, A.K.; Bhattacharyya, B.K. Supercapacitor module solely powering DC–DC buck converter for hand-held applications. *IETE J. Res.* **2020**, *66*, 115–123.
283. Sambit, S.; Sanchali, D.; Bidyut, K.B. How and where to use super-capacitors effectively, an integration of review of past and new characterization works on super-capacitors. *J. Energy Storage* **2020**, *27*, 101044.
284. Ahsan, E.; Arslan, A.A.; Umar, T.S.; Muhammad, T.U.; Muhammad, S.I. Efficient wireless charging system for supercapacitor-based electric vehicle using inductive coupling power transfer technique. *Adv. Mech. Eng.* **2019**, *11*, 168781401988696.

285. Ali, N.; Liu, Z.; Armghan, H.; Ahmad, I.; Hou, Y. LCC-S-based integral terminal sliding mode controller for a hybrid energy storage system using a wireless power system. *Energies* **2021**, *14*, 1693. [CrossRef]
286. Gao, C.; Huang, J.; Xiao, Y. A seamlessly integrated device of micro- supercapacitor and wireless charging with ultrahigh energy density and capacitance. *Nat. Commun.* **2021**, *12*, 2647. [CrossRef]
287. López Antón, R.; Andres, J.P.; Gonzalez, J.A.; Garcia-Gomez, A.; Zhukova, V.; Chizhik, A.; Salaheldeen, M.; Zhukov, A. Tuning of magnetic properties and Giant Magnetoimpedance effect in multilayered microwires. *J. Sci. Adv. Mater. Devices* **2024**, *9*, 100821. [CrossRef]
288. Felsmann, B.; Mezősi, A.; Szabó, L. Market versus Bureaucracy—Price Regulation in the Electricity Retail Sector. In Proceedings of the Importance of Kornai's Research Today, Budapest, Hungary, 21–22 February 2018.
289. Tawalbeh, M.; Murtaza, S.Z.; Al-Othman, A.; Alami, A.H.; Singh, K.; Olabi, A.G. Ammonia: A versatile candidate for the use in energy storage systems. *Renew. Energy* **2022**, *194*, 955–977. [CrossRef]
290. Chakraborty, M.R.; Dawn, S.; Saha, P.K.; Basu, J.B.; Ustun, T.S. A Comparative Review on Energy Storage Systems and Their Application in Deregulated Systems. *Batteries* **2022**, *8*, 124. [CrossRef]
291. Asri, L.I.M.; Ariffin, W.N.S.F.W.; Zain, A.S.M.; Nordin, J.; Saad, N.S.C. Comparative Study of Energy Storage Systems (ESSs). *J. Phys. Conf. Ser.* **2021**, *1962*, 012035. [CrossRef]
292. Hossain, E.; Faruque, H.M.R.; Sunny, M.S.H.; Mohammad, N.; Nawar, N. A Comprehensive Review on Energy Storage Systems: Types, Comparison, Current Scenario, Applications, Barriers, and Potential Solutions, Policies, and Future Prospects. *Energies* **2020**, *13*, 3651. [CrossRef]
293. David, B.R.; Spencer, S.; Miller, J.; Almahmoud, S.; Jouhara, H. Comparative environmental life cycle assessment of conventional energy storage system and innovative thermal energy storage system. *Int. J. Thermofluids* **2021**, *12*, 100116. [CrossRef]
294. Christen, T.; Carlen, M.W. Theory of Ragone plots. *J. Power Sources* **2000**, *91*, 210–216. [CrossRef]
295. Schneuwly, A.; Gallay, R. Properties and applications of supercapacitors from the state-of-the-art to future trends. In *Proceeding PCIM*; Mesago PCIM GmbH: Nuremberg, Germany, 2000.
296. Kosala, G. 4-Capacitors as energy storage devices—Simple basics to current commercial families. In *Energy Storage Devices for Electronic Systems*; Kularatna, N., Ed.; Academic Press: Cambridge, MA, USA, 2015; pp. 137–148. ISBN 9780124079472.
297. Inga, B.; Astrid, B.; Hanke-Rauschenbach, R. Ragone plots revisited: A review of methodology and application across energy storage technologies. *J. Energy Storage* **2023**, *73*, 109097.
298. Shifei, H.; Samrat, S.; Yufeng, Z. Challenges and opportunities for supercapacitors. *APL Mater.* **2019**, *7*, 100901.
299. Sayed, Y.A.; Saad, G.M.; Yosry, F.B.; Hamdy, H.H.; Zoubi, W.A. Supercapacitor electrode materials: Addressing challenges in mechanism and charge storage. *Rev. Inorg. Chem.* **2021**, *42*, 53–88.
300. Wu, Y.; Cao, C. The way to improve the energy density of supercapacitors: Progress and perspective. *Sci. China Mater.* **2018**, *61*, 1517–1526. [CrossRef]
301. Yaseen, M.; Khattak, M.A.K.; Humayun, M.; Usman, M.; Shah, S.S.; Bibi, S.; Hasnain, B.S.U.; Ahmad, S.M.; Khan, A.; Shah, N. A review of supercapacitors: Materials design, modification, and applications. *Energies* **2021**, *14*, 7779. [CrossRef]
302. Aqib, M.; Basheer Ahamed, M.; Kalim, D.; Jagannathan, T. A review on recent advances in hybrid supercapacitors: Design, fabrication and applications. *Renew. Sustain. Energy Rev.* **2019**, *101*, 123–145.
303. Sandhiya, M.; Suresh Balaji, S.; Sathish, M. Boosting the energy density of flexible supercapacitors by redox-additive hydrogels. *Energy Fuels* **2020**, *34*, 11536–11546. [CrossRef]
304. Edurne, R.; Lewis, W.; Le, F.; Richard, F.; Rebecca, T.; Andrew, J.F.; Dryfe, R.A.W. Enhancing supercapacitor energy density by mass- balancing of graphene composite electrodes. *Electrochim. Acta* **2020**, *360*, 136957.
305. Rani, J.R.; Thangavel, R.; Oh, S.I.; Lee, Y.S.; Jang, J.H. An ultra-high-energy density supercapacitor; fabrication based on thiol-functionalized graphene oxide scrolls. *Nanomaterials* **2019**, *9*, 148. [CrossRef]
306. Hérou, S.; Josh, J.B.; Matt, K.; Philipp, S.; Rhodri, J.; Dan, J.L.B.; Paul, R.S.; Maria, C.R.; Magdalena, T. High- density lignin-derived carbon nanofiber supercapacitors with enhanced volumetric energy density. *Adv. Sci.* **2021**, *8*, 2100016. [CrossRef]
307. Katkar, P.K.; Naikwade, M.B.; Patil, S.A.; Lee, S.W. Self-assembly of $Sr_2P_2O_7$@2D rGO nano/micro-architecture for highly durable and bendable solid-state supercapattery. *Mater. Today Phys.* **2024**, *46*, 101510. [CrossRef]
308. Ren, K.; Liu, Z.; Wei, T. Recent developments of transition metal compounds- carbon hybrid electrodes for high energy/power supercapacitors. *Nano-Micro Lett.* **2021**, *13*, 129. [CrossRef]
309. Jiandong, D.; Minghang, D.; Zhang, K.; Zhengkai, L.; Sun, L. Research on voltage equalization among multiple supercapacitor modules based on multiwinding transformer. *Int. J. Electr. Power Energy Syst.* **2020**, *120*, 106031.
310. Yu, Y.; Saasaa, R.; Khan, A.A.; Eberle, W. A series resonant energy storage cell voltage balancing circuit. *IEEE J. Emerg. Sel. Top. Power Electron.* **2020**, *8*, 3151–3161. [CrossRef]
311. Liao, W.; Chen, Y.; Zeng, J.; Lai, Z.; Liu, J. Topology, analysis and modeling of voltage equalizers based on reutilization technique for supercapacitor storage system. *IEEE Trans. Transp. Electrif.* 2021; *early access*.
312. Jiang, F. Consensus-Based Cell Balancing of Reconfigurable Supercapacitors. *IEEE Trans. Ind. Appl.* **2020**, *56*, 4146–4154. [CrossRef]

313. Zhang, L.; Xiaosong, L.; Zhenpo, W.; Fengchun, S.; David, G.D. A review of supercapacitor modeling, estimation, and applications: A control/management perspective. *Renew. Sustain. Energy Rev.* **2018**, *81*, 1868–1878. [CrossRef]
314. Helseth, L.E. Modelling supercapacitors using a dynamic equivalent circuit with a distribution of relaxation times. *J. Energy Storage* **2019**, *25*, 100912. [CrossRef]
315. Miniguano, H.; Barrado, A.; Fernández, C.; Zumel, P.; Lázaro, A. A general parameter identification procedure used for the comparative study of supercapacitors models. *Energies* **2019**, *12*, 1776. [CrossRef]
316. Zhao, Y.; Xie, W.; Fang, Z.; Liu, S. A parameters identification method of the equivalent circuit model of the supercapacitor cell module based on segmentation optimization. *IEEE Access* **2020**, *8*, 92895–92906. [CrossRef]
317. Khaled, L.; Antonio, J.M.C. A review of supercapacitors modeling, SoH, and SoE estimation methods: Issues and challenges. *Int. J. Energy Res.* **2021**, *45*, 18424–18440.
318. Ankaj, S.; Satadru, D.; Munmun, K. Second-life applications of supercapacitors: Effective capacitance prognosis and aging. *J. Power Sources* **2021**, *496*, 229824.
319. Zhao, J.; Burke, A. Review on supercapacitors: Technologies and performance evaluation. *J. Energy Chem.* **2020**, *59*, 276–291. [CrossRef]
320. Riaz, A.; Sarker, M.R.; Saad, M.H.M.; Mohamed, R. Review on comparison of different energy storage technologies used in micro-energy HARVESTING, WSNs, Low-Cost microelectronic devices: Challenges and recommendations. *Sensors* **2021**, *21*, 5041. [CrossRef]
321. Ebadi, R.; Sadeghi, Y.A.; Kazemzadeh, R.; Mohammadi-Ivatloo, B. Techno-economic evaluation of transportable battery energy storage in robust day-ahead scheduling of integrated power and railway transportation networks. *Int. J. Electr. Power Energy Syst.* **2021**, *126*, 106606. [CrossRef]
322. Can Supercapacitors Surpass Batteries for Energy Storage? 2021. Available online: https://www.electronicdesign.com/power-management/article/21801779/can-supercapacitors-surpass-batteries-for-energy-storage (accessed on 20 April 2025).

batteries

MDPI

Review

Application of Defect Engineering via ALD in Supercapacitors

Tiange Gao [1], Xiaoyang Xiao [1], Zhenliang Dong [1], Xilong Lu [1], Liwen Mao [1], Jinzheng Wang [1], Yiming Liu [2,*], Qingmin Hu [1,*] and Jiaqiang Xu [1]

1 New Energy and Sensing Technology (NEST) Lab, Department of Chemistry, College of Sciences, Shanghai University, Shanghai 200444, China

2 Future Battery Research Center, Global Institute of Future Technology, Shanghai Jiaotong University, Shanghai 200240, China

* Correspondence: liuyiming-sjtu@sjtu.edu.cn (Y.L.); huqingmin@shu.edu.cn (Q.H.)

Abstract: Supercapacitors are a kind of energy storage device that lie between traditional capacitors and batteries, characterized by high power density, long cycle life, and rapid charging and discharging capabilities. The energy storage mechanism of supercapacitors mainly includes electrical double-layer capacitance and pseudocapacitance. In addition to constructing multi-level pore structures to increase the specific surface area of electrode materials, defect engineering is essential for enhancing electrochemical active sites and achieving additional extrinsic pseudocapacitance. Therefore, developing a simple and efficient method for defect engineering is essential. Atomic layer deposition (ALD) technology enables precise control over thin film thickness at the atomic level through layer-by-layer deposition. This capability allows the intentional introduction of defects, such as vacancies, heteroatom doping, or misalignment, at specific sites within the material. The ALD process can regulate the defects in materials without altering the overall structure, thereby optimizing both the electrochemical and physical properties of the materials. Its self-limiting surface reaction mechanism also ensures that defects and doping sites are introduced uniformly across the material surface. This uniform defect distribution is particularly profitable for high surface area electrodes in supercapacitor applications, as it promotes consistent performance across the entire electrode. This review systematically summarizes the latest advancements in defect engineering via ALD technology in supercapacitors, including the enhancement of conductivity and the increase of active sites in supercapacitor electrode materials through ALD, thereby improving specific capacitance and energy density of the supercapacitor device. Furthermore, we discuss the underlying mechanisms, advantages, and future directions for ALD in this field.

Keywords: supercapacitors; defect engineering; atomic layer deposition; oxygen vacancy; electrode materials; surface engineering

1. Introduction

The rise of the global economy, coupled with fossil resource depletion and environmental pollution, has led to a growing emphasis on sustainable development and renewable energy sources like solar, hydropower, and wind energy. However, despite their potential, the intermittent nature of these energy sources highlights the ongoing challenge of effective electricity storage and supply [1–3]. Among various energy storage devices, batteries and supercapacitors (SCs) are two advanced electrochemical energy storage technologies. High-energy-density lithium-ion batteries (LIBs) have been widely commercialized and applied in various electronic products and new energy vehicles. However, due to their high cost, low charge/discharge rates, low electron and ion transport, and dendrite problems during high-power operation, the applications of LIBs are limited. Supercapacitors can safely provide fast charging and an ultra-long cycle life (>100,000 cycles), making them capable of compensating for or even replacing lithium-ion batteries in certain fields [4–9]; the significance of supercapacitors is shown in Figure 1a. However, the energy density of supercapacitors is significantly lower than LIBs and fuel-cells, as shown in Figure 1b [10]. Therefore, increasing the energy density of the supercapacitor without sacrificing the power density and cycle stability is the key to its further application.

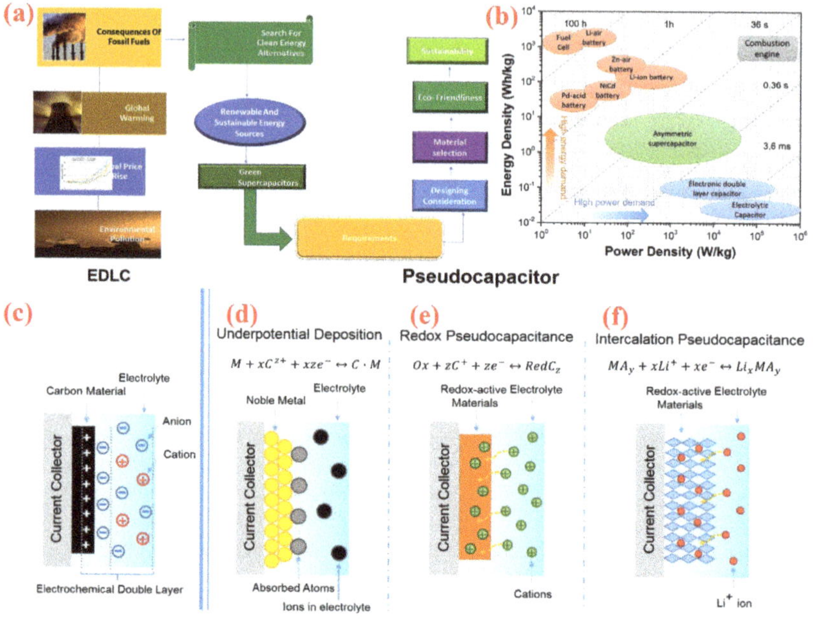

Figure 1. (**a**) Graphical illustration portraying the requisites for green supercapacitors and aspects associated with their development. Reproduced with permission. Copyright 2024, Elsevier Publishing [4]. (**b**) Ragone plot illustrating the performances of specific power vs. specific energy for different electrical energy-storage technologies. Schematics of charge-storage mechanisms for (**c**) an EDLC. Reproduced with permission. Copyright 2018, ACS Publishing [10]. (**d**–**f**) Different types of pseudocapacitive electrodes: (**d**) underpotential deposition, (**e**) redox pseudocapacitor, and (**f**) ion intercalation pseudocapacitor. Reproduced with permission. Copyright 2014 Royal Society of Chemistry [11].

The energy storage mechanisms of supercapacitors can be categorized into two types, as illustrated in Figure 1c–f [10,11]: electrical double-layer capacitance (EDLC) and pseudocapacitance. When electrode materials are immersed in an electrolyte, due to electrostatic interaction, charges spontaneously migrate in a directional manner towards the two electrodes, forming charge layers, as shown in Figure 1c. These charge layers, together with the electrode plates, constitute the electrical double layer. In this process, there is no charge transfer, and charge storage relies on electrostatic charge adsorption. The specific capacitance of EDLC primarily depends on the specific surface area and surface properties of the material. In contrast to EDLC, pseudocapacitive electrode materials store charges through rapid and reversible Faraday reactions occurring on or near the surface, relying on changes in elemental valence states, which result in electron transfer. The mechanisms of pseudocapacitance are primarily categorized into three types: underpotential deposition, redox reactions, and ion intercalation. Underpotential deposition is commonly observed in the adsorption of hydrogen atoms by noble metals with catalytic properties, as shown in Figure 1d. In Figure 1e, the pseudocapacitive electrode materials based on redox reactions exhibit electron transfer between oxidized and reduced states during energy storage. Specifically, the oxidized surface undergoes electrochemical adsorption of cations accompanied by rapid and reversible electron transfer. Pseudocapacitance also involves the insertion and extraction of ions, as shown in Figure 1f. Despite differing energy storage mechanisms, the above three pseudocapacitive systems exhibit similar electrochemical characteristics. This is attributed to the identical relationship between the potential and coverage of electrolyte ions on or near the material surface during charging and discharging, which results in a higher charge storage capacity on the material surface compared to that of EDLC.

Electrode materials are crucial for supercapacitors, but they are largely limited by poor reaction kinetics, resulting in suboptimal performance [12,13]. Therefore, developing electrode materials with robust reaction kinetics is of great significance. Currently, the main problems we are facing encompass two aspects. Firstly, metallic oxides lack an efficient electron transport path and electron/ion transfer interface, thus obtaining a slow rate capability [14,15]. In addition, the volume contraction/expansion during the charging and discharging process causes the electrode to shatter, resulting in decreased stability [16,17]. In the past few decades, people have designed and prepared electrode materials with high reaction kinetics using various methods, including integrating conductive substrates to improve conductivity [18,19], preparing porous structures to increase surface area [20,21], and constructing heterostructures and surfactant functional groups as electrochemical reaction sites. However, the improvement in reaction kinetics of supercapacitors is quite limited, as this "external" approach cannot eliminate the intrinsic property limitations of materials.

Among various design strategies, defect engineering, as an "intrinsic regulation" method, can effectively regulate the intrinsic properties of electrode materials to promote their reaction kinetics and thus achieve excellent electrochemical performance [22–24]. Taking energy density as an example, the introduction of defects, such as oxygen vacancies or metal doping, can increase the carrier concentration in materials, thereby enhancing their electrical conductivity. The improvement in electrical conductivity aids in reducing the internal resistance of supercapacitors during charging and discharging processes, thus elevating their energy density. Furthermore, the incorporation of defects creates additional ion diffusion pathways and electrochemical active sites. The former helps shorten the response time of supercapacitors during charging and discharging, while the latter facilitates charge storage and release, thereby further enhancing the energy density of supercapacitors. Many studies now aim to achieve faster reaction kinetics through defect engineering, thereby improving electrochemical performance [25–27]. For example, Fu et al. successfully constructed a yolk-shell structured MnO_2 microsphere with oxygen vacancies through a simple three-step method. The obtained MnO_2 microspheres possess superior conductivity, which can accelerate the diffusion of electrolyte ions, leading to higher specific capacitance (the specific capacitance of ov-MnO_2 is twice that of pure MnO_2) and high energy density (40.2 Wh kg^{-1}) [28]. Yang et al. proposed a strategy to generated oxygen vacancies in MoO_{3-x} to achieve larger interlayer spacing with improved charge storage kinetics. Oxygen vacancies can not only significantly increase the interlayer spacing and conductivity of MoO_3, but also greatly enhance its electrochemical activity; the fabricated asymmetric supercapacitor exhibits an ultrahigh energy density of 150 Wh kg^{-1} [29]. Li et al. have demonstrated that doping Cu into Ni_3S_2 could introduce defect energy levels near the fermi level to enhance intrinsic electronic conductivity and electrochemical activity, which helps to accelerate the kinetics of redox reactions, thereby exhibiting high energy density (33.7 Wh g^{-1} at 850.1 W kg^{-1}) and long cycle stability (the retention after 5000 cycles is 94.0%) [30].

Atomic layer deposition (ALD) technology is an advanced method that utilizes the principle of chemical vapor deposition to deposit materials layer by layer on a substrate surface through precise control of chemical reactions [31,32]. The core of this technology lies in its self-limiting growth characteristics, where each chemical reaction occurs only on a single layer of atoms/molecules, thereby achieving precise control on the thickness and composition of the as-prepared thin film [33,34]. This means that ALD can accurately introduce defects into different material systems, such as metals, semiconductors, insulators, organic materials, etc. This means that we can design a general defect engineering method based on ALD for the selection of the most suitable materials as needed to introduce or repair defects [35–37]. In addition, ALD technology has shown great potential in designing and engineering high-performance electrode materials. As a unique thin film deposition technology, ALD achieves precise thickness control and excellent deposition quality through self-terminating surface reactions, which is crucial for optimizing the performance of supercapacitor electrode materials [38–40].

In this review, we first introduced the history, principles, and wide applications of ALD in various fields. Subsequently, the application of ALD in defect engineering is reviewed, including the use of ALD technology to manufacture oxygen vacancies, surface defects, and passivation of defects. The unique advantages of ALD in defect engineering are analyzed. Finally, the application of ALD in supercapacitors is introduced, including the preparation of supercapacitor electrode materials and their defect engineering using ALD technology.

2. Introduction of ALD

Atomic layer deposition (ALD) is a high-precision thin film preparation technique. In the 1970s, ALD technology was first referred to as atomic layer epitaxy (ALE) when it was used to deposit ZnS thin films on flat display panels [41]. Since the beginning of the 21st century, continuous, self-limiting surface reactions have been introduced into the process of thin films, no longer requiring epitaxial growth to their underlying substrates [42]. Modern ALD technology has made considerable progress and became a mature scientific technology. ALD relies on continuous self-limiting surface reactions to deposit thin films layer by layer on the material surface, achieving atomic level control accuracy [36,43–46]. This technology allows the introduction of chemical precursors (mainly gaseous compounds) into the reaction chamber to interact with specific adsorption layers on the substrate surface, and then effectively remove unreacted precursors and reaction by-products through inert carrier gas blowing. This process will be repeated until the desired film thickness, composition, and structure is achieved [37,47–49].

The reaction mechanism of ALD technology is grounded in a self-limiting surface reaction [50,51]. In each reaction cycle, the precursor only reacts with the reactive sites available on the surface, forming a monolayer of material, after which the reaction ceases spontaneously until the subsequent precursor is introduced. This self-limiting attribute guarantees the uniform deposition of atoms layer by layer, thereby achieving precise control over the thickness of the thin film [52–54]. Figure 2 visually illustrates the ALD process of SnN_x; the deposition process is divided into two and a half reactions [55].

$$2NH^* + Sn(N(CH_3)_2)_4 \rightarrow N_2Sn(N(CH_3)_2)_2^* + 2HN(CH_3)_2 \quad (1)$$

$$N_2Sn(N(CH_3)_2)_2^* + NH_3 \rightarrow N_2SnNH_2^* + 2HN(CH_3)_2 + N_2 \quad (2)$$

Substances transferred or anchored onto substrates through chemical adsorption are marked with asterisks (*). Reactions (1) and (2) are both ammonia exchange reactions, where the surface material of NH_x^* is replaced by the surface material of $Sn(N(CH_3)_2)_y^*$.

The core advantage of ALD technology depend on its self-limiting growth characteristics [56–58]. This self-limiting property enables ALD to grow highly uniform conformal films on three-dimensional structures while achieving atomic level growth and regulation [59–61]. ALD possesses the properties of sequential, self-limiting, and controllable surface reactions, allowing thin films to grow conformally and atomically on various substrates. The key features of ALD include excellent thickness control, high uniformity, and the ability to coat complex geometries, which are attributed to the self-limiting growth mechanism where each cycle deposits a predictable amount of material [62]. The low temperature processing capabilities of ALD make it suitable for a wide range of materials [63], including thermally sensitive substrates, while the layer-by-layer growth enables the engineering of nanolaminates and multifunctional coatings with tunable properties. Moreover, the high aspect ratio gap filling is another notable advantage of ALD, which is crucial for advanced semiconductor devices and nanoelectronics [64,65]. These distinctive characteristics of ALD have positioned it as a pivotal technology in fields such as microelectronics, energy storage and conversion, and surface modification [66,67].

ALD has found widespread applications across various industries due to its ability to deposit high-quality, conformal, and ultrathin films with precise control over thickness and composition. ALD plays a pivotal role in the development of transparent conductive oxides (TCOs) for solar cells, where it contributes to enhanced light transmission and reduced resistance [43,68]. Moreover, in the realm of energy storage, ALD is employed for the coating of battery electrodes to improve cycling stability and capacitance, with applications ranging from lithium-ion batteries to emerging battery technologies like sodium-ion and solid-state batteries [69]. Additionally, the catalysis field benefits from ALD for the preparation of well-defined catalysts with enhanced activity and selectivity, which is crucial for chemical synthesis and environmental protection [70,71]. These applications underscore the versatility and importance of ALD in enabling advanced technologies across multiple sectors.

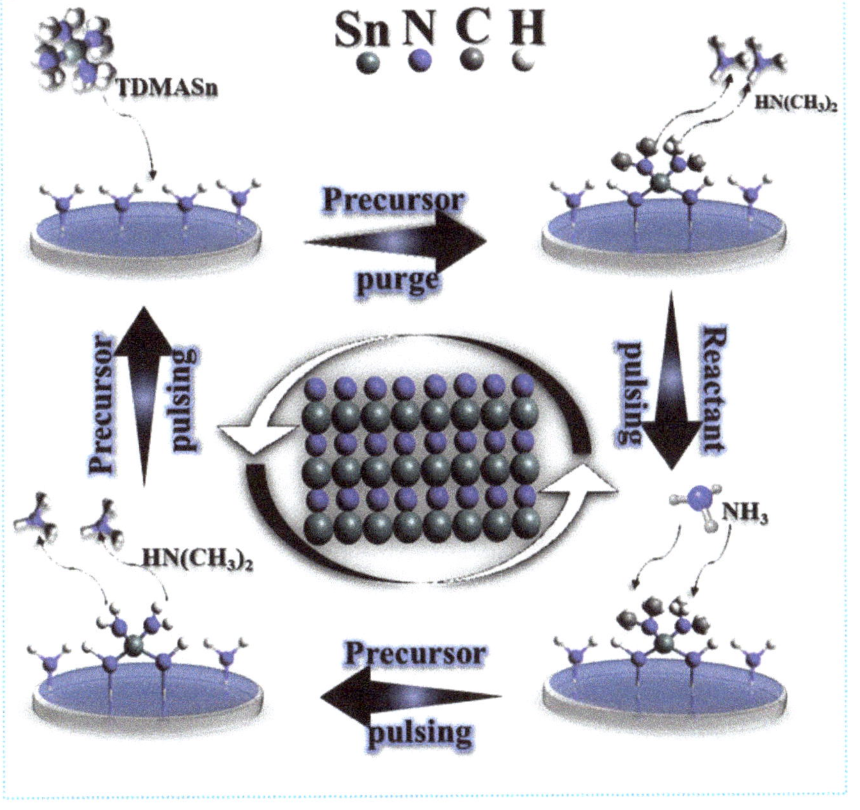

Figure 2. Schematic of ALD of tin nitride (SnN_x) process using TDMASn and NH_3 precursors as an example. Reproduced with permission. Copyright 2023, Wiley-VCH [55].

3. ALD for Defect Engineering

ALD technology plays a pivotal role in defect engineering. By precisely controlling the deposition process, it forms uniform and dense thin films on material surfaces at the atomic level, effectively modifying and reducing surface defects. This technology not only exhibits high step coverage, ensuring the formation of continuous films on complex or irregular surfaces, but also features low-temperature deposition, thereby avoiding the adverse effects of high-temperature processing on material properties. Compared with other technologies, ALD has unique advantages in defect engineering. Chemical vapor deposition (CVD) is a common synthesis method, whose principle primarily involves the utilization of gaseous substances to induce chemical reactions and transport reactions on solids at high temperatures, resulting in the formation of solid deposits. This technique is widely applied in various fields. However, in some reactions, excessively high temperatures can lead to the disappearance of defects [72]. In contrast, ALD is typically conducted at low temperatures, effectively avoiding such issues. Solvent-phase synthesis has been demonstrated to be capable of preparing electrode materials containing defects [73,74]. However, solvents may pose environmental and safety concerns, and separating and purifying products from solutions may require additional steps, such as evaporation, extraction, and crystallization, which can increase costs and time [75]. ALD, on the other hand, can be performed without the use of solvents and can achieve a more uniform distribution of defects.

3.1. *ALD for Oxygen Vacancy Engineering*

ALD technology alternately introduces precursor gas and reactant gas to perform chemical reactions on the substrate surface, forming a single-layer film. During the reaction process, the concentration and distribution of oxygen vacancies in the thin film can be controlled by adjusting type of precursor, temperature and gas partial pressure during the deposition process [76,77]. It possesses the advantages of high precision, good uniformity, and strong controllability. Thin films with specific oxygen vacancy concentrations prepared by ALD technology have shown excellent performance and broad application prospects in fields such as photocatalytic materials, perovskite solar cells, and supercapacitors [78–80].

3.1.1. ALD for Bulk Phase Oxygen Vacancy

Transition metal oxides with abundant oxygen vacancies can be grown via ALD on substrates by tuning deposition parameters. For example, Wang et al. synthesized corrosion-resistant thin films rich in oxygen vacancies on Al substrates by controlling the blowing time and pulse ratio of $TiCl_4/H_2O$ (Figure 3a). As shown in Figure 3b, The O/Ti ratio of TiO_2 would change accordingly through controlling the $TiCl_4/H_2O$ pulse ratio in order to regulate the content of oxygen vacancies in different TiO_2 samples. In corrosion testing, as shown in Figure 3c, the sample coated with a TiO_2 coating layer rich in oxygen vacancies has a higher corrosion barrier, meaning that the leaching of Al in the underlying matrix is reduced, and the corrosion resistance is the best [81]. Li et al. grew ZnO thin films directly on SiO_2/Si chips using ALD and then subjected the ZnO thin films to Ar treatment, as shown in Figure 3d. XPS analysis showed that the ZnO thin films grown directly using ALD contained abundant oxygen vacancies, and the concentration of oxygen vacancies could be further controlled through Ar treatment (Figure 3e). The atomic structure diagram of ZnO thin film is shown in Figure 3f. There are both oxygen vacancies and internal defects on the ZnO surface, and the content of oxygen vacancies is usually directly related to the gas sensing performance [82]. The test shown in Figure 3g also proves that the sample (noted as 700) with the highest concentration of oxygen vacancies exhibits the best gas sensing performance. Subsequently, the mechanism was explained in detail (Figure 3h). The abundant oxygen vacancies increase the concentration of adsorbed oxygen, allowing more gas molecules to adsorb on the surface to promote subsequent redox reactions. In addition, oxygen vacancies act as electron donors, increasing the charge density of the maximum valence band and minimum conduction band in the vicinity. This is due to the reduced bandgap of ZnO, which is conducive to the adsorption and activation of TEA. Therefore, it exhibits excellent sensing performance for TEA [83].

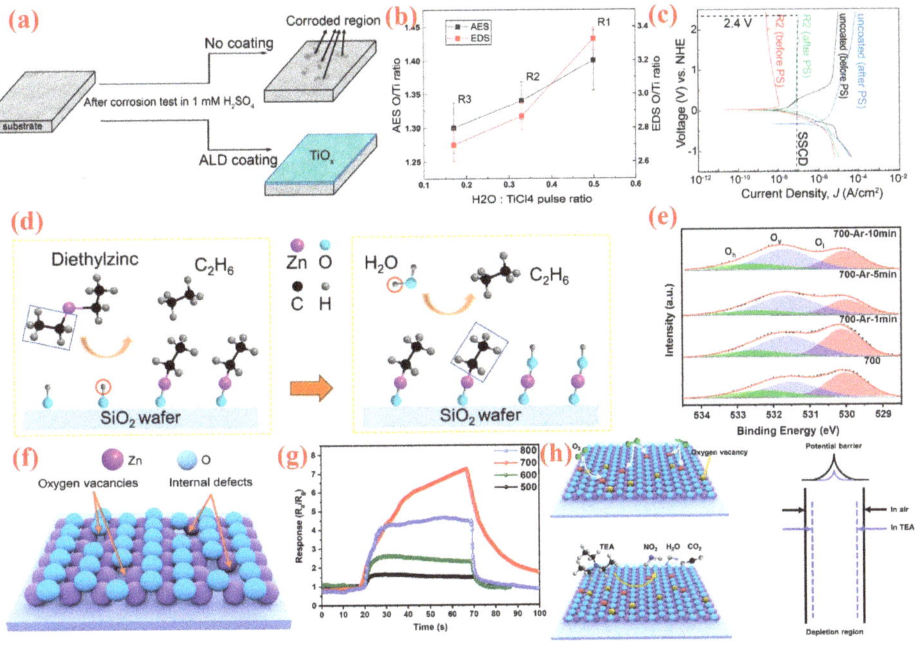

Figure 3. (**a**) Schematic outline of the corrosion protection of Al substrate achieved by optimized ALD TiOx protection layers; (**b**) O/Ti ratios for the ALD recipes; (**c**) polarization curves. Reproduced with permission. Copyright 2024, ACS Publishing [81]. (**d**) ALD deposition process of ZnO; (**e**) XPS survey spectra of ZnO films; (**f**) atomic structure of oxygen-deficient ZnO surface; (**g**) gas sensing test; (**h**) surface reaction mechanism. Reproduced with permission. Copyright 2021, Elsevier Publishing [83].

3.1.2. ALD for Interface/Surface Oxygen Vacancy

The ALD process is a technology from bottom to top that enables precise adjustment of surfaces and interfaces [84,85]. Therefore, in addition to directly generating metal oxides containing oxygen vacancies, oxygen vacancies can also be constructed at the interface between two thin films. For example, Hu et al. modified ZnO on HKUST-1 via ALD and calcined it to form Cu-doped ZnO (CZO-6) with a core–shell structure, as shown in Figure 4a. As shown in Figure 4b, the oxygen vacancies generated at the Zn-O-Cu interface can act as adsorption sites for hydrogen sulfide, further improving selectivity and sensitivity. This has also been demonstrated in the gas sensing testing (Figure 4c) [86]. Similarly, Hu et al. used ALD to modify the CoO_x surface of ZnO thin films and deposited ultra-thin $ZnCo_y$ gas sensitive films (Figure 4d). Due to the different deposition cycles, it was observed from Figure 4e,f that Co elements were uniformly distributed in the form of particles with a diameter of 5 nm on the surface of ZnO thin films. As shown in Figure 4g,h, XPS analysis showed that the addition of Co increased the oxygen vacancy content at the Co-O-Zn interface, and the synergistic effect of abundant oxygen vacancies and heterojunction interface sites increased the surface oxygen content and H_2S selective oxidation to S, which is consistent with the gas sensing test results, as shown in Figure 4i,j [87]. In addition, unlike traditional immersion or drop coating thin films, ALD ZnO thin films have a compact structure, which is not only beneficial for improving sensing performance but also plays an important role in the field of energy storage [88–91].

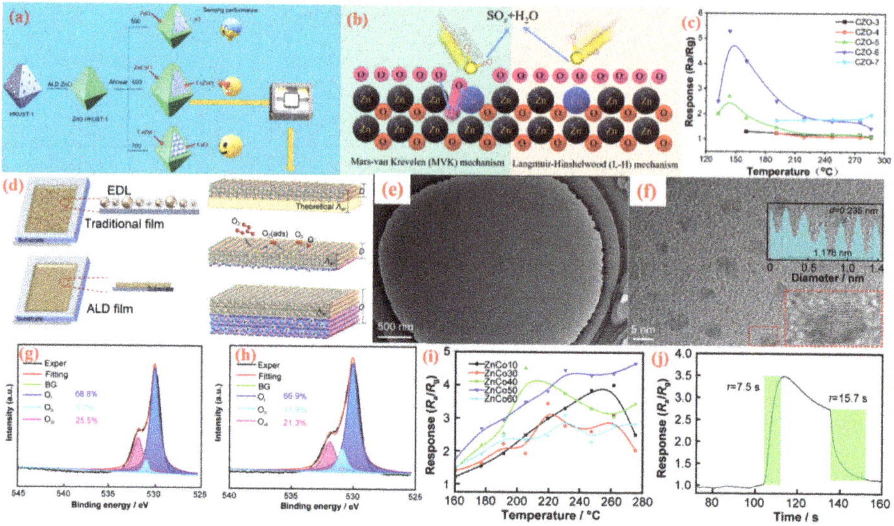

Figure 4. (**a**) Schematic diagram of the formation process of confined samples; (**b**) the schematic illustration for proposed sensing mechanism; (**c**) sensing performance. Reproduced with permission. Copyright 2024, Elsevier Publishing [86]. (**d**) Illustration of intergranular contact regions and relation between ALD ZnO thickness and length of space-charge layer; (**e**,**f**) TEM and HRTEM images. XPS survey spectra of (**g**) ZnO films and (**h**) ZnCo50 film; (**i**,**j**) gas sensing test. Reproduced with permission. Copyright 2023, Spinger Nature Publishing [87].

3.1.3. ALD for Inside Oxygen Vacancy

Oxygen vacancies in samples fabricated with ALD techniques are not only present at the interfaces between different oxide layers or on the surface of the materials but also can be generated inside the hollow structure. This usually requires the active species, such as zeolites [92], carbon nanotubes [93], and metal–organic frameworks (MOFs), to be confined in nanometer or subnanometer spaces [94] to create a unique microenvironment, thereby regulating the rate and path of chemical reactions. This is also known as spatial confinement and is widely used to improve the activity and selectivity of catalysts. Dong et al. deposited Pt nanoparticles into the pores of SnO_2 using the ALD method, as shown in Figure 5a. The closed separation of Pt sites accelerates the adsorption of acetone and the redox reaction between acetone and adsorbed oxygen, thus exhibiting excellent sensing performance (Figure 5b). XPS analysis also showed that the addition of Pt increased the concentration of oxygen vacancies, proving that oxygen vacancies were generated inside the material by the ALD method (Figure 5c). The DFT calculation shown in Figure 5d indicates that acetone molecules are more easily adsorbed on SnO_2-Pt2. The closed structure increased the concentration of acetone, and the synergistic effect of Pt separation catalysis and the restricted structure significantly improved the sensing performance, which is also reflected in Figure 5e [95]. Hu et al. used the sacrificial template method to deposit Ni and Sn species in different orders on the surface of carbon nanotubes, followed by calcination to remove carbon, resulting in NiO-SnO_2 nanocoils (NiSnNCs), as shown in Figure 5f. Depending on the location of Ni, they were labeled as outside NiSnNCs (O-NiSnNCs), inside NiSnNCs (I-NiSnNCs), and both outside and inside NiSnNCs (B-NiSnNCs). The position of Ni in different samples can be accurately observed form TEM images, shown in Figure 5g. The XPS results also indicate that the generation of oxygen vacancies can be controlled outside, inside, or both inside and outside the hollow tube, depending on the location of Ni, with I-NiSnNCs having the highest proportion of oxygen vacancies (Figure 5i–k). The high sensing performance of confined NiO in SnO_2 nanocoils is due to the overflow of dissociated H on NiO to the interface active sites, as well as the synergistic effect of

NiO and SnO_2, leading to the enrichment of hydrogen concentration. This phenomenon is called the funnel effect [96].

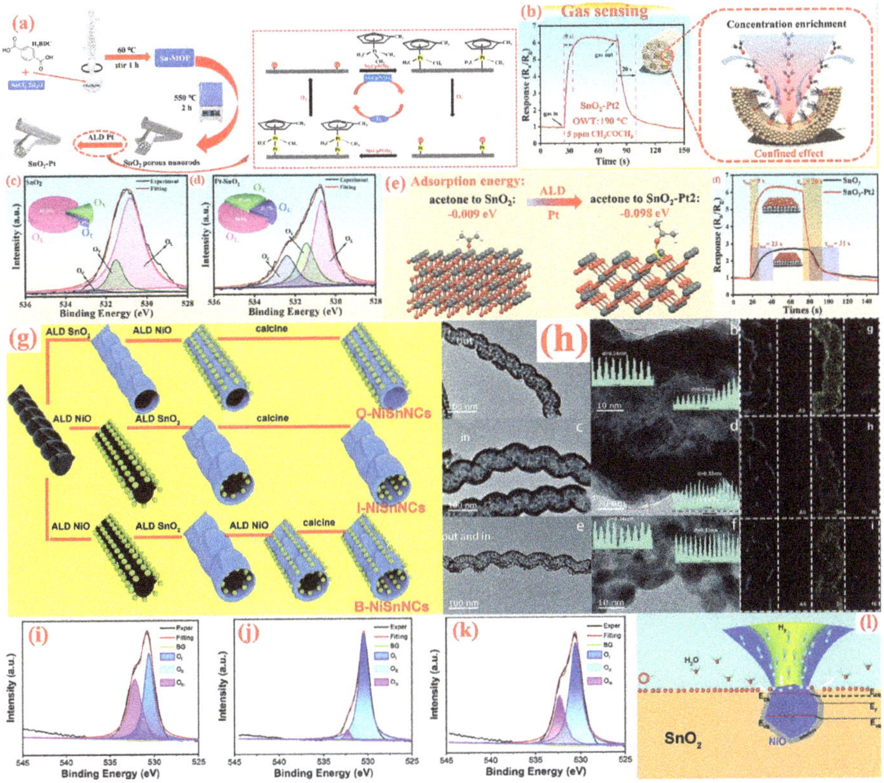

Figure 5. (**a**) Synthesis route of SnO_2 porous nanorods; (**b**) the scheme of detection of VOCs and the sensing performance; (**c**,**d**) XPS survey; (**e**) the adsorption of acetone molecules; (**f**) the response transient. Reproduced with permission. Copyright 2024, Elsevier Publishing [95]. (**g**) Schematic illustration of the fabrication of NiSnNC samples; (**h**) TEM and HRTEM images; (**i–k**) XPS survey spectra of O-NiSnNCs, I-NiSnNCs, and B-NiSnNCs. (**l**) The funnelling effects of the hydrogen concentration effects during the sensing performance. Reproduced with permission. Copyright 2022, RSC Publishing [96].

In summary, not only samples containing oxygen vacancies can be directly prepared, but oxygen vacancies can also be formed at the interface between layers through multiple deposition through ALD. In addition, for some specific situations, the generation location of oxygen vacancies can have a significant impact on the entire reaction. Through the controllability of ALD technology, it is possible to generate oxygen vacancies at specific locations.

3.2. ALD for Surface Defect

Surface defect engineering is a technique that introduces or controls defects on the surface of materials to improve their performance, significantly affecting the electronic structure and surface properties of the materials [97,98]. It demonstrates its unique value and potential in various fields, such as catalysis, energy conversion, electrochemical sensing, and environmental protection [99–101]. Surface defect engineering can be achieved through methods such as chemical vapor deposition (CVD), laser processing, and ion implantation [102–104]. ALD can adjust atomic stoichiometry and doping levels in self-limiting surface reactions, thereby obtaining unique conformal and pinhole free films, which can play an important role in surface defect engineering [105].

Yu et al. altered the density of Ni vacancy (V_{Ni}) defects in bulk NiO_x by adjusting the oxygen injection time (t_{OE}) during the ALD process, as shown in Figure 6a–d. The Ni-O bond is disrupted to a varying degree by O_2 plasma; the ratio of Ni^{3+}/Ni^{2+} and O/Ni increase with the increase of oxygen injection time, indicating that the content of surface Ni defects (V_{Ni}) is proportional to the oxygen injection time (Figure 6e,f). The alteration in Ni defects concentration fine-tunes the physical and chemical properties of NiO_x thin films, improving carrier transport as well as reducing adverse redox reactions at the PVK/NiO_x interface. And that also increased power conversion efficiency (PCE) from 18.19% to 19.86%, shown in Figure 6g,h [106]. Dai et al. deposited an additional TiO_2 layer on the surface of TiO_2 nanowire film (TNF) using ALD. The structure of ALD TiO_2 undergoes a transition from ordered to completely disordered. The presence of Ti^{3+} defects in the ALD TiO_2 layer can be confirmed from XPS results. The effective ion diffusion coefficient of the electrode can be enhanced by increasing the thickness of the ALD TiO_2 layer, and the electrochemical performance of ALD TiO_2 modified TNF is significantly improved. The deposition of an additional TiO_2 layer on crystalline TiO_2 nanowire thin film by the ALD modification method comprehensively improves electrochemical performance, as its surface structure continuously evolves, including interconnected domains, Ti^{3+} defects, and exchangeable hydroxyl groups, thereby comprehensively enhancing the electrochemical performance of TiO_2 nanowire film [107].

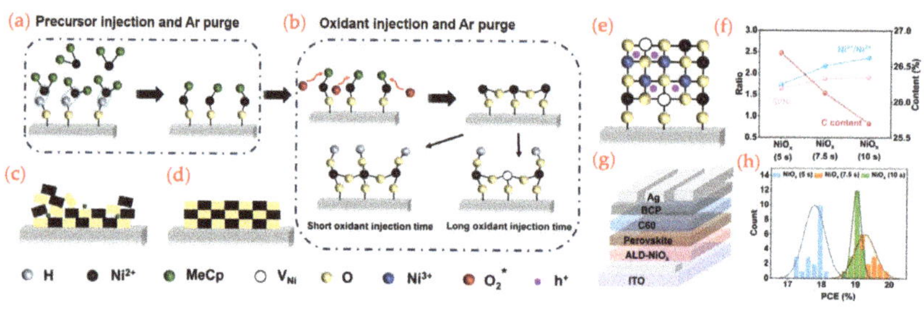

Figure 6. (**a–d**) Schematic illustration of the principle of modulating the valence state of the Ni element using ALD; (**e**) schematic NiO_x lattice structure with Ni vacancy defects; (**f**) surface O/Ni ratio, Ni^{3+}/Ni^{2+} ratio, and C content of different NiO_x; (**g**) schematic stack of the p-i-n structured PSCs; (**h**) PCE histograms of 20 devices. Reproduced with permission. Copyright 2024, ACS Publishing [106].

3.3. *ALD for Passivation of Defect*

In some fields, the appearance of defects often leads to a decline in material properties, and the formation of defects is inevitable in the material preparation process. Repairing or passivating defects can improve material properties. Defect passivation is a technique that introduces or regulates defects on the surface of materials to improve their performance [108,109]. This technology has demonstrated its unique value and potential in multiple fields such as catalysis, energy conversion, electrochemical sensing, and environmental protection [110–113]. Taking perovskite solar cells as an example, the presence of defects can lead to energy loss, ion migration, operational instability, and hysteresis, seriously hindering their potential for industrial applications. Therefore, researchers have developed various defect passivation strategies, including using additives to passivate defects in perovskite materials [114]. These strategies involve surface modification of materials as well as control of internal defects to reduce nonradiative recombination, charge capture and scattering, ion migration, and other phenomena.

ALD has unique advantages in defect passivation, especially in improving material performance and stability. ALD technology can perform thin film growth at lower temperatures, which helps to protect the substrate material from thermal damage and reduce the occurrence of thermally induced defects [115]. And ALD technology can also deposit various materials, including metals, semiconductors, insulators, and organic materials. This multifunctionality enables ALD technology to select suitable materials for surface defect passivation according to different application requirements [116]. For example, Ng et al. reported the use of the ALD method to coat 3D-printed nanocarbon surfaces, as shown in Figure 7a. Figure 7b displays the 3D electrode surface morphology of an 80 ALD cycle Al_2O_3 coating at different magnifications. Owing to the extremely thin nanoscale coating, scanning electron microscopy (STEM) imaging of the carbon surface of Al_2O_3 coating was performed using a high angle annular dark field (HAADF) detector with 20 and 80 ALD cycles. The initial growth of Al_2O_3 in the 20th cycle is characterized by unevenly sized islands dispersed on the surface. As the ALD process continues and the number of cycles increases, the deposited islands become additional nucleation sites for subsequent growth. These islands combine to form larger islands and expand into adjacent uninhabited spaces. Therefore, 80 cycles of Al_2O_3 produced larger and thicker island layers, gradually forming a more continuous coating, which confines the initial nucleation at the defect sites and edges as anchoring sites. Figure 7c reveals SECM images of electrocatalytic oxidation for blank and 80 cycle Al_2O_3-coated 3D-printed nanocarbon electrodes. At the same voltage, 80 cycle Al_2O_3 exhibits a stronger current, and the Al_2O_3 coating can catalyze the oxidation of catechol, which is further confirmed by Figure 7d,e. Al_2O_3-coated 3D-printed nanocarbon electrodes exhibit superior electrochemical performance. The thin Al_2O_3 film deposited on the electrode can not only effectively serve as an active layer for adsorbing catechins but also serve as a surface passivation layer to reduce the defect density inside the carbon fiber electrode [117].

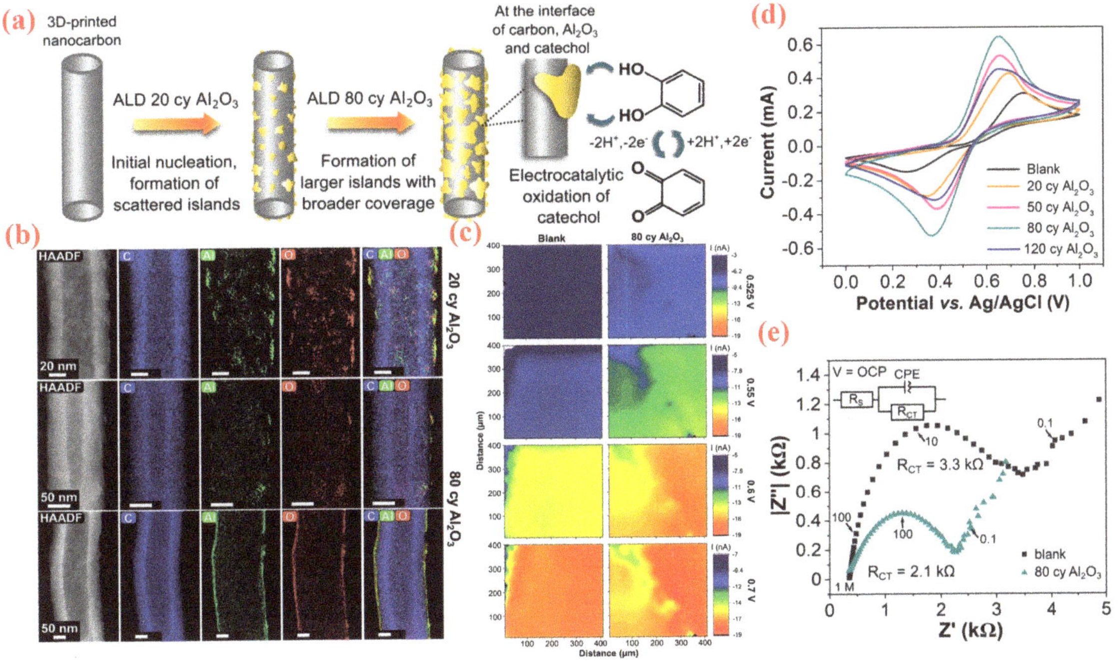

Figure 7. (**a**) Mechanism of the electrocatalytic oxidation of catechol to o-benzoquinone; (**b**) STEM-HAADF images; (**c**) SECM images of electrocatalytic oxidation acquired via substrate generation/tip collection (SG/TC) mode for blank and 80 cycle Al_2O_3-coated 3D-printed nanocarbon electrodes; (**d**,**e**) electrochemical characterization. Reproduced with permission. Copyright 2021, ACS Publishing [117].

4. Defect Engineering via ALD for Supercapacitors

4.1. ALD for Supercapacitors

Electrochemical supercapacitors have become one of the most promising strategies for the next generation of energy storage systems due to their high-power density, fast charging and discharging rates, significant cycling stability, and low maintenance costs [118–121]. Generally speaking, capacitance electrodes with high power capacity require high surface area and good conductivity to form double layers and undergo redox reactions. Therefore, suitable production techniques are needed to produce supercapacitor electrode materials with ideal geometric shapes, appropriate charge transport behavior, controlled thickness, and surface chemical properties [122–125]. ALD, with its unique processing advantages, has become the primary technology for developing advanced generation supercapacitor electrode materials with superior macroscopic/nanoscale structures and surface properties [126–130]. The advantages of ALD, such as superior thickness control of the deposition layer, uniform deposition on large surface areas, consistency in deposition on complex features, and ease of manufacturing composite nanostructures, make it particularly suitable for application in energy storage devices [131–134].

4.1.1. ALD for Thin Film Electrode Materials

The uniform thickness of thin film electrode materials for supercapacitors is crucial for achieving rapid electrochemical response, as it helps to shorten the diffusion length of charges and ions [135,136]. In addition, thin film electrode materials directly affect the performance of supercapacitors, including energy density, power density, cycle stability, and charge and discharge rates [137]. Thin film electrodes have a thin and uniform thickness, which not only ensures rapid electrochemical response during charging and discharging processes, but also allows for the use of thin film electrodes in portable and/or miniaturized devices [138]. The research on thin film electrode materials includes the optimization of the mechanical strength and electrochemical properties of the materials. For example, by combining two-dimensional materials such as MXenes with one-dimensional materials such as nanocellulose, researchers have developed supercapacitor electrode materials with both high mechanical strength and excellent electrochemical performance. This hybrid material not only maintains high conductivity but also exhibits excellent mechanical strength, allowing supercapacitors to bend and roll without sacrificing performance [139].

ALD technology possesses a suite of significant advantages in the fabrication of thin-film electrode materials, rendering it an essential tool in the development of energy storage devices such as supercapacitors. Through precise control of chemical reactions, ALD enables the layer-by-layer deposition of films on material surfaces, achieving atomic-level control over film growth. This technique can introduce ultra-thin and uniform coatings on material surfaces, effectively enhancing interfacial compatibility and improving performance and safety [140–142]. Lopa et al. utilized ALD technology to fabricate WO_3-MoO_3 nanocomposite electrodes with a thickness of less than 10.0 nm for supercapacitor applications, as depicted in Figure 8a–c. The thickness of the WO_3 and MoO_3 layers exhibited a good linear relationship with the number of ALD cycles, demonstrating the precise control of film thickness in thin-film electrode materials achievable through ALD. The thickness of the WO_3 film deposited via ALD was approximately 3.95 nm, while that of the MoO_3 film was about 3.93 nm, as shown in Figure 8d–g. Through meticulous optimization and control, the prepared electrodes featured uniform thickness, high purity, and good interfacial contact. The fabricated heterostructure electrodes displayed excellent pseudocapacitive behavior, highlighting their suitability for energy storage applications (Figure 8h,i). At a high current density of 19.23 A g^{-1}, the specific capacitance (Cs) reached a value of 803.10 F g^{-1}. Furthermore, the electrode exhibited excellent cyclic stability and low Cs loss after 5000 GCD cycles, indicating its superior durability and long-term reliability for various applications. Additionally, they reported the fabrication of different film electrode materials for supercapacitors using ALD technology. The prepared SnO_2-ZnO (Figure 8j–m) and SnO_2-In_2O_3 (Figure 8n–p) thin-film electrode materials, with their uniform thickness, high purity, and good crystallinity, demonstrated the versatility of ALD in depositing uniform thin-film electrode materials for supercapacitors. The excellent performance in long-term cyclic testing further confirmed the electrochemical stability of ALD-deposited thin-film electrode materials [143–145].

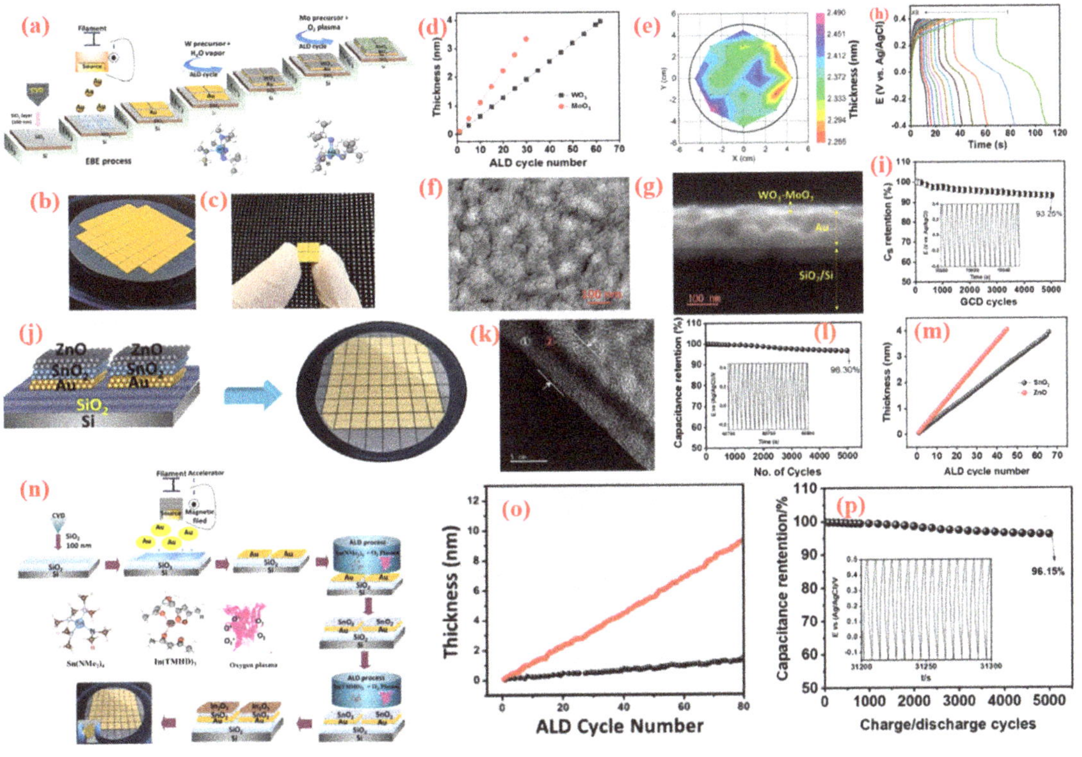

Figure 8. (**a**–**c**) Schematic representation of the fabrication steps involved in the production of WO_3-MoO_3 electrodes; (**d**) thickness vs. ALD cycle number; (**e**) wafer scale spectroscopic ellipsometry measurements; (**f**) to-view and (**g**) cross-sectional FE-SEM images of the WO_3-MoO_3 film; (**h**,**i**) electrochemical characterization. Reproduced with permission. Copyright 2024, Elsevier Publishing [143]. (**j**) Schematic illustration of 2D SnO_2-ZnO heterojunction electrodes; (**k**) TEM images; (**l**) long-term cycle test of 2D SnO_2-ZnO heterojunction electrodes; (**m**) film thickness vs. ALD cycle number monitored by an in situ sensing electrode for SnO_2 and ZnO. Reproduced with permission. Copyright 2023, ACS Publishing [144]. (**n**) Schematic illustration of 2D SnO_2-In_2O_3 electrodes; (**o**) film thickness vs. ALD cycle number for SnO_2 and In_2O_3; (**p**) long-term cycle test. Reproduced with permission. Copyright 2022, Elsevier Publishing [145].

4.1.2. ALD for Surface Coating of Electrode Materials

The application of coating layers in supercapacitor electrode materials plays a multifaceted role in enhancing the performance and stability of the electrodes. Coating layers provide a protective barrier for electrode materials, which reduces side reactions between active substances and the electrolyte, thereby extending the cycle life of the supercapacitor [146,147]. Coating layers also improve the interfacial compatibility of electrode materials, facilitating the transport of electrons and ions. In some cases, coating layers can act as a medium to increase the specific surface area of electrode materials, offering more charge storage sites [148]. The optimization of structure helps to increase the power density and energy density of supercapacitors [149,150]. Furthermore, coating layers enhance the structural stability of electrode materials, particularly during volume expansion or contraction in the charging and discharging processes. This protective effect is conducive to maintaining the integrity of the electrode materials, reducing structural degradation, and thus improving the long-term stability and reliability of supercapacitors [151,152].

ALD plays a significant role in the construction of coating layers for supercapacitor electrode materials. ALD enables precise control over chemical reactions, allowing for the layer-by-layer deposition of thin films on material surfaces and achieving atomic-level thin-film growth. This effectively enhances the material's interfacial compatibility [153]. Films deposited by ALD, such as aluminum oxide, exhibit excellent thermal stability. Electrodes coated with these films can maintain better performance when operating at high temperatures, reducing the risk of thermal runaway [154], further enhancing the performance and safety of supercapacitors. Adhikari et al. successfully constructed freestanding core–shell NiO/Co_3O_4 electrodes on nickel foam by growing cobalt nanocone structures through a simple hydrothermal method, followed by annealing to form Co_3O_4. The advanced atomic layer deposition (ALD) technique was then employed to produce a shell of NiO on the Co_3O_4 nanocone structures. The Co_3O_4 structure serves as a layered substrate for the NiO atoms and conforms to the shape of the substrate, forming a core–shell structure. The layer-by-layer precipitation simplifies the process for controlling the thickness and uniformity of the Co_3O_4@NF. The deposited NiO adopts the primary nanoparticle shape with a nanocone structure. The optimized NiO/Co_3O_4@NF exhibits a specific capacitance of 1242 C g^{-1} (2760 F g^{-1}) at a current density of 2 A g^{-1} and retains a capacitance of 959.8 C g^{-1} (2133 F g^{-1}) at a current density of 10 A g^{-1}, demonstrating superior electrochemical performance. After 12,000 charge–discharge cycles, the material's capacitance retention is about 95.5%, which is exceptionally high compared with the retention capability of Co_3O_4@NF. Impedance studies confirm that the NiO/Co_3O_4@NF nanocone structure has low charge transfer resistance, and the better interaction between the core and shell facilitates faster transport phenomena. The excellent electrochemical performance of NiO/Co_3O_4@NF is not only attributed to the synergistic effect of the core Co_3O_4 and the atomic layer-deposited NiO shell but also to the effective stabilization of the Co_3O_4 structure against degradation by the ALD NiO protective shell, while maintaining superior electrochemical properties [155].

In addition to stabilizing the structure, ALD coatings play a significant role in enhancing electrochemical performance by increasing the specific surface area of electrode materials and the electronic synergy between layers. For instance, Pawar et al. reported the excellent performance of hierarchical nanosheets modified with an ultrathin molybdenum disulfide (MoS_2) layer on $NiCo_2O_4$ electrodes for supercapacitor applications. As shown in Figure 9a, an ultrathin MoS_2 layer was synthesized on the $NiCo_2O_4$ surface using the atomic layer deposition (ALD) method. From the TSEM images shown in Figure 9b,c, the $NiCo_2O_4$ nanosheets are much thinner than the Co_3O_4 hexagonal nanosheets, which may positively impact achieving a larger surface area and more active sites, conducive to higher capacitance values. At the same time, the ALD-MoS_2 layer has a rough surface at the atomic scale, which also increases the total surface area of the electrode, significantly improving the charge storage capacitance [156,157]. The thinner and well-connected $NiCo_2O_4$ nanosheets provide more interfaces between the electrode and electrolyte, while the optimal thickness of MoS_2 contributes to maximizing device performance. According to CV measurements, shown in Figure 9d, a significant portion of the capacitance is transmitted through Faradaic reactions. The high surface capacitance (2445 mF cm^2) and enhanced rate performance of the optimized $NiCo_2O_4$- MoS_2 composite material are attributed to the synergistic effects of the high specific surface area of the oxide nanosheets and the better electronic conductivity of the sulfide (Figure 9e–g) [158].

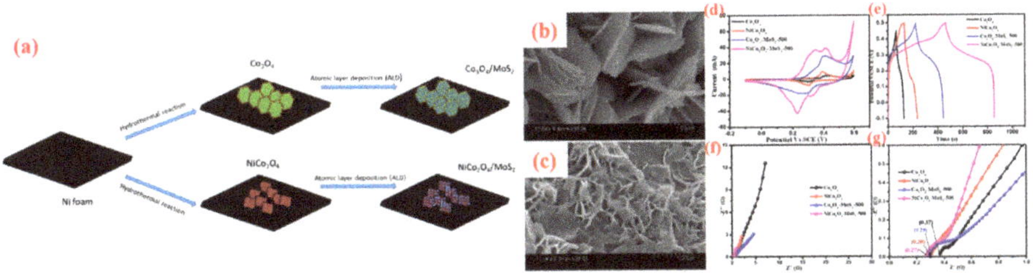

Figure 9. (**a**) Schematic representation of the synthesis of MoS_2-coated Co_3O_4 and $NiCo_2O_4$ nanostructures; (**b**,**c**) FESEM images of Co_3O_4-MoS_2 and $NiCo_2O_4$-MoS_2; (**d**–**g**) electrochemical behaviors of $NiCo_2O_4$-MoS_2. Reproduced with permission. Copyright 2022, Elsevier Publishing [158].

4.1.3. ALD for Loaded Nanoparticle

The application of nanoparticles in the electrode materials of supercapacitors has significantly enhanced their electrochemical performance. These nanoparticles can serve as dopants to modify the surface of electrode materials or embed in composite materials to effectively reduce resistance and increase specific capacitance [159–161]. For instance, low-crystallinity iron oxyhydroxide nanoparticles, as anode materials for supercapacitors, have demonstrated excellent electrochemical energy storage characteristics. Their design and assembly in aqueous electrolytes can yield supercapacitor energy storage devices with high energy density, high power density, and long cycle life [162]. Nanoparticles can be directly used as electrode materials, and although they are susceptible to oxidation, their stability and performance can be significantly improved through appropriate protective layers or by compounding with other materials. Metal nanoparticles can even act as current collectors, especially precious metals like gold or platinum, due to their excellent stability and high conductivity [163,164]. Furthermore, the introduction of nanoparticles can increase the specific surface area of electrode materials, thereby providing more charge storage sites [165].

As depicted in Figure 10a, Hai et al. employed the ALD method to deposit WO_3 and TiO_2 on a SiO_2/Si substrate, with a subsequent annealing process leading to the aggregation of TiO_2 films into nanoparticles, thereby functionalizing the WO_3 electrode material with TiO_2 nanoparticles. It is clearly observable from Figure 10b–d that the top blank electrode exhibits a bright yellow color, while both the two-dimensional WO_3 electrode on the left and the two-dimensional TiO_2NP-WO_3 electrode on the right display a deep yellow, indicating the deposition of WO_3 and TiO_2NP-WO_3 nanofilms. In contrast to the pure two-dimensional WO_3 film (Figure 10e), TiO_2 after annealing is uniformly distributed on the WO_3 surface in the form of small nanoparticles, with the ultrathin TiO_2 layer gradually condensing into small nanoparticles in air at 380 °C (Figure 10f,g), resulting in an increased surface area. The GCD test results reveal that the surface functionalization of the two-dimensional WO_3 electrode with TiO_2 nanoparticles shifts the capacitive behavior from double-layer capacitance to pseudocapacitance, with a significant increase in specific capacitance of approximately 1.5 times (Figure 10h,i). CV and rate capability tests also demonstrate the enhanced electrochemical performance of the 2D TiO_2NP-WO_3 electrode (Figure 10j,k) [166].

In summary, atomic layer deposition (ALD) technology exhibits notable advantages in the preparation of electrode materials for supercapacitors. The high precision and uniformity ensure the superior performance of electrode materials, with the capability of achieving monolayer deposition that ensures uniform coverage even on complex shapes and structures. Table 1 compares the electrochemical performance of supercapacitor electrodes prepared by ALD with those prepared by other methods. It can be observed that the electrochemical performance of the electrodes deposited via ALD is significantly higher than that of electrodes synthesized by other techniques. ALD technology holds broad application prospects and significant research value in the field of electrode material preparation for supercapacitors.

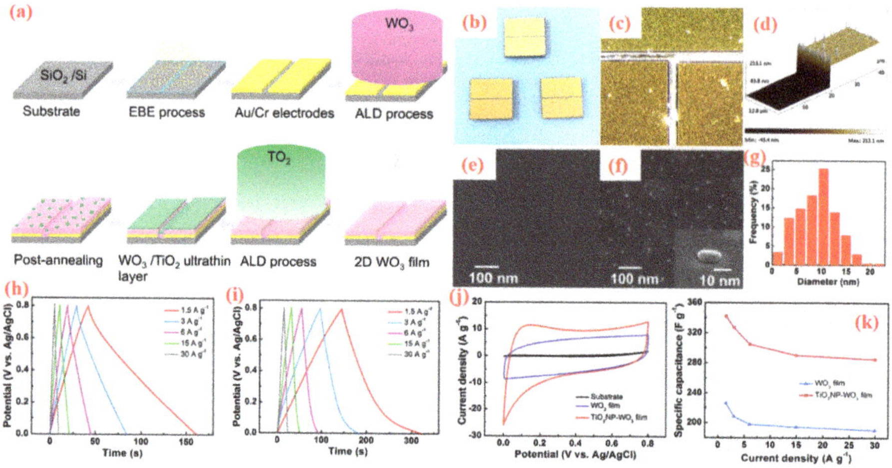

Figure 10. (**a**) Schematic illustration of TiO_2NP-WO_3 film; (**b**) optical image of the samples; (**c**) high-resolution optical image of the samples; (**d**) AFM image; (**e**) 2D WO_3 film and (**f**) 2D TiO_2NP-WO_3 film after post-annealing; (**g**) size distribution histogram of TiO_2 nanoparticles; (**h–k**) electrochemical behaviors. Reproduced with permission. Copyright 2017, Elsevier Publishing [166].

Table 1. Summary of electrochemical performance of ALD-based electrodes in comparison to other preparation methods.

Electrodes Prepared via ALD	Electrochemical Performance	Electrodes Prepared by Other Methods	Electrochemical Performance	Ref.
MoS_2	C_{sp} = 3400 mF cm^{-2} at 3 mA cm^{-2}, C_R = 82% after 4000 cycles	MoS_2	C_{sp} = 403 F g^{-1} at 1 mV s^{-1}, C = 80% after 2000 cycles at 150 mV s^{-1}	[167–169]
		MoS_2 quantum sheets	C_{sp} = 140 F g^{-1} at 5 mV s^{-1}, C_R = 84% after 5000 cycles at 7.5 mA s^{-1}	
Co_9S_8	C_{sp} = 1309 F g^{-1} at 45 A g^{-1}, C_R = 94.4% after 2000 cycles at 45 A g^{-1}	Co_9S_8/CNF	C_{sp} = 718 F g^{-1} at 1 A g^{-1}, C_R = 83.1% after 5000 cycles at 10 A g^{-1}	[170–172]
		Co_9S_8@C	C_{sp} = 514 F g^{-1} at 1 A g^{-1}, C_R = 88% after 1000 cycles at 8 A g^{-1}	
TiO_2	C_{sp} = 2332 F g^{-1} at 1 A g^{-1}	TiO_2-rGO	C_{sp} = 338 F g^{-1} at 5 mV s^{-1}, C_R = 93.3% after 3000 cycles at 2.5 A g^{-1}	[173–175]
		TiO_2-NF	C_{sp} = 1052 F g^{-1} at 25 mV s^{-1}, C_R = 80% after 5000 cycles at 7 A g^{-1}	
V_2O_5	C_{sp} = 540 F g^{-1}, C_R = 89% after 10,000 cycles at 5 A g^{-1}	V_2O_5	C_{sp} = 204 C g^{-1} at 1 A g^{-1}	[176–178]
		V_2O_5	C_{sp} = 437 F g^{-1} at 1 A g^{-1}, C_R = 91% after 10,000 cycles	

Table 1. *Cont.*

Electrodes Prepared via ALD	Electrochemical Performance	Electrodes Prepared by Other Methods	Electrochemical Performance	Ref.
RuO_2	C_{sp} = 644 F g^{-1}, C_R = 117% after 2000 cycles	RuO_2/PCs	C_{sp} = 539.6 F g^{-1} at 5 mV s^{-1}, C_R = 98% after 1000 cycles at 5 A g^{-1}	[179–181]
		RuO_2	C_{sp} = 209 F g^{-1} at 45 A g^{-1}, C_R = 94.4% after 2000 cycles at 45 A g^{-1}	
NiO/CNC	C_{sp} = 430 C g^{-1} at 1 A g^{-1}, C_R = 95.5% after 12,000 cycles at 5 A g^{-1}	C/NiO	C_{sp} = 585 F g^{-1} at 1 A g^{-1}, C_R = 100% after 6000 cycles at 5 A g^{-1}	[182–184]
		NiO/rGO	C_{sp} = 435.25 F g^{-1} at 1 A g^{-1}	
NiO/Co_3O_4	C_{sp} = 2760 F g^{-1} at 2 A g^{-1}, C_R = 95.5% after 12,000 cycles at 5 A g^{-1}	NiO/Co_3O_4	C_{sp} = 405 F g^{-1} at 1 A g^{-1}, C_R = 97.4% after 1000 cycles at 15 A g^{-1}	[155,185, 186]
		NiO/Co_3O_4	C_{sp} = 1303.9 F g^{-1} at 1 A g^{-1}, C_R = 88.5% after 10,000 cycles at 10 A g^{-1}	
Ni_3C/CNT	C_{sp} = 1850 F g^{-1} at 2 mA cm^{-2}, C_R = 98.5% after 5000 cycles at 20 mA cm^{-2}	g-C_3N_4/Ni_3C	C_{sp} = 1137.3 F g^{-1} at 1 A g^{-1}	[187–189]
		Ni_3C	C_{sp} = 390 C g^{-1} at 0.5 A g^{-1}, C_R = 86.9% after 3000 cycles at 2 A g^{-1}	
$NiCo_2O_4$/NiO	C_{sp} = 2439 F g^{-1} at 45 A g^{-1}, C_R = 94.2% after 20,000 cycles at 4 A g^{-1}	NiO/$NiCo_2O_4$	C_{sp} = 866 F g^{-1} at 5 mV s^{-1}, C_R = 85% after 5000 cycles at 5 A g^{-1}	[129,190, 191]
		NiO/$NiCo_2O_4$	C_{sp} = 1623 F g^{-1} at 2 A g^{-1}, C_R = 90% after 10,000 cycles at 10 A g^{-1}	

4.2. *Defect Engineering via ALD for Supercapacitors*

In the previous content, the role of ALD in defect engineering and electrode material preparation for supercapacitors was reviewed. The application of defect engineering in supercapacitors through ALD is a promising technical strategy. By precisely controlling the types and distribution of defects in electrode materials, the electrochemical performance of supercapacitors can be significantly improved.

4.2.1. Oxygen Vacancies

Oxygen vacancies play a pivotal role in supercapacitors, enhancing the electrochemical performance of electrode materials through multifaceted mechanisms. The introduction of oxygen vacancies can increase the number of active sites on the surface of electrode materials, thereby improving the charge transfer efficiency between the electrode and the electrolyte [192,193]. Additionally, the incorporation of oxygen vacancies can enhance electronic transport properties, reduce charge transfer resistance, and result in a faster kinetic response of the electrode material during charging and discharging processes [194,195]. The regulatory effect of oxygen vacancies can also optimize the surface properties of electrode materials, such as hydrophilicity and electrochemical stability, thereby enhancing the wettability and long-term stability of the materials in the electrolyte [196,197]. The microstructure of the material can be further optimized through the introduction of oxygen vacancies, such as pore size distribution and specific surface area, providing more storage space for electrolyte ions and consequently increasing the specific capacitance of the electrode material [198,199]. Moreover, oxygen vacancies can promote pseudocapacitive behavior, enhancing the energy storage performance of supercapacitors through surface adsorption and redox reactions [200,201]. For instance, Wang et al. introduced a dual-defect strategy to synthesize Mo-doped $NiCo_2O_4$ (R-Mo-$NiCo_2O_4$) with abundant oxygen vacancies. By employing $NaBH_4$ reduction, oxygen vacancy defects were generated within the $NiCo_2O_4$ lattice, resulting in an increased carrier concentration and a substantially enhanced conductivity. This, in turn, facilitated the involvement of more active materials in electrochemical redox reactions, resulting in its significantly enhanced specific capacity, capacity retention, and cycling stability [202]. Additionally, Wang et al. modeled the bulk oxygen substitution and surface oxygen vacancies in $ZnCo_2O_4$ nanowires (F-$ZnCo_2O_4$-x) grown on nickel foam. The increase in oxygen vacancy concentration enhances the conductivity and electrochemical activity of the electrode. The surface O vacancies could lower the conduction band, serving as active sites for electron trapping, Therefore, the obtained F-$ZnCo_2O_4$-x electrode achieves a high specific capacity of 664 $mAh{\cdot}g^{-1}$ at 1 $A{\cdot}g^{-1}$ [203].

ALD can directly obtain supercapacitor electrode materials with oxygen vacancies. Kao et al. utilized atomic layer deposition (ALD) to deposit a conformal coating of titanium nitride (TiN) on the surface of a carbon nanotube forest. The increased surface oxygen vacancies in titanium nitride (TiN) facilitated Faradaic redox surface reactions, as depicted in Figure 11a,b. Figure 11c,d reveal the SEM images of a vertically aligned carbon nanotube forest coated with ALD TiN, where a forest approximately 30 μm in height was successfully grown. From the TEM images in Figure 11e,f, the exposed carbon nanotubes exhibit a hollow core and a multi-walled structure with a uniform coating thickness. The thickness of the TiN coating, corresponding to 400 ALD cycles, is approximately 20 nm, demonstrating a good uniformity. XPS measurements (Figure 11g) indicate a relatively high oxygen content on the electrode surface, suggesting a high concentration of oxygen in the native oxide layer. Furthermore, nitrogen doping in titanium dioxide materials has been shown to increase oxygen vacancies, thereby enhancing the potential for charge insertion and storage [204]. The capability to store more charge stems from the recombination of electrons associated with oxygen deficiency. Chemical testing (Figure 11h–j) demonstrates that the capacitance of the TiN-CNT devices increased by over 500% compared with bare carbon nanotubes (TiN-CNT at 81 mF/cm^2, CNT electrode at 14 mF/cm^2), while the capacitance of the bare carbon nanotubes was nearly 400 times higher than that of the planar TiN electrode (0.2 mF/cm^2) [80].

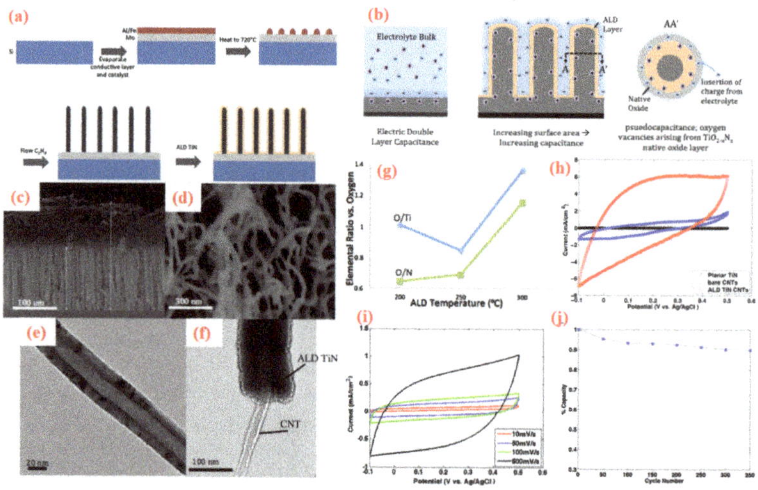

Figure 11. (**a**) Schematic presentation for TiN-CNT architecture; (**b**) conceptual illustration of increasing capacitance through increased surface area and the pseudo capacitive effect; (**c**) SEM image of

uncycled ALD TiN coated onto vertically aligned CNT forest; (**d**) close-up SEM image of uncycled ALD TiN-coated CNTs; (**e**) TEM image of single multiwalled carbon nanotube; (**f**) TEM image of ALD TiN-coated multiwalled carbon nanotube; (**g**) XPS characterization of ALD TiN; (**h**–**j**) electrochemical behaviors of ALD TiN-CNTs. Reproduced with permission. Copyright 2016, Elsevier Publishing [80].

4.2.2. Interface Defects

Interface defects play a critical role in supercapacitors, significantly enhancing the electrochemical performance of electrode materials through various mechanisms. The introduction of interface defects can increase the number of active sites on the surface of electrode materials, thereby improving the charge transfer efficiency between the electrode and the electrolyte [205,206]. Additionally, the integration of interface defects can enhance electron transport properties, reduce charge transfer resistance, and enable a faster dynamic response of the electrode material during charging and discharging processes [207,208]. The regulatory effect of interface defects can also optimize the surface properties of electrode materials, such as hydrophilicity and electrochemical stability, thereby modifying the material's wettability and long-term stability in the electrolyte [209,210]. Furthermore, interface defects can promote pseudocapacitive behavior, enhancing the energy storage performance of supercapacitors through surface adsorption and redox reactions [211,212]. For instance, Zhou et al. fabricated dual-transition metal oxide heterojunction (V_o-ZnO/CoO) nanowires with oxygen vacancies through hydrothermal and thermal treatments. The formation of the heterojunction led to the redistribution of interfacial charges between ZnO and CoO, generating an internal electric field that accelerated electron transfer. Due to the abundant electron density on the surface and unimpeded electron transfer within the bulk, the ZnO/CoO electrode exhibited higher capacity and superior cycling stability [213].

In addition to directly preparing electrode materials containing oxygen vacancies, oxygen vacancies can also be formed at the interface of multi-layer materials deposited by ALD. Zhang et al. employed plasma treatment and ALD technology to fabricate Al_2O_3 ALD coatings on activated carbon (AC) electrodes (Figure 12a). As shown in Figure 12b–e, the surface structure of the AC gradually became destroyed and roughened with the extension of plasma treatment time compared with the bare AC. Clearly, the AC structure could no longer maintain an intact block structure after 30 min. This may be attributed to the breakdown of original C-C bonds under the action of plasma energy, leading to the formation of new functional group bonds and defect structures. With the prolongation of processing time, the graphitization degree of the samples (named as OP-x, x = 5, 15 and 30) decreased, indicating that the introduction of oxygen led to an increase of defects. The ID/IG ratio of the four samples slightly increased with the extension of treatment time, indicating an increase in defects (Figure 12f). Impedance test results (Figure 12g) showed that the charge transfer resistances (R_{ct}) for the bare AC, OP-5, OP-15, and OP-30 were 17.5, 6.72, 6.62, and 6.98, respectively. The equivalent series resistance Rs values were 1.5, 0.945, 0.851, and 0.825, respectively; the induced oxygen doping led to more defects and a richer array of ion adsorption active sites. Additional oxygen was introduced into the AC electrodes through plasma treatment to increase the active sites and defects. This plasma-induced oxygen doping reduced the charge transfer resistance, improved the kinetics, and was beneficial for enhancing the capacitance and ion transport. It also resulted in superior electrochemical performance (Figure 12h,i). Moreover, The ALD coating layer can be observed more clearly from the TEM images, shown in Figure 12j–m. Compared with the exposed AC and OP-15 electrodes (Figure 12j,k), a roughly 2 nm thick Al_2O_3 layer can be detected on the carbon surface, as shown in Figure 12l,m. After 5000 retention test cycles, four battery samples were dissected to further inspect the degradation. The relatively intact thin layer maintained good stability, demonstrating the highly stability of the ALD-deposited film [214].

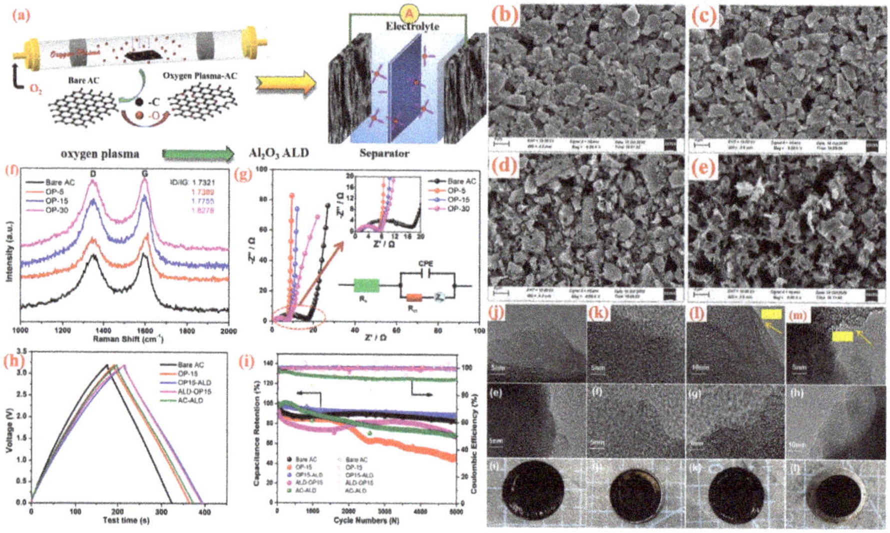

Figure 12. (**a**) Illustration for the function of oxygen plasma and Al_2O_3 ALD coating on the AC electrode. SEM image: (**b**) bare AC; (**c**) oxygen-plasma treatment at 5 min (OP-5); (**d**) oxygen-plasma treatment at 15 min (OP-15); (**e**) oxygen-plasma treatment at 30 min (OP-30). (**f**) Raman spectra; (**g–i**) Electrochemical behaviors. TEM image of fresh electrode: (**j**) bare AC; (**k**) OP-15; (**l**) ALD-OP15; (**m**) OP15-ALD. Reproduced with permission. Copyright 2024, Frontiers Publishing [214].

4.2.3. Passivation of Defects

Passivation of defect can significantly enhance electrochemical performance and stability of supercapacitors through the reduction of defects within electrode materials, such as vacancies, interstitials, and grain boundaries [215–217]. Defects act as traps for charge carriers, leading to hindered charge transport and consequently decreased capacitance and power density of the electrode materials [218,219]. Passivation techniques, including surface modification, doping, atomic layer deposition (ALD), and plasma treatment, can effectively fill these defects, thereby reducing charge transport resistance and improving the specific capacitance and cycling stability of electrode materials [220,221].

Graphene is considered an exciting electrode material for supercapacitors due to its excellent conductivity and large specific surface area [222–226]. However, the inherent defects in the oxidized/reduced graphene oxide structure reduce its conductivity and electrochemical performance, limiting its application in its original form. ALD provides a unique convenience for accurately functionalizing graphene oxide/reducing graphene oxide defects [227,228]. Yang et al. reported an optimization method for electrode materials utilizing a combined system of RuO_2 nanomaterials and graphene oxide. Initially, graphene oxide sheets were immobilized on carbon cloth via an electrochemical process. Subsequently, RuO_2 nanoparticles were grown on the immobilized graphene oxide using ALD with $Ru(EtCp)_2$ [bis(ethylcyclopentadienyl)ruthenium] and O_2 as precursors (Figure 13a). As the number of deposition cycles increased, the size of the RuO_2 nanoparticles also enlarged (Figure 13b). The ALD precursors require a functional site for chemisorption during the first half of the reaction. Functional sites typically exist at defects. SEM and STEM images in Figure 13c–h revealed that individual carbon fibers were fully covered with graphene oxide sheets containing one or more graphene layers. This process tightly bound the graphene oxide to the carbon fibers. After undergoing the ALD step, randomly distributed RuO_2 nanoparticles were observed to grow on the reduced graphene oxide (rGO). As shown in Figure 13i–l, the synthesized composite exhibited a specific capacitance of 1132 Fg^{-1} at a scan rate of 50 mV s^{-1}, approaching the theoretical limit. Traditional two-electrode configurations indicated that the specific capacitance of the graphene oxide electrode increased by approximately 40% when the optimal Ru loading was about 9.3 wt%. RuO_2 selectively chemically bound to defect sites on the graphene oxide substrate, where C-O sites were initially replaced by C-O-Ru bonds, altering the chemical nature of these defects and exhibiting excellent electrochemical performance [229].

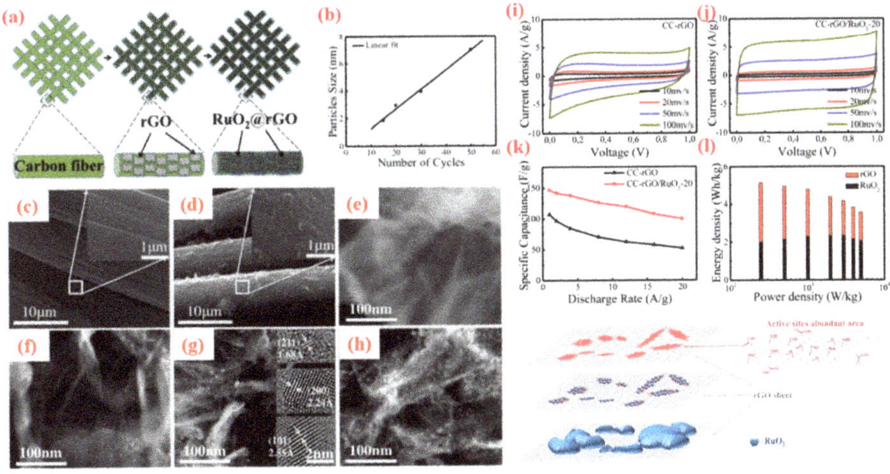

Figure 13. (**a**) Schematic of fabrication process of rGO/RuO_2; (**b**) average size dependence of RuO_2 nanoparticles on number of ALD cycles; (**c**,**d**) SEM images of (**c**) carbon fiber and (**d**) carbon fiber covered with rGO; (**e**–**h**) STEM images of (**e**) rGO, (**f**) rGO/RuO_2-15, (**g**) rGO/RuO_2-20, and (**h**) rGO/RuO_2-30; (**i**–**l**) electrochemical behaviors. Reproduced with permission. Copyright 2015, ACS Publishing [229].

5. Conclusions

This article provides an overview of ALD application in supercapacitors. It can be divided into several parts: ALD for defect engineering, ALD for the preparation of supercapacitor electrode materials, and defect engineering of electrode materials through ALD. The summary is as follows:

(1) ALD technology is effective in defect engineering, allowing for the direct deposition of thin films by adjusting deposition parameters or introducing defects at the interfaces between film layers through controlled deposition sequences and cycles. In terms of defect passivation, ALD can precisely fill and cover defects to enhance the stability of materials, influencing the electrical, optical, and mechanical properties of nanomaterials, optoelectronic devices, and catalysts.

(2) ALD technology enables precise control over the composition and structure of electrode materials, directly impacting their electrochemical performance in supercapacitors. By selecting appropriate precursors and deposition conditions, ALD can not only directly fabricate thin-film electrodes with excellent electrochemical properties for supercapacitors but also increase the specific surface area and stability of electrode materials through the deposition of covering layers, while optimizing these conditions further refines the morphology and structure of the thin films.

(3) ALD technology is also a potential tool for defect engineering, significantly enhancing the electrochemical performance of supercapacitor electrode materials. Electrode materials prepared containing oxygen vacancies through ALD can promote Faradaic oxidation-reduction surface reactions, enhancing the ability of electrodes to store electrical charge. The formation of interface defects using ALD can enhance electron transport performance and reduce charge transfer resistance, as well as make the

dynamic response of electrode materials faster during charging and discharging processes. Furthermore, with the advantage of atomic-level control, ALD can precisely repair defects on the surface of electrode materials, improving inherent electrical conductivity, thus enhancing electrochemical performance.

In addition, there are also some challenges in applying defect engineering via ALD for supercapacitors.

(1) ALD demonstrates significant potential in the fabrication of supercapacitor electrode materials, with unique advantages including precise thickness control and excellent deposition quality, which play a crucial role in the field of energy storage, particularly in the construction and of supercapacitor electrode materials. ALD can optimize the electrochemical performance of electrode materials, such as specific surface area and conductivity, thereby enhancing the specific capacitance of supercapacitors and achieving high energy and power density. However, ALD technology also faces challenges in the preparation of supercapacitor electrode materials, including the complex charge storage issues of high-loading electrodes, performance requirements under unconventional/extreme conditions, and the understanding of the structure–property relationship. These challenges necessitate further research and exploration by scientific researchers to achieve the commercial application of supercapacitor electrode materials.

(2) ALD with precise control over thickness and exceptional deposition quality offers unique advantages in the defect engineering for supercapacitor electrode materials. Through precisely controlling defects, including the type, quantity, and distribution, the electrochemical performance can be optimized, such as specific surface area and conductivity, thereby enhancing the energy density, power density, and cyclic stability of supercapacitors. However, implementing defect engineering requires precise control over deposition conditions, reaction mechanisms, and surface chemical processes, increasing the technical difficulty and complexity.

(3) While atomic layer deposition (ALD) has demonstrated remarkable potential in a wide array of applications, transitioning this advanced technology from the laboratory to commercial-scale production presents a multitude of challenges. Among these, maintaining precise process control, ensuring material quality and uniformity across larger substrates, and managing the economic viability of ALD systems stand out as critical areas requiring further attention. Additionally, the scalability of ALD infrastructure to meet industrial demands, as well as the need for continuous innovation in materials and processes to address market trends and emerging applications, pose significant hurdles. As such, future research should focus on developing scalable and cost-effective ALD technologies, exploring new materials and deposition strategies, and optimizing process parameters to enhance productivity and material performance. Addressing these challenges will be instrumental in unlocking the full commercial potential of ALD, paving the way for its broader adoption and integration into industrial manufacturing processes.

(4) ALD, besides demonstrating immense potential in the preparation of electrode materials for supercapacitors, also holds significant promise for other sustainable and scalable energy applications. However, it still faces numerous challenges. For instance, ALD technology can be employed to construct protective layers on lithium metal surfaces to suppress the growth of lithium dendrites. Yet, research is still required to ensure the stability and long-term effectiveness of these protective layers. In lithium-sulfur batteries, ALD technology can be utilized to build barrier layers on cathodes or separators to inhibit the shuttle of polysulfides. Nevertheless, further investigation is needed to guarantee the permeability and ionic conductivity of these barrier layers. In sodium-ion batteries, where sodium exhibits higher chemical reactivity than lithium, ALD coatings need to provide superior interface stability to prevent the decomposition of electrolytes.

Author Contributions: Writing—original draft preparation, T.G.; Conceptualization, X.X. and Z.D.; Writing—review and editing, X.L., L.M. and J.W.; Project administration, Y.L.; funding acquisition, Q.H. and J.X. All authors have read and agreed to the published version of the manuscript.

Funding: This work was supported by National Key R&D Program of China (Grant No. 2022YFC3320700), National Natural Science Foundation of China (Grant No. 22302118), Beijing Nova Program (20230484316) and China Postdoctoral Science Foundation (No. 2022M712019).

Data Availability Statement: All collected data are presented in the manuscript.

Conflicts of Interest: The authors declare no conflict of interest.

References

1. Zong, W.; Lai, F.; He, G.; Feng, J.; Wang, W.; Lian, R.; Miao, Y.E.; Wang, G.C.; Parkin, I.P.; Liu, T. Sulfur-Deficient Bismuth Sulfide/Nitrogen-Doped Carbon Nanofibers as Advanced Free-Standing Electrode for Asymmetric Supercapacitors. *Small* **2018**, *14*, 1801562. [CrossRef] [PubMed]
2. Zhong, C.; Deng, Y.; Hu, W.; Qiao, J.; Zhang, L.; Zhang, J. A review of electrolyte materials and compositions for electrochemical supercapacitors. *Chem. Soc. Rev.* **2015**, *44*, 7484–7539. [CrossRef] [PubMed]
3. Zhang, Y.; Liu, L.; Zhang, L.; Yu, Y.; Lv, H.; Chen, A. Template-free method for fabricating carbon nanotube combined with thin N-doped porous carbon composite for supercapacitor. *J. Mater. Sci.* **2019**, *54*, 6451–6460. [CrossRef]
4. Muzaffar, A.; Ahamed, M.B.; Hussain, C.M. Green supercapacitors: Latest developments and perspectives in the pursuit of sustainability. *Renew. Sustain. Energy Rev.* **2024**, *195*, 114324. [CrossRef]
5. Padalkar, N.S.; Sadavar, S.V.; Shinde, R.B.; Patil, A.S.; Patil, U.M.; Dhawale, D.S.; Bulakhe, R.N.; Kim, H.; Im, H.; Vinu, A.; et al. Layer-by-layer nanohybrids of Ni-Cr-LDH intercalated with 0D polyoxotungstate for highly efficient hybrid supercapacitor. *J. Colloid Interface Sci.* **2022**, *616*, 548–559. [CrossRef]
6. Cai, B.; Li, J.; Wang, L.; Li, X.; Yang, X.; Lü, W. Enhanced performance of asymmetric supercapacitor based on NiZn-LDH@NiCoSe_2 electrode materials. *Nanotechnology* **2022**, *33*, 295402. [CrossRef]
7. Zhong, M.; Zhang, M.; Li, X. Carbon nanomaterials and their composites for supercapacitors. *Carbon Energy* **2022**, *4*, 950–985. [CrossRef]
8. Xiong, Y.; Zhang, Y.-F.; Zhu, C.-L.; Yang, L.; Liang, H.-Y.; Shi, J.; Chen, J.-W.; Tian, W.-Q.; Liu, S.; Li, Z.; et al. Carbon cathode with heteroatom doping and ultrahigh surface area enabling enhanced capacitive behavior for potassium-ion hybrid capacitors. *Rare Met.* **2024**, *43*, 2136–2149. [CrossRef]
9. Wang, L.; Zhang, X.; Kong, Y.-Y.; Li, C.; An, Y.-B.; Sun, X.-Z.; Wang, K.; Ma, Y.-W. Metal–organic framework-derived $CoSe_2$@N-doped carbon nanocubes for high-performance lithium-ion capacitors. *Rare Met.* **2024**, *43*, 2150–2160. [CrossRef]
10. Shao, Y.; El-Kady, M.F.; Sun, J.; Li, Y.; Zhang, Q.; Zhu, M.; Wang, H.; Dunn, B.; Kaner, R.B. Design and Mechanisms of Asymmetric Supercapacitors. *Chem. Rev.* **2018**, *118*, 9233–9280. [CrossRef]
11. Augustyn, V.; Simon, P.; Dunn, B. Pseudocapacitive oxide materials for high-rate electrochemical energy storage. *Energy Environ. Sci.* **2014**, *7*, 1597. [CrossRef]
12. Tang, C.; Wang, X.; Ma, M.; Wang, Z.; Li, Y.; Li, H.; Li, B.; Zhang, Y.; Zhu, X. Optimizing the electrons/ions diffusion kinetics in δ-MnO_2 for realizing an ultra-high rate-capability supercapacitor. *Chem. Eng. J.* **2023**, *471*, 144784. [CrossRef]
13. Song, Z.; Bi, R.; Li, J.; He, Y.; Rao, F.; Chen, X.; Wang, J.; Wang, Z.; Yu, R.; Wang, D. Metallic 1T-MoS_2 boosts the kinetics for NiS_2-based hybrid supercapacitors with superb rate performance. *J. Mater. Sci. Technol.* **2024**. [CrossRef]
14. Shinde, P.A.; Olabi, A.G.; Chodankar, N.R.; Patil, S.J.; Hwang, S.-K.; Abdelkareem, M.A. Realizing superior redox kinetics of metal-metal carbides/carbon coordination supported heterointerface for stable solid-state hybrid supercapacitor. *Chem. Eng. J.* **2023**, *454*, 140246. [CrossRef]
15. Thalji, M.R.; Ali, G.A.M.; Shim, J.-J.; Chong, K.F. Cobalt-doped tungsten suboxides for supercapacitor applications. *Chem. Eng. J.* **2023**, *473*, 140246. [CrossRef]
16. Beknalkar, S.A.; Teli, A.M.; Satale, V.V.; Amate, R.U.; Morankar, P.J.; Yewale, M.A.; Shin, J.C. A critical review of recent advancements in high-temperature supercapacitors: Thermal kinetics, interfacial dynamics, employed strategies, and prospective trajectories. *Energy Storage Mater.* **2024**, *66*, 103217. [CrossRef]
17. Malavekar, D.; Pujari, S.; Jang, S.; Bachankar, S.; Kim, J.H. Recent Development on Transition Metal Oxides-Based Core–Shell Structures for Boosted Energy Density Supercapacitors. *Small* **2024**, *20*, 202312179. [CrossRef]
18. Pu, S.; Wang, Z.; Xie, Y.; Fan, J.; Xu, Z.; Wang, Y.; He, H.; Zhang, X.; Yang, W.; Zhang, H. Origin and Regulation of Self-Discharge in MXene Supercapacitors. *Adv. Funct. Mater.* **2022**, *33*, 202208715. [CrossRef]
19. Lan, S.; Yu, C.; Yu, J.; Zhang, X.; Liu, Y.; Xie, Y.; Wang, J.; Qiu, J. Recent Advances in Low-Temperature Liquid Electrolyte for Supercapacitors. *Small* **2024**, 202309286. [CrossRef]
20. Lyu, S.; Chang, H.; Zhang, L.; Wang, S.; Li, S.; Lu, Y.; Li, S. High specific surface area MXene/SWCNT/cellulose nanofiber aerogel film as an electrode for flexible supercapacitors. *Compos. Part B: Eng.* **2023**, *264*, 110888. [CrossRef]
21. Wang, T.; Liu, Z.; Li, P.; Wei, H.; Wei, K.; Chen, X. Lignin-derived carbon aerogels with high surface area for supercapacitor applications. *Chem. Eng. J.* **2023**, *466*, 143118. [CrossRef]

22. Chen, S.; Xu, Q.; Chen, L.; Hu, Y.; Jiang, H.; Li, C. Engineering V_2O_3 nanoarrays with abundant localized defects towards high-voltage aqueous supercapacitors. *J. Mater. Chem. A* **2022**, *10*, 4825–4832. [CrossRef]
23. Ji, Z.; Tang, G.; Chen, L.; Zhong, J.; Chen, Y.; Zhu, G.; Chuan, X.; Zhang, J.; Shen, X. Defect engineering of hollow porous N, S co-doped carbon spheres-derived materials for high-performance hybrid supercapacitors. *Chem. Eng. J.* **2024**, *480*, 148213. [CrossRef]
24. Sun, Y.; Li, T.; Liu, X.; Han, Y.; Liu, Y.; Zada, A.; Deng, W.; Yuan, Z.; Dang, A. Modulating oxygen vacancies in MXene/MoO_{3-x} smart fiber by defect engineering for ultrahigh volumetric energy density supercapacitors and wearable SERS sensors. *Chem. Eng. J.* **2024**, *494*, 152911. [CrossRef]
25. Lu, W.; Yan, L.; Ye, W.; Ning, J.; Zhong, Y.; Hu, Y. Defect engineering of electrode materials towards superior reaction kinetics for high-performance supercapacitors. *J. Mater. Chem. A* **2022**, *10*, 15267–15296. [CrossRef]
26. Wang, Q.; Qu, Z.; Chen, S.; Zhang, D. Metal organic framework derived P-doping CoS@C with sulfide defect to boost high-performance asymmetric supercapacitors. *J. Colloid Interface Sci.* **2022**, *624*, 385–393. [CrossRef]
27. Peng, Z.; Li, S.; Huang, Y.; Guo, J.; Tan, L.; Chen, Y. Sodium-Intercalated Manganese Oxides for Achieving Ultra-Stable and Fast Charge Storage Kinetics in Wide-Voltage Aqueous Supercapacitors. *Adv. Funct. Mater.* **2022**, *32*, 202206539. [CrossRef]
28. Fu, Y.; Gao, X.; Zha, D.; Zhu, J.; Ouyang, X.; Wang, X. Yolk–shell-structured MnO_2microspheres with oxygen vacancies for high-performance supercapacitors. *J. Mater. Chem. A* **2018**, *6*, 1601–1611. [CrossRef]
29. Yang, J.; Xiao, X.; Chen, P.; Zhu, K.; Cheng, K.; Ye, K.; Wang, G.; Cao, D.; Yan, J. Creating oxygen-vacancies in MoO_3- nanobelts toward high volumetric energy-density asymmetric supercapacitors with long lifespan. *Nano Energy* **2019**, *58*, 455–465. [CrossRef]
30. Li, G.; Cui, X.; Song, B.; Ouyang, H.; Wang, K.; Sun, Y.; Wang, Y. One-pot synthesis of Cu-doped Ni_3S_2 nano-sheet/rod nanoarray for high performance supercapacitors. *Chem. Eng. J.* **2020**, *388*, 124319. [CrossRef]
31. Ziegler, M.; Dathe, A.; Pollok, K.; Langenhorst, F.; Hübner, U.; Wang, D.; Schaaf, P. Metastable Atomic Layer Deposition: 3D Self-Assembly toward Ultradark Materials. *ACS Nano* **2020**, *14*, 15023–15031. [CrossRef] [PubMed]
32. Olowoyo, J.O.; Gharahshiran, V.S.; Zeng, Y.; Zhao, Y.; Zheng, Y. Atomic/molecular layer deposition strategies for enhanced CO_2 capture, utilisation and storage materials. *Chem. Soc. Rev.* **2024**, *53*, 5428–5488. [CrossRef] [PubMed]
33. Hui, L.; Nixon, R.; Tolman, N.; Mukai, J.; Bai, R.; Wang, R.; Liu, H. Area-Selective Atomic Layer Deposition of Metal Oxides on DNA Nanostructures and Its Applications. *ACS Nano* **2020**, *14*, 13047–13055. [CrossRef] [PubMed]
34. Yang, H.; Chen, Y.; Qin, Y. Application of atomic layer deposition in fabricating high-efficiency electrocatalysts. *Chin. J. Catal.* **2020**, *41*, 227–241. [CrossRef]
35. Yoo, C.; Jeon, J.W.; Yoon, S.; Cheng, Y.; Han, G.; Choi, W.; Park, B.; Jeon, G.; Jeon, S.; Kim, W.; et al. Atomic Layer Deposition of Sb_2Te_3/GeTe Superlattice Film and Its Melt-Quenching-Free Phase-Transition Mechanism for Phase-Change Memory. *Adv. Mater.* **2022**, *34*, 202207143. [CrossRef]
36. Plutnar, J.; Pumera, M. Applications of Atomic Layer Deposition in Design of Systems for Energy Conversion. *Small* **2021**, *17*, 202102088. [CrossRef]
37. Fonseca, J.; Lu, J. Single-Atom Catalysts Designed and Prepared by the Atomic Layer Deposition Technique. *ACS Catal.* **2021**, *11*, 7018–7059. [CrossRef]
38. Seenivasan, S.; Shim, K.I.; Lim, C.; Kavinkumar, T.; Sivagurunathan, A.T.; Han, J.W.; Kim, D.-H. Boosting Pseudocapacitive Behavior of Supercapattery Electrodes by Incorporating a Schottky Junction for Ultrahigh Energy Density. *Nano-Micro Lett.* **2023**, *15*, 62. [CrossRef]
39. Gupta, B.; Hossain, M.A.; Riaz, A.; Sharma, A.; Zhang, D.; Tan, H.H.; Jagadish, C.; Catchpole, K.; Hoex, B.; Karuturi, S. Recent Advances in Materials Design Using Atomic Layer Deposition for Energy Applications. *Adv. Funct. Mater.* **2021**, *32*, 202109105. [CrossRef]
40. Du, C.; Shi, S.; Chen, G.; Zhang, Y.; Wei, Q.; Li, L.; Wan, G.; Deng, Z.; Wu, Y.; Su, Y.; et al. Structurally stable Fe–Ni–Co phosphide/graphene and nitrogen-doped carbon/graphene constructed by atomic layer deposition for high-performance hybrid supercapacitor. *Mater. Today Energy* **2023**, *34*, 101287. [CrossRef]
41. Nishizawa, J.-I. Molecular Layer Epitaxy. *Appl. Surf. Sci.* **1994**, *79–80*, 1–11. [CrossRef]
42. George, S.M. Atomic Layer Deposition: An Overview. *Chem. Rev.* **2010**, *110*, 111–131. [CrossRef] [PubMed]
43. Macco, B.; Kessels, W.M.M. Atomic layer deposition of conductive and semiconductive oxides. *Appl. Phys. Rev.* **2022**, *9*, 116732. [CrossRef]
44. Kolhep, M.; Pantle, F.; Karlinger, M.; Wang, D.; Scherer, T.; Kübel, C.; Stutzmann, M.; Zacharias, M. Atomic Layer Deposition and Strain Analysis of Epitaxial GaN-ZnO Core–Shell Nanowires. *Nano Lett.* **2023**, *23*, 6920–6926. [CrossRef] [PubMed]
45. Zuo, Y.; Wang, Z.; Zhao, H.; Zhao, L.; Zhang, L.; Yi, B.; Bao, W.; Zhang, Y.; Su, L.; Yu, Y.; et al. Synthesis of a Spatially Confined, Highly Durable, and Fully Exposed Pd Cluster Catalyst via Sequential Site-Selective Atomic Layer Deposition. *ACS Appl. Mater. Interfaces* **2022**, *14*, 14466–14473. [CrossRef]
46. Jin, Y.; Yu, H.; Liang, X. Understanding the roles of atomic layer deposition in improving the electrochemical performance of lithium-ion batteries. *Appl. Phys. Rev.* **2021**, *8*, 031301. [CrossRef]
47. Jia, J.; Jiang, Z.; Ma, S.; Guo, S.; Wu, J.; Zhang, Y.; Cao, B.; Dong, J. Novel Strategy for High Efficient and Stable Perovskite Solar Cells through Atomic Layer Deposition. *ACS Appl. Mater. Interfaces* **2024**, *16*, 3576–3585. [CrossRef]
48. Wang, D.; Huang, Q.; Shi, W.; You, W.; Meyer, T.J. Application of Atomic Layer Deposition in Dye-Sensitized Photoelectrosynthesis Cells. *Trends Chem.* **2020**, *3*, 59–71. [CrossRef]

49. McNeary, W.W.; Tacey, S.A.; Lahti, G.D.; Conklin, D.R.; Unocic, K.A.; Tan, E.C.D.; Wegener, E.C.; Erden, T.E.; Moulton, S.; Gump, C.; et al. Atomic Layer Deposition with TiO_2 for Enhanced Reactivity and Stability of Aromatic Hydrogenation Catalysts. *ACS Catal.* **2021**, *11*, 8538–8549. [CrossRef]
50. Chen, R.; Cao, K.; Wen, Y.; Yang, F.; Wang, J.; Liu, X.; Shan, B. Atomic layer deposition in advanced display technologies: From photoluminescence to encapsulation. *Int. J. Extrem. Manuf.* **2024**, *6*, 022003. [CrossRef]
51. Guo, H.J.; Sun, Y.; Zhao, Y.; Liu, G.X.; Song, Y.X.; Wan, J.; Jiang, K.C.; Guo, Y.G.; Sun, X.; Wen, R. Surface Degradation of Single-crystalline Ni-rich Cathode and Regulation Mechanism by Atomic Layer Deposition in Solid-State Lithium Batteries. *Angew. Chem. Int. Ed.* **2022**, *61*, 202211626. [CrossRef] [PubMed]
52. Karimzadeh, S.; Safaei, B.; Yuan, C.; Jen, T.-C. Emerging Atomic Layer Deposition for the Development of High-Performance Lithium-Ion Batteries. *Electrochem. Energy Rev.* **2023**, *6*, 24. [CrossRef]
53. Cai, J.; Han, X.; Wang, X.; Meng, X. Atomic Layer Deposition of Two-Dimensional Layered Materials: Processes, Growth Mechanisms, and Characteristics. *Matter* **2020**, *2*, 587–630. [CrossRef]
54. Ghosh, S.; Pariari, D.; Behera, T.; Boix, P.P.; Ganesh, N.; Basak, S.; Vidhan, A.; Sarda, N.; Mora-Seró, I.; Chowdhury, A.; et al. Buried Interface Passivation of Perovskite Solar Cells by Atomic Layer Deposition of Al_2O_3. *ACS Energy Lett.* **2023**, *8*, 2058–2065. [CrossRef]
55. Ansari, M.Z.; Hussain, I.; Mohapatra, D.; Ansari, S.A.; Rahighi, R.; Nandi, D.K.; Song, W.; Kim, S.H. Atomic Layer Deposition—A Versatile Toolbox for Designing/Engineering Electrodes for Advanced Supercapacitors. *Adv. Sci.* **2023**, *11*, 202303055. [CrossRef]
56. Sivagurunathan, A.T.; Adhikari, S.; Kim, D.-H. Strategies and implications of atomic layer deposition in photoelectrochemical water splitting: Recent advances and prospects. *Nano Energy* **2021**, *83*, 105802. [CrossRef]
57. Gao, Z.; Zhang, C.; Gao, J.; Wang, Q.; Xu, G.; Cao, H.; Zhang, H. Nucleation and growth of low resistivity copper thin films on polyimide substrates by low-temperature atomic layer deposition. *Appl. Surf. Sci.* **2023**, *638*, 158072. [CrossRef]
58. Henning, A.; Bartl, J.D.; Zeidler, A.; Qian, S.; Bienek, O.; Jiang, C.M.; Paulus, C.; Rieger, B.; Stutzmann, M.; Sharp, I.D. Aluminum Oxide at the Monolayer Limit via Oxidant-Free Plasma-Assisted Atomic Layer Deposition on GaN. *Adv. Funct. Mater.* **2021**, *31*, 202101441. [CrossRef]
59. Asundi, A.S.; Raiford, J.A.; Bent, S.F. Opportunities for Atomic Layer Deposition in Emerging Energy Technologies. *ACS Energy Lett.* **2019**, *4*, 908–925. [CrossRef]
60. Xu, S.; Wang, Z.; Dull, S.; Liu, Y.; Lee, D.U.; Lezama Pacheco, J.S.; Orazov, M.; Vullum, P.E.; Dadlani, A.L.; Vinogradova, O.; et al. Direct Integration of Strained-Pt Catalysts into Proton-Exchange-Membrane Fuel Cells with Atomic Layer Deposition. *Adv. Mater.* **2021**, *33*, 202007885. [CrossRef]
61. Liu, X.; Yang, B.; Zhou, X.; Wu, M.; Spiecker, E.; Bachmann, J.; Hauke, F.; Hirsch, A.; Wei, T. Synergistic Combination of Reductive Covalent Functionalization and Atomic Layer Deposition—Towards Spatially Defined Graphene-Organic-Inorganic Heterostructures. *Angew. Chem. Int. Ed.* **2023**, *62*, 202314183. [CrossRef] [PubMed]
62. Yang, X.; Sun, P.; Zhang, H.; Xia, Z.; Waldman, R.Z.; Mane, A.U.; Elam, J.W.; Shao, L.; Darling, S.B. Polyphenol-Sensitized Atomic Layer Deposition for Membrane Interface Hydrophilization. *Adv. Funct. Mater.* **2020**, *30*, 201910062. [CrossRef]
63. Popov, G.; Bačić, G.; Mattinen, M.; Manner, T.; Lindström, H.; Seppänen, H.; Suihkonen, S.; Vehkamäki, M.; Kemell, M.; Jalkanen, P.; et al. Atomic Layer Deposition of PbS Thin Films at Low Temperatures. *Chem. Mater.* **2020**, *32*, 8216–8228. [CrossRef]
64. Yang, X.; Sun, P.; Wen, Y.; Mane, A.U.; Elam, J.W.; Ma, J.; Liu, S.; Darling, S.B.; Shao, L. Protein-activated atomic layer deposition for robust crude-oil-repellent hierarchical nano-armored membranes. *Sci. Bull.* **2024**, *69*, 218–226. [CrossRef]
65. Zhang, H.; Marshall, C.L. Atomic layer deposition: Catalytic preparation and modification technique for the next generation. *Chin. J. Catal.* **2019**, *40*, 1311–1323. [CrossRef]
66. Guo, D.; Wan, Z.; Li, Y.; Xi, B.; Wang, C. TiN@$Co_{5.47}N$ Composite Material Constructed by Atomic Layer Deposition as Reliable Electrocatalyst for Oxygen Evolution Reaction. *Adv. Funct. Mater.* **2020**, *31*, 202008511. [CrossRef]
67. Zhang, J.; Zhang, G.; Chen, Z.; Dai, H.; Hu, Q.; Liao, S.; Sun, S. Emerging applications of atomic layer deposition for lithium-sulfur and sodium-sulfur batteries. *Energy Storage Mater.* **2020**, *26*, 513–533. [CrossRef]
68. Coutancier, D.; Zhang, S.-T.; Bernardini, S.; Fournier, O.; Mathieu-Pennober, T.; Donsanti, F.; Tchernycheva, M.; Foldyna, M.; Schneider, N. ALD of ZnO:Ti: Growth Mechanism and Application as an Efficient Transparent Conductive Oxide in Silicon Nanowire Solar Cells. *ACS Appl. Mater. Interfaces* **2020**, *12*, 21036–21044. [CrossRef]
69. Yu, F.; Du, L.; Zhang, G.; Su, F.; Wang, W.; Sun, S. Electrode Engineering by Atomic Layer Deposition for Sodium-Ion Batteries: From Traditional to Advanced Batteries. *Adv. Funct. Mater.* **2019**, *30*, 201906890. [CrossRef]
70. Yang, X.; Martinson, A.B.F.; Elam, J.W.; Shao, L.; Darling, S.B. Water treatment based on atomically engineered materials: Atomic layer deposition and beyond. *Matter* **2021**, *4*, 3515–3548. [CrossRef]
71. McNeary, W.W.; Zaccarine, S.F.; Lai, A.; Linico, A.E.; Pylypenko, S.; Weimer, A.W. Improved durability and activity of Pt/C catalysts through atomic layer deposition of tungsten nitride and subsequent thermal treatment. *Appl. Catal. B: Environ.* **2019**, *254*, 587–593. [CrossRef]
72. Ngambou, M.W.N.; Pellet-Mary, C.; Brinza, O.; Antonino, A.; Hetet, G.; Tallaire, A.; Bénédic, F.; Achard, J. Improving NV centre density during diamond growth by CVD process using N_2O gas. *Diam. Relat. Mater.* **2022**, *123*, 108884. [CrossRef]
73. Mendhe, A.C.; Dhas, S.; Kim, Y.; Kim, D. Hierarchically structured $Cu_2P_2O_7$ nanoflakes as a binder-free electrodes for high-performance supercapacitors. *Chem. Eng. J.* **2024**, *496*, 153857. [CrossRef]

74. Zhou, H.; Zhu, G.; Dong, S.; Liu, P.; Lu, Y.; Zhou, Z.; Cao, S.; Zhang, Y.; Pang, H. Ethanol-Induced Ni^{2+}-Intercalated Cobalt Organic Frameworks on Vanadium Pentoxide for Synergistically Enhancing the Performance of 3D-Printed Micro-Supercapacitors. *Adv. Mater.* **2023**, *35*, 2211523. [CrossRef] [PubMed]
75. Anuar, N.; Yusop, S.N.; Roberts, K.J. Crystallisation of organic materials from the solution phase: A molecular, synthonic and crystallographic perspective. *Crystallogr. Rev.* **2022**, *28*, 97–215. [CrossRef]
76. Martínez-Puente, M.A.; Horley, P.; Aguirre-Tostado, F.S.; López-Medina, J.; Borbón-Nuñez, H.A.; Tiznado, H.; Susarrey-Arce, A.; Martínez-Guerra, E. ALD and PEALD deposition of HfO_2 and its effects on the nature of oxygen vacancies. *Mater. Sci. Eng. B* **2022**, *285*, 115964. [CrossRef]
77. Pan, H.; Wang, C.; Zhang, Z.; Li, Y.; Hou, X.; Zheng, W.; Liu, X.; Wan, Y.; Zhang, J. Oxygen vacancy-enriched ALD NiO sub-50 nm thin films for enhanced triethylamine detection. *Appl. Phys. Lett.* **2022**, *121*, 111603. [CrossRef]
78. Hsu, C.-H.; Chen, K.-T.; Lin, L.-Y.; Wu, W.-Y.; Liang, L.-S.; Gao, P.; Qiu, Y.; Zhang, X.-Y.; Huang, P.-H.; Lien, S.-Y.; et al. Tantalum-Doped TiO_2 Prepared by Atomic Layer Deposition and Its Application in Perovskite Solar Cells. *Nanomaterials* **2021**, *11*, 1504. [CrossRef]
79. Yang, F.; Yang, R.; Yan, L.; Liu, X.; Luo, X.; Zhang, L. In situ growth of porous TiO_2 with controllable oxygen vacancies on an atomic scale for highly efficient photocatalytic water splitting. *Catal. Sci. Technol.* **2020**, *10*, 5659–5665. [CrossRef]
80. Kao, E.; Yang, C.; Warren, R.; Kozinda, A.; Lin, L. ALD titanium nitride on vertically aligned carbon nanotube forests for electrochemical supercapacitors. *Sens. Actuators A: Phys.* **2016**, *240*, 160–166. [CrossRef]
81. Wang, K.; Paxson, A.; Valdez, T.I.; Erlat, A.; Lee, P.-C.; Yun, S.; Khajanji, P.; Zhang, Z.; Kummel, A.C.; Bandaru, P. Enhanced Corrosion Resistance in Aluminum-Based Electrolyzer Components via Stoichiometry Tuned Atomic Layer-Deposited TiOx Films. *ACS Appl. Mater. Interfaces* **2024**, *16*, 35043–35052. [CrossRef] [PubMed]
82. Wang, J.; Ni, Y.; Li, X.; Jiang, X.; Jin, Q.; Zhang, X.; Wei, S.; Jin, H.; Yuan, T.; Zou, J.; et al. Enhancement of nitric oxide sensing performance via oxygen vacancy promotion on strontium-doped $LaFeO_3$ perovskites. *Sens. Actuators B: Chem.* **2024**, *417*, 136157. [CrossRef]
83. Li, Z.; Liu, X.; Zhou, M.; Zhang, S.; Cao, S.; Lei, G.; Lou, C.; Zhang, J. Plasma-induced oxygen vacancies enabled ultrathin ZnO films for highly sensitive detection of triethylamine. *J. Hazard. Mater.* **2021**, *415*, 125757. [CrossRef] [PubMed]
84. Hu, Q.; Wang, S.; Gao, Z.; Li, Y.; Zhang, Q.; Xiang, Q.; Qin, Y. The precise decoration of Pt nanoparticles with Fe oxide by atomic layer deposition for the selective hydrogenation of cinnamaldehyde. *Appl. Catal. B: Environ.* **2017**, *218*, 591–599. [CrossRef]
85. Cao, L.; Liu, W.; Luo, Q.; Yin, R.; Wang, B.; Weissenrieder, J.; Soldemo, M.; Yan, H.; Lin, Y.; Sun, Z.; et al. Atomically dispersed iron hydroxide anchored on Pt for preferential oxidation of CO in H_2. *Nature* **2019**, *565*, 631–635. [CrossRef]
86. Hu, Q.; Lu, X.; Gao, T.; Zhang, J.; Dong, Z.; Ma, Z.; Wang, X.; Lu, B.; Lv, Z.; Xu, J. A novel Cu-doped ZnO confined structure: Precisely preparation, and sensitization mechanism for ppb-level H_2S gas detection. *Sens. Actuators B: Chem.* **2024**, *414*, 135852. [CrossRef]
87. Hu, Q.-M.; Dong, Z.; Zhang, G.-X.; Li, Y.-X.; Xing, S.-F.; Ma, Z.-H.; Dong, B.-Y.; Lu, B.; Sun, S.-H.; Xu, J.-Q. Ultra-thin ALD CoO_x-ZnO heterogenous films as highly sensitive and environmentally friendly H2S sensor. *Rare Met.* **2023**, *42*, 3054–3063. [CrossRef]
88. Jiang, Y.; Wang, Z.; Xu, C.; Li, W.; Li, Y.; Huang, S.; Chen, Z.; Zhao, B.; Sun, X.; Wilkinson, D.P.; et al. Atomic layer deposition for improved lithiophilicity and solid electrolyte interface stability during lithium plating. *Energy Storage Mater.* **2020**, *28*, 17–26. [CrossRef]
89. Guo, Y.; Wang, S.; Du, X.; Liang, S.; Huang, S.; Peng, S.; Xie, Y.; Ma, M.; Xiong, L. Construction of ultrahigh capacity density carbon nanotube based MIM capacitor. *Energy Storage Mater.* **2023**, *63*, 103064. [CrossRef]
90. Hossain, M.I.; Saleque, A.M.; Ahmed, S.; Saidjafarzoda, I.; Shahiduzzaman, M.; Qarony, W.; Knipp, D.; Biyikli, N.; Tsang, Y.H. Perovskite/perovskite planar tandem solar cells: A comprehensive guideline for reaching energy conversion efficiency beyond 30%. *Nano Energy* **2021**, *79*, 105400. [CrossRef]
91. Widiapradja, L.J.; Hong, S.; Kim, K.-T.; Bae, H.; Im, S. Van der Waals crystal radio with Pt/$MoSe_2$ Schottky diode and h-BN capacitor for RF energy harvesting. *Nano Energy* **2022**, *92*, 106771. [CrossRef]
92. Wu, S.M.; Yang, X.Y.; Janiak, C. Confinement Effects in Zeolite-Confined Noble Metals. *Angew. Chem. Int. Ed.* **2019**, *58*, 12340–12354. [CrossRef]
93. Pan, X.; Bao, X. The Effects of Confinement inside Carbon Nanotubes on Catalysis. *Acc. Chem. Res.* **2011**, *44*, 553–562. [CrossRef] [PubMed]
94. Easun, T.L.; Moreau, F.; Yan, Y.; Yang, S.; Schröder, M. Structural and dynamic studies of substrate binding in porous metal–organic frameworks. *Chem. Soc. Rev.* **2017**, *46*, 239–274. [CrossRef] [PubMed]
95. Dong, Z.; Hu, Q.; Gao, T.; Wang, J.; Kong, J.; Li, J.; Mao, L.; Wu, C.; Xu, J. Confining isolated Pt atoms into the porous SnO_2 for efficient gas sensing. *Ceram. Int.* **2024**, *50*, 22220–22231. [CrossRef]
96. Hu, Q.; Wu, C.; Dong, Z.; Zhang, G.; Ma, Z.; Wang, X.; Sun, S.; Xu, J. Direct confirmation of confinement effects by NiO confined in helical SnO_2 nanocoils and its application in sensors. *J. Mater. Chem. A* **2022**, *10*, 2786–2794. [CrossRef]
97. Li, W.; Wang, D.; Zhang, Y.; Tao, L.; Wang, T.; Zou, Y.; Wang, Y.; Chen, R.; Wang, S. Defect Engineering for Fuel-Cell Electrocatalysts. *Adv. Mater.* **2020**, *32*, 201907879. [CrossRef]
98. Bai, S.; Zhang, N.; Gao, C.; Xiong, Y. Defect engineering in photocatalytic materials. *Nano Energy* **2018**, *53*, 296–336. [CrossRef]

99. Park, K.; Lee, J.-H.; Lee, J.-W. Surface Defect Engineering of Metal Halide Perovskites for Photovoltaic Applications. *ACS Energy Lett.* **2022**, *7*, 1230–1239. [CrossRef]
100. Ortiz-Medina, J.; Wang, Z.; Cruz-Silva, R.; Morelos-Gomez, A.; Wang, F.; Yao, X.; Terrones, M.; Endo, M. Defect Engineering and Surface Functionalization of Nanocarbons for Metal-Free Catalysis. *Adv. Mater.* **2019**, *31*, 201805717. [CrossRef]
101. Zeng, H.; Chen, H.; Yang, B.; Zeng, J.; Meng, L.; Shi, D.; Chen, L.; Huang, Y. Highly-oxidizing Au@MnO_{2-X} nanozymes mediated homogeneous electrochemical detection of organophosphorus independent of dissolved oxygen. *J. Hazard. Mater.* **2023**, *459*, 132116. [CrossRef] [PubMed]
102. Ma, Z.; Martins, S.; Tan, Z.; Chen, S.; Monteblanco, E.; Liedke, M.O.; Butterling, M.; Attallah, A.G.; Hirschmann, E.; Wagner, A.; et al. Controlling Magneto-Ionics by Defect Engineering Through Light Ion Implantation. *Adv. Funct. Mater.* **2024**, *34*, 202312827. [CrossRef]
103. Wu, G.; Wang, Q.; Wu, Y.; Sun, X.; Liao, J.; Pan, J.; Chen, M.; Kasu, M.; Liu, S. Evolution of defects, morphologies and fundamental growth characteristics of CVD diamond films induced by nitrogen addition. *Mater. Today Commun.* **2020**, *25*, 101504. [CrossRef]
104. Ling, W.; Wang, X.; Wang, L.; Wang, X.; Zhan, X. Defect formation mechanism of laser additive manufacturing joint for large-thickness Ti_6Al_4V titanium alloy with Y_2O_3 nanoparticles. *Opt. Laser Technol.* **2023**, *157*, 108648. [CrossRef]
105. Meng, L.; Tian, W.; Wu, F.; Cao, F.; Li, L. TiO_2 ALD decorated CuO/$BiVO_4$ p-n heterojunction for improved photoelectrochemical water splitting. *J. Mater. Sci. Technol.* **2019**, *35*, 1740–1746. [CrossRef]
106. Yu, X.; Liu, C.; Li, C.; Wang, C.; Li, Y.; Liang, L.; Yu, W.; Wang, Y.; Liu, C.; Liu, Y.; et al. Controlled NiO_x Defect Engineering to Harnessing Redox Reactions in Perovskite Photovoltaic Cells via Atomic Layer Deposition. *ACS Appl. Mater. Interfaces* **2024**, *16*, 31114–31125. [CrossRef]
107. Dai, B.; Wu, C.; Xie, Y. Boosting the electrochromic performance of TiO_2 nanowire film via successively evolving surface structure. *Sci. China Chem.* **2021**, *64*, 745–752. [CrossRef]
108. Xiong, S.; Hou, Z.; Zou, S.; Lu, X.; Yang, J.; Hao, T.; Zhou, Z.; Xu, J.; Zeng, Y.; Xiao, W.; et al. Direct Observation on p- to n-Type Transformation of Perovskite Surface Region during Defect Passivation Driving High Photovoltaic Efficiency. *Joule* **2021**, *5*, 467–480. [CrossRef]
109. Zhou, L.; Zhang, L.; Li, H.; Shen, W.; Li, M.; He, R. Defect Passivation in Air-Stable Tin(IV)-Halide Single Crystal for Emissive Self-Trapped Excitons. *Adv. Funct. Mater.* **2021**, *31*, 202108561. [CrossRef]
110. Feng, W.; Zhao, Y.; Lin, K.; Lu, J.; Liang, Y.; Liu, K.; Xie, L.; Tian, C.; Lyu, T.; Wei, Z. Polymer-Assisted Crystal Growth Regulation and Defect Passivation for Efficient Perovskite Light-Emitting Diodes. *Adv. Funct. Mater.* **2022**, *32*, 202203371. [CrossRef]
111. Xu, T.; Xie, Y.; Qi, S.; Zhang, H.; Ma, W.; Wang, J.; Gao, Y.; Wang, L.; Zong, X. Simultaneous Defect Passivation and Co-Catalyst Engineering Leads to Superior Photocatalytic Hydrogen Evolution on Metal Halide Perovskites. *Angew. Chem. Int. Ed.* **2024**, *63*, 202409945. [CrossRef] [PubMed]
112. Jung, S.M.; Kang, H.L.; Won, J.K.; Kim, J.; Hwang, C.; Ahn, K.; Chung, I.; Ju, B.-K.; Kim, M.-G.; Park, S.K. High-Performance Quantum Dot Thin-Film Transistors with Environmentally Benign Surface Functionalization and Robust Defect Passivation. *ACS Appl. Mater. Interfaces* **2018**, *10*, 3739–3749. [CrossRef] [PubMed]
113. Jesna, G.K.; Halali, V.V.; Sanjayan, C.G.; Suvina, V.; Sakar, M.; Balakrishna, R.G. Perovskite nanomaterials as optical and electrochemical sensors. *Inorg. Chem. Front.* **2020**, *7*, 2702–2725. [CrossRef]
114. Li, Y.; Wu, F.; Han, M.; Li, Z.; Zhu, L.; Li, Z. Merocyanine with Hole-Transporting Ability and Efficient Defect Passivation Effect for Perovskite Solar Cells. *ACS Energy Lett.* **2021**, *6*, 869–876. [CrossRef]
115. Nam, T.; Lee, H.; Seo, S.; Cho, S.M.; Shong, B.; Lee, H.-B.-R.; Kim, H. Moisture barrier properties of low-temperature atomic layer deposited Al_2O_3 using various oxidants. *Ceram. Int.* **2019**, *45*, 19105–19112. [CrossRef]
116. Kim, S.; Lee, S.-H.; Jo, I.H.; Park, T.J.; Kim, J.H. Reduced leakage current in atomic-layer-deposited HfO_2 thin films deposited at low temperature by in-situ defect passivation. *Appl. Surf. Sci.* **2024**, *645*, 158790. [CrossRef]
117. Ng, S.; Iffelsberger, C.; Michalička, J.; Pumera, M. Atomic Layer Deposition of Electrocatalytic Insulator Al_2O_3 on Three-Dimensional Printed Nanocarbons. *ACS Nano* **2021**, *15*, 686–697. [CrossRef]
118. Jayaraman, S.; Rawson, T.J.; Belyustina, M.A. Designing supercapacitor electrolyte via ion counting. *Energy Environ. Sci.* **2022**, *15*, 2948–2957. [CrossRef]
119. Tang, Y.; Guo, W.; Zou, R. Nickel-based bimetallic battery-type materials for asymmetric supercapacitors. *Coord. Chem. Rev.* **2022**, *451*, 214242. [CrossRef]
120. Zhou, Y.; Qi, H.; Yang, J.; Bo, Z.; Huang, F.; Islam, M.S.; Lu, X.; Dai, L.; Amal, R.; Wang, C.H.; et al. Two-birds-one-stone: Multifunctional supercapacitors beyond traditional energy storage. *Energy Environ. Sci.* **2021**, *14*, 1854–1896. [CrossRef]
121. Patil, S.J.; Chodankar, N.R.; Hwang, S.-K.; Raju, G.S.R.; Ranjith, K.S.; Huh, Y.S.; Han, Y.-K. Ultra-stable flexible Zn-ion capacitor with pseudocapacitive 2D layered niobium oxyphosphides. *Energy Storage Mater.* **2022**, *45*, 1040–1051. [CrossRef]
122. Zhang, Y.; Mei, H.-X.; Cao, Y.; Yan, X.-H.; Yan, J.; Gao, H.-L.; Luo, H.-W.; Wang, S.-W.; Jia, X.-D.; Kachalova, L.; et al. Recent advances and challenges of electrode materials for flexible supercapacitors. *Coord. Chem. Rev.* **2021**, *438*, 213910. [CrossRef]
123. Xu, M.; Huang, Y.; Chen, R.; Huang, Q.; Yang, Y.; Zhong, L.; Ren, J.; Wang, X. Green conversion of Ganoderma lucidum residues to electrode materials for supercapacitors. *Adv. Compos. Hybrid Mater.* **2021**, *4*, 1270–1280. [CrossRef]
124. Ma, Y.; Xie, X.; Yang, W.; Yu, Z.; Sun, X.; Zhang, Y.; Yang, X.; Kimura, H.; Hou, C.; Guo, Z.; et al. Recent advances in transition metal oxides with different dimensions as electrodes for high-performance supercapacitors. *Adv. Compos. Hybrid Mater.* **2021**, *4*, 906–924. [CrossRef]

125. Sowbakkiyavathi, E.S.; Arunachala Kumar, S.P.; Maurya, D.K.; Balakrishnan, B.; Guo, J.Z.; Subramania, A. Research progress in the development of transition metal chalcogenides and their composite-based electrode materials for supercapacitors. *Adv. Compos. Hybrid Mater.* **2024**, *7*, 130. [CrossRef]
126. Kavinkumar, T.; Seenivasan, S.; Sivagurunathan, A.T.; Kwon, Y.; Kim, D.-H. Three-Dimensional Hierarchical Core/shell Electrodes Using Highly Conformal TiO_2 and Co_3O_4 Thin Films for High-Performance Supercapattery Devices. *ACS Appl. Mater. Interfaces* **2021**, *13*, 29058–29069. [CrossRef]
127. Naeem, F.; Naeem, S.; Zhao, Z.; Shu, G.-Q.; Zhang, J.; Mei, Y.; Huang, G. Atomic layer deposition synthesized ZnO nanomembranes: A facile route towards stable supercapacitor electrode for high capacitance. *J. Power Sources* **2020**, *451*, 227740. [CrossRef]
128. Li, Z.; Su, J.; Wang, X. Atomic layer deposition in the development of supercapacitor and lithium-ion battery devices. *Carbon* **2021**, *179*, 299–326. [CrossRef]
129. Chodankar, N.R.; Selvaraj, S.; Ji, S.H.; Kwon, Y.; Kim, D.H. Interface-Engineered Nickel Cobaltite Nanowires through NiO Atomic Layer Deposition and Nitrogen Plasma for High-Energy, Long-Cycle-Life Foldable All-Solid-State Supercapacitors. *Small* **2018**, *15*, 1803716. [CrossRef]
130. Jayaraman, V.; Sivagurunathan, A.T.; Adhikari, S.; Kim, D.H. CoO_x@NiMoN/$Ti_3C_2T_x$ Interface for Stable High-Performance Electrochemical Energy Storage Devices. *Small* **2023**, *20*, 2305868. [CrossRef]
131. Saha, A.; Shalev, O.; Maiti, S.; Wang, L.; Akella, S.H.; Schmerling, B.; Targin, S.; Tkachev, M.; Fan, X.; Noked, M. Evolution of ternary $Li_xSn_yO_z$ artificial cathode-electrolyte interphase (ACEI) through ALD: A surface strengthened NCM811 with enhanced electrochemical performances for Li-ion batteries. *Mater. Today Energy* **2023**, *31*, 101207. [CrossRef]
132. Zyulkov, I.; Madhiwala, V.; Voronina, E.; Snelgrove, M.; Bogan, J.; O'Connor, R.; De Gendt, S.; Armini, S. Area-Selective ALD of Ru on Nanometer-Scale Cu Lines through Dimerization of Amino-Functionalized Alkoxy Silane Passivation Films. *ACS Appl. Mater. Interfaces* **2020**, *12*, 4678–4688. [CrossRef] [PubMed]
133. Multia, J.; Heiska, J.; Khayyami, A.; Karppinen, M. Electrochemically Active In Situ Crystalline Lithium-Organic Thin Films by ALD/MLD. *ACS Appl. Mater. Interfaces* **2020**, *12*, 41557–41566. [CrossRef] [PubMed]
134. Azaceta, E.; García, S.; Leonet, O.; Beltrán, M.; Gómez, I.; Chuvilin, A.; Mainar, A.R.; Blazquez, J.A.; Knez, M. Particle atomic layer deposition as an effective way to enhance Li-S battery energy density. *Mater. Today Energy* **2020**, *18*, 100567. [CrossRef]
135. Spurling, D.; Krüger, H.; Kohlmann, N.; Rasch, F.; Kremer, M.P.; Kienle, L.; Adelung, R.; Nicolosi, V.; Schütt, F. 3D networked MXene thin films for high performance supercapacitors. *Energy Storage Mater.* **2024**, *65*, 103148. [CrossRef]
136. Lin, D.; Qian, O.; Huo, D.; Pan, Q.; Zhang, S.; Wang, Z.; Han, F.; Wei, B. Alternately stacked thin film electrodes for high-performance compact energy storage. *Nano Energy* **2020**, *78*, 105323. [CrossRef]
137. Li, H.; Liu, Y.; Lin, S.; Li, H.; Wu, Z.; Zhu, L.; Li, C.; Wang, X.; Zhu, X.; Sun, Y. Laser crystallized sandwich-like MXene/Fe_3O_4/MXene thin film electrodes for flexible supercapacitors. *J. Power Sources* **2021**, *497*, 229882. [CrossRef]
138. Bhosale, R.P.; Kumbhar, S.S.; Bhosale, S.B.; Salunkhe, R.R.; Kadam, V.A.; Pardhi, S.P.; Gholap, S.S.; Lokhande, C.D.; Jamadade, V.S. Morphology modulation of $MnFe_2O_4$ thin film electrode for enhanced performance of hybrid supercapacitor. *J. Energy Storage* **2024**, *86*, 111146. [CrossRef]
139. Liang, Q.; Wang, Y.; Yang, Y.; Xu, T.; Xu, Y.; Zhao, Q.; Heo, S.-H.; Kim, M.-S.; Jeong, Y.-H.; Yao, S.; et al. Nanocellulose/two dimensional nanomaterials composites for advanced supercapacitor electrodes. *Front. Bioeng. Biotechnol.* **2022**, *10*, 1024453. [CrossRef]
140. Kim, M.-J.; Bae, J.-S.; Jung, M.-J.; Jeon, E.; Park, Y.; Khan, H.; Kwon, S.-H. Atomic Layer Deposition of Defective Amorphous TiO_x Thin Films with Improved Photoelectrochemical Performance. *ACS Appl. Mater. Interfaces* **2023**, *15*, 45732–45744. [CrossRef]
141. Giraldo, B.; Fryauf, D.M.; Cheney, B.; Sands, J.H.; Kobayashi, N.P. Demonstration of Sputtering Atomic Layer Augmented Deposition: Aluminum Oxide–Copper Dielectric–Metal Nanocomposite Thin Films. *ACS Appl. Mater. Interfaces* **2020**, *12*, 14280–14288. [CrossRef] [PubMed]
142. Park, H.; Hwang, J.H.; Oh, S.H.; Ryu, J.J.; Jeon, K.; Kang, M.; Chai, H.-J.; Ham, A.; Kim, G.H.; Kang, K.; et al. Direct Growth of Bi_2SeO_5 Thin Films for High-k Dielectrics via Atomic Layer Deposition. *ACS Nano* **2024**, *18*, 22071–22079. [CrossRef] [PubMed]
143. Lopa, N.S.; Akbari, M.K.; Lu, H.L.; Zhuiykov, S. ALD-enabled WO_3-MoO_3 nanohybrid heterostructure for high-performance electrochemical supercapacitors. *J. Energy Storage* **2024**, *84*, 110777. [CrossRef]
144. Lopa, N.S.; Akbari, M.K.; Wu, D.; Verpoort, F.; Zhuiykov, S. Two-Dimensional SnO_2-ZnO Nanohybrid Electrode Fabricated via Atomic Layer Deposition for Electrochemical Supercapacitors. *Energy Fuels* **2023**, *37*, 3142–3151. [CrossRef]
145. Lopa, N.S.; Akbari, M.K.; Zhuiykov, S. ALD-fabricated two-dimensional SnO_2-In_2O_3 n-n nanohybrid electrode for electrochemical supercapacitors. *Electrochim. Acta* **2022**, *434*, 141322. [CrossRef]
146. Ye, F.; Xu, B.; Chen, R.; Li, R.; Chang, G. A high performance flexible cotton-based supercapacitor prepared by in-situ polyaniline and MXene coating. *J. Energy Storage* **2023**, *62*, 106803. [CrossRef]
147. Dharmasiri, B.; Usman, K.A.S.; Qin, S.A.; Razal, J.M.; Tran, N.T.; Coia, P.; Harte, T.; Henderson, L.C. $Ti_3C_2T_x$ MXene coated carbon fibre electrodes for high performance structural supercapacitors. *Chem. Eng. J.* **2023**, *476*, 146739. [CrossRef]
148. Baig, M.M.; Khan, M.A.; Gul, I.H.; Rehman, S.U.; Shahid, M.; Javaid, S.; Baig, S.M. A Review of Advanced Electrode Materials for Supercapacitors: Challenges and Opportunities. *J. Electron. Mater.* **2023**, *52*, 5775–5794. [CrossRef]
149. Etman, A.S.; Halim, J.; Rosen, J. MXene-based Zn-ion hybrid supercapacitors: Effects of anion carriers and MXene surface coatings on the capacities and life span. *J. Energy Storage* **2022**, *52*, 104823. [CrossRef]

150. Dong, W.; Lin, T.; Huang, J.; Wang, Y.; Zhang, Z.; Wang, X.; Yuan, X.; Lin, J.; Chen, I.W.; Huang, F. Electrodes with Electrodeposited Water-excluding Polymer Coating Enable High-Voltage Aqueous Supercapacitors. *Research* **2020**, *2020*, 4178179. [CrossRef]
151. Kim, H.; Maity, C.K.; Venkateswarlu, S.; Kim, M.J. Alternative inner filling and outer surface coating of BNNT by Tungsten(VI) oxide for supercapacitor electrode. *Compos. Part B: Eng.* **2024**, *279*, 111436. [CrossRef]
152. Liu, X.; Li, J.; Liu, Y.; Zhou, L. Achieving enhanced multifunctional performance for structural composite supercapacitors by reinforcing interfaces with polymer coating. *J. Colloid Interface Sci.* **2024**, *665*, 603–612. [CrossRef] [PubMed]
153. Hsain, H.A.; Lee, Y.; Lancaster, S.; Materano, M.; Alcala, R.; Xu, B.; Mikolajick, T.; Schroeder, U.; Parsons, G.N.; Jones, J.L. Role of Oxygen Source on Buried Interfaces in Atomic-Layer-Deposited Ferroelectric Hafnia–Zirconia Thin Films. *ACS Appl. Mater. Interfaces* **2022**, *14*, 42232–42244. [CrossRef] [PubMed]
154. Ahaliabadeh, Z.; Miikkulainen, V.; Mäntymäki, M.; Colalongo, M.; Mousavihashemi, S.; Yao, L.; Jiang, H.; Lahtinen, J.; Kankaanpää, T.; Kallio, T. Stabilized Nickel-Rich-Layered Oxide Electrodes for High-Performance Lithium-Ion Batteries. *Energy Environ. Mater.* **2024**, *7*, 12741. [CrossRef]
155. Adhikari, S.; Selvaraj, S.; Ji, S.H.; Kim, D.H. Encapsulation of Co_3O_4 Nanocone Arrays via Ultrathin NiO for Superior Performance Asymmetric Supercapacitors. *Small* **2020**, *16*, 202005414. [CrossRef]
156. Wang, H.; Niu, H.; Wang, H.; Wang, W.; Jin, X.; Wang, H.; Zhou, H.; Lin, T. Micro-meso porous structured carbon nanofibers with ultra-high surface area and large supercapacitor electrode capacitance. *J. Power Sources* **2021**, *482*, 228986. [CrossRef]
157. Bian, Z.; Wang, Y.; Zhu, B.; Ye, R.; Yuan, F.; Zhang, Y.; Liu, T.; Zhang, P.; Chen, C.; Zhu, G.; et al. Converting Hippophae rhamnoides fruit extraction residue into porous carbons with ultra-high surface area for superior-performance supercapacitors. *J. Energy Storage* **2024**, *98*, 113058. [CrossRef]
158. Pawar, S.A.; Patil, D.S.; Nandi, D.K.; Monirul Islam, M.; Sakurai, T.; Kim, S.-H.; Shin, J.C. Cobalt-based metal oxide coated with ultrathin ALD-MoS_2 as an electrode material for supercapacitors. *Chem. Eng. J.* **2022**, *435*, 135066. [CrossRef]
159. Sahoo, B.B.; Kumar, N.; Panda, H.S.; Panigrahy, B.; Sahoo, N.K.; Soam, A.; Mahanto, B.S.; Sahoo, P.K. Self-assembled 3D graphene-based aerogel with Au nanoparticles as high-performance supercapacitor electrode. *J. Energy Storage* **2021**, *43*, 103157. [CrossRef]
160. Zhao, Y.; Wang, S.; Yuan, M.; Chen, Y.; Huang, Y.; Lian, J.; Yang, S.; Li, H.; Wu, L. Amorphous MoS nanoparticles grown on cobalt-iron-based needle-like array for high-performance flexible asymmetric supercapacitor. *Chem. Eng. J.* **2021**, *417*, 127927. [CrossRef]
161. Karim, M.R.; Mohammad, A.; Mukta, C.B.; Lee, J.; Yoon, T. Biofilm-engineered fabrication of Ag nanoparticles with modified ZIF-8-derived ZnO for a high-performance supercapacitor. *J. Energy Storage* **2024**, *75*, 109646. [CrossRef]
162. Albashir, A.I.M.; Shang, W.; Hadi, M.K.; Zhang, J.; Zhang, T.; Ran, F. Straightforward Solution Polymerization Synthesis of Porous Carbon@Gold Nanoparticles Electrode for High-Performance Supercapacitor. *J. Energy Storage* **2021**, *33*, 102041. [CrossRef]
163. Zhang, Y.; Cao, J.; Yuan, Z.; Zhao, L.; Wang, L.; Han, W. Assembling Co_3O_4 Nanoparticles into MXene with Enhanced electrochemical performance for advanced asymmetric supercapacitors. *J. Colloid Interface Sci.* **2021**, *599*, 109–118. [CrossRef] [PubMed]
164. Ma, Q.; Liu, M.; Cui, F.; Zhang, J.; Cui, T. Fabrication of 3D graphene microstructures with uniform metal oxide nanoparticles via molecular self-assembly strategy and their supercapacitor performance. *Carbon* **2023**, *204*, 336–345. [CrossRef]
165. Jiang, H.; Yan, X.; Miao, J.; You, M.; Zhu, Y.; Pan, J.; Wang, L.; Cheng, X. Super-conductive silver nanoparticles functioned three-dimensional Cu_xO foams as a high-pseudocapacitive electrode for flexible asymmetric supercapacitors. *J. Mater.* **2021**, *7*, 156–165. [CrossRef]
166. Hai, Z.; Karbalaei Akbari, M.; Wei, Z.; Xue, C.; Xu, H.; Hu, J.; Hyde, L.; Zhuiykov, S. TiO_2 nanoparticles-functionalized two-dimensional WO_3 for high-performance supercapacitors developed by facile two-step ALD process. *Mater. Today Commun.* **2017**, *12*, 55–62. [CrossRef]
167. Nandi, D.K.; Sahoo, S.; Sinha, S.; Yeo, S.; Kim, H.; Bulakhe, R.N.; Heo, J.; Shim, J.-J.; Kim, S.-H. Highly Uniform Atomic Layer-Deposited MoS_2@3D-Ni-Foam: A Novel Approach To Prepare an Electrode for Supercapacitors. *ACS Appl. Mater. Interfaces* **2017**, *9*, 40252–40264. [CrossRef]
168. Ramadoss, A.; Kim, T.; Kim, G.-S.; Kim, S.J. Enhanced activity of a hydrothermally synthesized mesoporous MoS_2 nanostructure for high performance supercapacitor applications. *New J. Chem.* **2014**, *38*, 2379–2385. [CrossRef]
169. Nardekar, S.S.; Krishnamoorthy, K.; Pazhamalai, P.; Sahoo, S.; Mariappan, V.K.; Kim, S.-J. Exceptional interfacial electrochemistry of few-layered 2D MoS_2 quantum sheets for high performance flexible solid-state supercapacitors. *J. Mater. Chem. A* **2020**, *8*, 13121–13131. [CrossRef]
170. Li, H.; Gao, Y.; Shao, Y.; Su, Y.; Wang, X. Vapor-Phase Atomic Layer Deposition of Co_9S_8 and Its Application for Supercapacitors. *Nano Lett.* **2015**, *15*, 6689–6695. [CrossRef]
171. Zhang, N.; Wang, W.; Teng, C.; Wu, Z.; Ye, Z.; Zhi, M.; Hong, Z. Co_9S_8nanoparticle-decorated carbon nanofibers as high-performance supercapacitor electrodes. *RSC Adv.* **2018**, *8*, 27574–27579. [CrossRef] [PubMed]
172. Lin, T.W.; Tsai, H.C.; Chen, T.Y.; Shao, L.D. Facile and Controllable One-Pot Synthesis of Hierarchical Co_9S_8 Hollow Microspheres as High-Performance Electroactive Materials for Energy Storage and Conversion. *ChemElectroChem* **2017**, *5*, 137–143. [CrossRef]
173. Naeem, F.; Naeem, S.; Zhao, Y.; Wang, D.; Zhang, J.; Mei, Y.; Huang, G. TiO_2 Nanomembranes Fabricated by Atomic Layer Deposition for Supercapacitor Electrode with Enhanced Capacitance. *Nanoscale Res. Lett.* **2019**, *14*, 92. [CrossRef]

174. Zhang, Y.; Wang, F.; Zhu, H.; Zhang, D.; Chen, J. Elongated TiO_2 nanotubes directly grown on graphene nanosheets as an efficient material for supercapacitors and absorbents. *Compos. Part A: Appl. Sci. Manuf.* **2017**, *101*, 297–305. [CrossRef]
175. Ambade, R.B.; Ambade, S.B.; Shrestha, N.K.; Salunkhe, R.R.; Lee, W.; Bagde, S.S.; Kim, J.H.; Stadler, F.J.; Yamauchi, Y.; Lee, S.-H. Controlled growth of polythiophene nanofibers in TiO_2 nanotube arrays for supercapacitor applications. *J. Mater. Chem. A* **2017**, *5*, 172–180. [CrossRef]
176. Daubert, J.S.; Lewis, N.P.; Gotsch, H.N.; Mundy, J.Z.; Monroe, D.N.; Dickey, E.C.; Losego, M.D.; Parsons, G.N. Effect of Meso- and Micro-Porosity in Carbon Electrodes on Atomic Layer Deposition of Pseudocapacitive V_2O_5 for High Performance Supercapacitors. *Chem. Mater.* **2015**, *27*, 6524–6534. [CrossRef]
177. Prakash, H.C.; Kumar, M.S.; Lin, T.-W.; Batabyal, S.K. Photo-assisted capacitive performance of V_2O_5 supercapacitor. *Electrochim. Acta* **2023**, *469*, 143229. [CrossRef]
178. Karade, S.S.; Lalwani, S.; Eum, J.-H.; Kim, H. Coin cell fabricated symmetric supercapacitor device of two-steps synthesized V_2O_5 Nanorods. *J. Electroanal. Chem.* **2020**, *864*, 114080. [CrossRef]
179. Warren, R.; Sammoura, F.; Tounsi, F.; Sanghadasa, M.; Lin, L. Highly active ruthenium oxide coating via ALD and electrochemical activation in supercapacitor applications. *J. Mater. Chem. A* **2015**, *3*, 15568–15575. [CrossRef]
180. Zhang, Q.; Hu, Z.; Yang, Y.; Zhang, Z.; Wang, X.; Yang, X.; An, Y.; Guo, B. Metal organic frameworks-derived porous carbons/ruthenium oxide composite and its application in supercapacitor. *J. Alloys Compd.* **2018**, *735*, 1673–1681. [CrossRef]
181. B, N.; Y, V.; Abdul Razack, S. Enhanced formation of ruthenium oxide nanoparticles through green synthesis for highly efficient supercapacitor applications. *Adv. Powder Technol.* **2020**, *31*, 1001–1006. [CrossRef]
182. Yu, L.; Wan, G.; Qin, Y.; Wang, G. Atomic layer deposition assisted fabrication of high-purity carbon nanocoil for electrochemical energy storage. *Electrochim. Acta* **2018**, *268*, 283–294. [CrossRef]
183. Liu, T.; Jiang, C.; Cheng, B.; You, W.; Yu, J. Hierarchical flower-like C/NiO composite hollow microspheres and its excellent supercapacitor performance. *J. Power Sources* **2017**, *359*, 371–378. [CrossRef]
184. Li, X.; Li, J.; Zhang, Y.; Zhao, P. Synthesis of Ni-MOF derived NiO/rGO composites as novel electrode materials for high performance supercapacitors. *Colloids Surf. A: Physicochem. Eng. Asp.* **2021**, *622*, 126653. [CrossRef]
185. Qiu, J.; Bai, Z.; Dai, E.; Liu, S.; Liu, Y. NiO/Co_3O_4 nanoheterostructure derived from nickelocene filled ZIF-67 for supercapacitors. *J. Alloys Compd.* **2018**, *763*, 966–974. [CrossRef]
186. Li, H.; Mi, W.; Zhang, F.; Song, Y.; Zhao, J. Microwave-assisted hydrothermal synthesis of NiO/Co_3O_4 nanosheets decorated on three-dimensional porous graphene for supercapacitors. *J. Phys. Chem. Solids* **2024**, *190*, 112025. [CrossRef]
187. Xiong, W.; Guo, Q.; Guo, Z.; Li, H.; Zhao, R.; Chen, Q.; Liu, Z.; Wang, X. Atomic layer deposition of nickel carbide for supercapacitors and electrocatalytic hydrogen evolution. *J. Mater. Chem. A* **2018**, *6*, 4297–4304. [CrossRef]
188. Liu, M.; Niu, B.; Guo, H.; Ying, S.; Chen, Z. Simple preparation of g-C_3N_4@Ni_3C nanosheets and its application in supercapacitor electrode materials, hydrogengeneration via $NaBH_4$ hydrolysis and reduction of p–nitrophenol. *Inorg. Chem. Commun.* **2021**, *130*, 108687. [CrossRef]
189. Liu, M.-C.; Hu, Y.-M.; An, W.-Y.; Kong, L.-B.; Kang, L. Liquid phase synthesis of dendritic nickel carbide alloy with high conductivity for advanced energy storage. *J. Energy Chem.* **2017**, *26*, 750–756. [CrossRef]
190. Ramachandran, T.; Hamed, F.; Raji, R.K.; Majhi, S.M.; Barik, D.; Kumar, Y.A.; Jauhar, R.M.; Pachamuthu, M.P.; Vijayalakshmi, L.; Ansar, S. Enhancing asymmetric supercapacitor performance with $NiCo_2O_4$–NiO hybrid electrode fabrication. *J. Phys. Chem. Solids* **2023**, *180*, 111467. [CrossRef]
191. Yao, D.; Ouyang, Y.; Jiao, X.; Ye, H.; Lei, W.; Xia, X.; Lu, L.; Hao, Q. Hierarchical NiO@$NiCo_2O_4$ Core–shell Nanosheet Arrays on Ni Foam for High-Performance Electrochemical Supercapacitors. *Ind. Eng. Chem. Res.* **2018**, *57*, 6246–6256. [CrossRef]
192. Xiang, K.; Wu, D.; Fan, Y.; You, W.; Zhang, D.; Luo, J.-L.; Fu, X.-Z. Enhancing bifunctional electrodes of oxygen vacancy abundant $ZnCo_2O_4$ nanosheets for supercapacitor and oxygen evolution. *Chem. Eng. J.* **2021**, *425*, 130583. [CrossRef]
193. Li, T.; Hu, Y.; Zhang, J.; Li, H.; Fang, K.; Wang, J.; Wang, Z.; Xu, M.; Zhao, B. Doping effect and oxygen vacancy engineering in nickel-manganese layered double hydroxides for high-performance supercapacitors. *Nano Energy* **2024**, *126*, 109690. [CrossRef]
194. Yue, J.; Lu, G.; Zhang, P.; Wu, Y.; Cheng, Z.; Kang, X. Oxygen vacancies modulation in graphene/MnO_x composite for high performance supercapacitors. *Colloids Surf. A: Physicochem. Eng. Asp.* **2019**, *569*, 10–17. [CrossRef]
195. Xu, L.; Xi, Y.; Huang, C.; Zhang, J.; Hua, Z.; Zhou, J.-J.; Yin, J.; Zhang, L.; Li, W.; Wang, J.; et al. Superior electronic/ionic transport dynamics of Zn-Co-OH/MnO_2 heterointerface containing oxygen vacancies for pseudocapacitive storage. *Chem. Eng. J.* **2023**, *468*, 143551. [CrossRef]
196. Feng, Y.; Liu, W.; Wang, Y.; Gao, W.; Li, J.; Liu, K.; Wang, X.; Jiang, J. Oxygen vacancies enhance supercapacitive performance of $CuCo_2O_4$ in high-energy-density asymmetric supercapacitors. *J. Power Sources* **2020**, *458*, 228005. [CrossRef]
197. Zhu, Z.; Tian, W.; Lv, X.; Wang, F.; Hu, Z.; Ma, K.; Wang, C.; Yang, T.; Ji, J. P-doped cobalt carbonate hydroxide@$NiMoO_4$ double-shelled hierarchical nanoarrays anchored on nickel foam as a bi-functional electrode for energy storage and conversion. *J. Colloid Interface Sci.* **2021**, *587*, 855–863. [CrossRef] [PubMed]
198. Zhang, H.; Bai, Y.; Chen, H.; Wu, J.; Li, C.M.; Su, X.; Zhang, L. Oxygen-defect-rich 3D porous cobalt-gallium layered double hydroxide for high-performance supercapacitor application. *J. Colloid Interface Sci.* **2022**, *608*, 1837–1845. [CrossRef]
199. Ramadan, A.; Ramadan, W. Superior performance of asymmetric supercapacitor based on Cu doped bismuth ferrite electrode. *J. Energy Storage* **2024**, *100*, 113702. [CrossRef]

200. Chen, Y.; Zhou, C.; Liu, G.; Kang, C.; Ma, L.; Liu, Q. Hydroxide ion dependent α-MnO_2 enhanced via oxygen vacancies as the negative electrode for high-performance supercapacitors. *J. Mater. Chem. A* **2021**, *9*, 2872–2887. [CrossRef]
201. Su, J.; Yang, L.; Yu, D.; Dong, S. Exploring the potential of constructing a hybrid supercapacitor with FeCoNi-LDH porous material containing oxygen vacancies for high-performance energy storage applications. *Nano Res.* **2024**, *17*, 8134–8144. [CrossRef]
202. Wang, C.; Wu, X.; Qin, Y.; Kong, Y. Reduced Mo-doped $NiCo_2O_4$ with rich oxygen vacancies as an advanced electrode material in supercapacitors. *Chem. Commun.* **2022**, *58*, 5120–5123. [CrossRef]
203. Wang, S.; Ding, N.; Han, D.; Wang, P.; Dang, Y.; Xu, P.; Sui, Y.; Wei, Y. Synergistic effect of fluorine doping and oxygen vacancies on electrochemical performance of $ZnCo_2O_4$ for advanced supercapacitors and Zn-ion batteries. *Acta Mater.* **2023**, *257*, 119190. [CrossRef]
204. Pacchioni, G. Numerical Simulations of Defective Structures: The Nature of Oxygen Vacancy in Non-reducible (MgO, SiO_2, ZrO_2) and Reducible (TiO_2, NiO, WO_3) Oxides. In *Defects at Oxide Surfaces*; Springer International Publishing: Cham, Switzerland, 2015; Volume 58, pp. 1–28. [CrossRef]
205. He, J.; Hu, Z.; Deng, K.; Zhao, R.; Lv, X.; Tian, W.; Zhang, Y.X.; Ji, J. A triple-layered PPy@NiCo LDH/$FeCo_2O_4$ hybrid crystalline structure with high electron conductivity and abundant interfaces for supercapacitors and oxygen evolution. *CrystEngComm* **2021**, *23*, 2262–2268. [CrossRef]
206. Zheng, J.; Zhang, Y.; Meng, C.; Wang, X.; Liu, C.; Bo, M.; Pei, X.; Wei, Y.; Lv, T.; Cao, G. V_2O_3/C nanocomposites with interface defects for enhanced intercalation pseudocapacitance. *Electrochim. Acta* **2019**, *318*, 635–643. [CrossRef]
207. Wang, P.; Ding, X.; Zhe, R.; Zhu, T.; Qing, C.; Liu, Y.; Wang, H.-E. Synchronous Defect and Interface Engineering of $NiMoO_4$ Nanowire Arrays for High-Performance Supercapacitors. *Nanomaterials* **2022**, *12*, 1094. [CrossRef] [PubMed]
208. Ranjith, K.S.; Lee, S.-Y.; Ghoreishian, S.M.; Chodankar, N.R.; Rama Raju, G.S.; Patil, S.J.; Huh, Y.S.; Park, S.-J.; Han, Y.-K. Defect and interface engineering of MXene-tagged N,F-doped carbon-$CoSe_2$ heterostructure for superior hydrogen evolution reactions and supercapacitors. *Carbon* **2023**, *206*, 246–259. [CrossRef]
209. Maria Mahimai, B.; Li, E.; Pang, J.; Zhang, J.; Zhang, J. Interface engineering in conducting polymers-based supercapacitor. *J. Energy Storage* **2024**, *96*, 112598. [CrossRef]
210. Zheng, X.; Jiang, J.; Bi, T.; Jin, F.; Li, M. $FeCo_2S_4$@Ni@graphene Nanocomposites with Rich Defects Induced by Heterointerface Engineering for High-Performance Supercapacitors. *ACS Appl. Energy Mater.* **2021**, *4*, 3288–3296. [CrossRef]
211. Wu, J.; Huang, F.; Lee, T.; Yan, Y.; Pei, X.; Wang, M.; Gao, S.; Guo, S.; Pan, X.; Wang, P. Interface-Guided Formation of 2D Ultrathin MnO_2 Nanosheets with Abundant Oxygen Defects for High Performance Supercapacitors. *ACS Appl. Energy Mater.* **2022**, *5*, 6962–6969. [CrossRef]
212. Chen, Y.; Yin, Z.; Huang, D.; Lei, L.; Chen, S.; Yan, M.; Du, L.; Xiao, R.; Cheng, M. Uniform polypyrrole electrodeposition triggered by phytic acid-guided interface engineering for high energy density flexible supercapacitor. *J. Colloid Interface Sci.* **2022**, *611*, 356–365. [CrossRef] [PubMed]
213. Zhou, X.; Yue, X.; Dong, Y.; Zheng, Q.; Lin, D.; Du, X.; Qu, G. Enhancing electrochemical performance of electrode material via combining defect and heterojunction engineering for supercapacitors. *J. Colloid Interface Sci.* **2021**, *599*, 68–78. [CrossRef] [PubMed]
214. Zhang, F.; Song, G.; Gandla, D.; Ein-Eli, Y.; Tan, D.Q. Synergy of Oxygen Plasma and Al_2O_3 Atomic Layer Deposition on Improved Electrochemical Stability of Activated Carbon-Based Supercapacitor. *Front. Energy Res.* **2021**, *9*, 653203. [CrossRef]
215. Özge Alaş Çolak, M.; Güngör, A.; Akturk, M.B.; Erdem, E.; Genç, R. Unlocking the full potential of citric acid-synthesized carbon dots as a supercapacitor electrode material via surface functionalization. *Nanoscale* **2024**, *16*, 719–733. [CrossRef]
216. Yang, L.; Zhang, L.; Jiao, X.; Qiu, Y.; Xu, W. The electrochemical performance of reduced graphene oxide prepared from different types of natural graphites. *RSC Adv.* **2021**, *11*, 4042–4052. [CrossRef]
217. Wang, J.; Zhang, C.; Chen, Z.; Dong, J.; Chang, P.; Yang, T.; Zhang, J.; Zhao, P.-Y. An environment-friendly method to prepare fulvic acid-based porous carbon for high energy density supercapacitors. *Diam. Relat. Mater.* **2024**, *149*, 111587. [CrossRef]
218. Zhang, S.; Wang, J.; Wu, R.; Liu, L.; Pan, B.; Liu, C. Structural and corrosion resistance properties of sputtered zirconium nitride thin films as electrode material for supercapacitor. *J. Alloys Compd.* **2022**, *900*, 163506. [CrossRef]
219. Xu, L.; Zhang, Y.; Zhou, W.; Jiang, F.; Zhang, H.; Jiang, Q.; Jia, Y.; Wang, R.; Liang, A.; Xu, J.; et al. Fused Heterocyclic Molecule-Functionalized N-Doped Reduced Graphene Oxide by Non-Covalent Bonds for High-Performance Supercapacitors. *ACS Appl. Mater. Interfaces* **2020**, *12*, 45202–45213. [CrossRef]
220. Fan, Y.-F.; Yi, Z.-L.; Song, G.; Wang, Z.-F.; Chen, C.-J.; Xie, L.-J.; Sun, G.-H.; Su, F.-Y.; Chen, C.-M. Self-standing graphitized hybrid Nanocarbon electrodes towards high-frequency supercapacitors. *Carbon* **2021**, *185*, 630–640. [CrossRef]
221. Garcés, L.; López-Mireles, D.A.; Padmasree, K.P.; Mtz-Enriquez, A.I.; Encinas, A.; Oliva, J. Stretchable supercapacitors made with plasticine/graphene/$CeNdNiO_2$ composite electrodes. *Mater. Chem. Phys.* **2024**, *313*, 128721. [CrossRef]
222. Lim, T.; Ho, B.T.; Suk, J.W. High-performance and thermostable wire supercapacitors using mesoporous activated graphene deposited on continuous multilayer graphene. *J. Mater. Chem. A* **2021**, *9*, 4800–4809. [CrossRef]
223. Meng, Q.; Du, C.; Xu, Z.; Nie, J.; Hong, M.; Zhang, X.; Chen, J. Siloxene-reduced graphene oxide composite hydrogel for supercapacitors. *Chem. Eng. J.* **2020**, *393*, 124684. [CrossRef]
224. Que, L.; Zhang, L.; Wu, C.; Zhang, Y.; Pei, C.; Nie, F. Pt-decorated graphene network materials for supercapacitors with enhanced power density. *Carbon* **2019**, *145*, 281–289. [CrossRef]

225. Zhang, H.; Yang, D.; Lau, A.; Ma, T.; Lin, H.; Jia, B. Hybridized Graphene for Supercapacitors: Beyond the Limitation of Pure Graphene. *Small* **2021**, *17*, 202007311. [CrossRef]
226. Mohamed, N.B.; El-Kady, M.F.; Kaner, R.B. Macroporous Graphene Frameworks for Sensing and Supercapacitor Applications. *Adv. Funct. Mater.* **2022**, *32*, 202203101. [CrossRef]
227. Kim, K.H.; Yang, M.; Cho, K.M.; Jun, Y.-S.; Lee, S.B.; Jung, H.-T. High quality reduced graphene oxide through repairing with multi-layered graphene ball nanostructures. *Sci. Rep.* **2013**, *3*, 3251. [CrossRef]
228. Chen, K.; Li, Z.; Liu, Q.; Xu, M.; Jiang, Q.; Dong, Z.; Wang, X.; Chen, X.; Li, S.; Gao, F. Graphene/GaN ultraviolet photodetector performance regulated by a HfO_2 insulating layer. *Appl. Phys. Lett.* **2024**, *124*, 052103. [CrossRef]
229. Yang, F.; Zhang, L.; Zuzuarregui, A.; Gregorczyk, K.; Li, L.; Beltrán, M.; Tollan, C.; Brede, J.; Rogero, C.; Chuvilin, A.; et al. Functionalization of Defect Sites in Graphene with RuO_2 for High Capacitive Performance. *ACS Appl. Mater. Interfaces* **2015**, *7*, 20513–20519. [CrossRef]

MDPI AG
Grosspeteranlage 5
4052 Basel
Switzerland
Tel.: +41 61 683 77 34

Batteries Editorial Office
E-mail: batteries@mdpi.com
www.mdpi.com/journal/batteries

www.ingramcontent.com/pod-product-compliance
Lightning Source LLC
Chambersburg PA
CBHW040315270926
51850CB00007BC/1493